Genetics—Principles and Perspectives: a series of texts

General Editors: Dr. K. R. Lewis, Professor Bernard John

Volumes already published

1 **The Differentiation of Cells**   Norman Maclean
2 **The Genetics of Recombination** D. G. Catcheside, F.R.S.
3 **Chromosome Genetics**   H. Rees, F.R.S. and R. N. Jones

Chromosome Genetics

H. Rees, D.Sc., F.R.S.,
R. N. Jones, Ph.D.

Department of Agricultural Botany, University College of Wales,
Aberystwyth.

University Park Press

© H. Rees and R. N. Jones 1977

First published 1977
by Edward Arnold (Publishers) Limited
25 Hill Street, London W1X 8LL

First published in the USA in 1977 by
University Park Press
233 East Redwood Street
Baltimore, Maryland 21202

Library of Congress Cataloging in Publication Data

Rees, Hubert, 1923–
  Chromosome genetics.

  (Genetics, principles and perspectives; 3)
  Includes bibliographical references and index.
  1. Chromosomes. 2. Genetics. I. Jones, Robert
Neil, 1939– joint author. II. Title. III. Series.
[DNLM: 1. Chromosomes. 2. Cytogenetics. QH605 R328c]
QH600.R43 1978   574.8'732   77-16211
ISBN 0–8391–1195–9

Printed in Great Britain

3-1303-00019-7096

Preface

The aim of this book is to present a brief but comprehensive account of the structure and organization of chromosomes, their behaviour during cell division, their variation within and among species, and to explain how the chromosome mechanism regulates the distribution of genetic information during growth and reproduction. During recent years many new facts have come to light about the molecular composition and fine structure of the chromosomes, but we are still a very long way from appreciating what these facts imply in terms of genetics and genetic systems. We have attempted to relate the implications of new knowledge and discoveries at the molecular level to established concepts of the chromosome as a whole.

In a small book of this kind we have been obliged to be selective. Some subjects, not extensively covered in other textbooks, are however presented at some length, *B*-chromosomes for example.

The book is intended for students with an elementary knowledge of genetics. We hope the facts and principles will be of value not only to cytologists and geneticists but to biologists generally.

We are grateful to our colleagues Dr. A. Durrant, Dr. G. M. Evans and Dr. R. K. J. Narayan for helpful comments on the manuscript. We also thank Miss Elizabeth Phillips for typing the script and Mr. E. Wintle for developing and printing some of the photographs. Finally, we thank those who have kindly allowed us to reproduce figures and tables from their published work.

Contents

1

The physical basis of heredity

1.1 Information

Genetics is concerned with the nature and utilization of the heritable information which controls the development of living organisms and with the distribution of this information during growth and reproduction. The information is located mainly within the chromosomes which, in most organisms, are organized within the cell nucleus. The location of heritable information within the chromosomes is directly established by two kinds of experiment: reciprocal grafts and reciprocal crosses.

Reciprocal grafts

Algae of the genus *Acetabularia* are unicellular with a single nucleus in the rhizoid of the filamentous cell. The shape of the cap at the top of the filament varies between species. In *A. mediterranea* the cap is round and smooth edged, in *A. crenulata* it is frilled. When the filament of *A. mediterranea* is grafted on to the nucleated rhizoid of *A. crenulata* the new cap which develops is of the *crenulata* type, and vice versa (Fig. 1.1). Information about cap development is evidently located in the chromosomes, which make up the nucleus, and not in the cytoplasm.

More sophisticated grafts can be achieved by nuclear transplantation (Danielli, 1958). By microdissection it is possible in *Amoeba* to replace the nucleus of one cell by that of another. When the nuclei from one strain are displaced by those of another observations on the nucleus/cytoplasm 'hybrids' and their progenies show some characters to be controlled by nuclear genes, others by genes located in the cytoplasm. Equally important the experiments show instances of interaction between nuclear and cytoplasmic determinants.

Reciprocal crosses

The contribution of material to the zygote by the female gamete in higher plants and animals is vastly greater than by the male gamete. The human egg has a diameter of 140 μm and weighs about 0.0015 mg. The sperm head by comparison has a length of only 4 μm and weighs 0.000 000 005 mg. The mass of the latter is made up almost entirely of the chromosomes. In addition to chromosomes the egg contains a

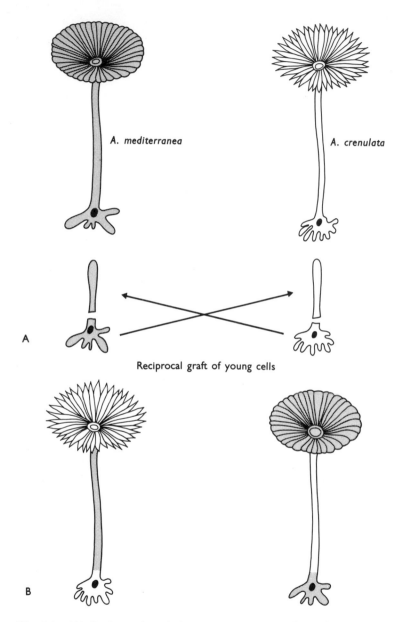

Fig. 1.1   (A) Reciprocal grafts between young cells of two *Acetabularia* species combining, in the main, the cytoplasm of one species of *Acetabularia* and the nucleus of another. (B) The adults display cap phenotypes determined by the nucleus, not the cytoplasm.

massive quantity of extranuclear, cytoplasmic material. A comparable situation applies in higher plants. Yet for the vast majority of characters the results of reciprocal crosses are identical. When Mendel crossed Tall peas with Dwarf, the F_1 and F_2 progenies were similar whether Tall plants were used as male or female parents. The inference is clear, namely that the genetic determinants are located in the chromosomes within the nucleus, not in the cytoplasm. Yet, as with the exceptions in Danielli's grafts in *Amoeba*, the results of reciprocal crosses are sometimes different. Control over certain aspects of the phenotype may be exercised by cytoplasmic 'genes' and by their interaction with the genes in the nucleus.

That the genetic information resides mainly within the chromosomes was established in the 1920s, and in the 1930s the Chromosome Theory of Heredity was elaborated, mainly on the basis of cytological and linkage analyses in *Drosophila* and maize. It was confirmed that the number of linkage groups corresponded with the basic chromosome number; that linkage maps, as one might suspect from the thread-like organization of chromosomes, were linear; and that structural rearrangements within or between chromosomes were accompanied by changes in linkage relations. This information, however, gave no indication of how or where the information was carried by the chromosomes which, in higher plants and animals, are organelles of considerable complexity in both chemical and physical terms. Certain key experiments served to locate the information precisely within the nucleic acid component of the chromosomes.

Virus fractionation and reconstitution
In the small plant viruses, such as Tobacco Mosaic Virus (TMV), individual particles comprise two components, a protein sheath embracing a single strand of ribonucleic acid (RNA). Together they make up a cylinder 300 nm in length and 15 nm in diameter. The disease symptoms exhibited by plants infected by virus are direct consequences of the physiological activities of these virus particles. Symptoms caused by different strains of the virus may differ sharply from one another. The information which determines the differing properties of different strains may be located unambiguously in the nucleic acid component of each particle. The experimental evidence is presented in Fig. 1.2. It will be observed that the nucleic acid component alone is sufficient to induce symptoms of virus disease; also that virus particles reconstituted from two different strains cause disease symptoms characteristic of the strain from which the nucleic acid is derived, and do not show symptoms of the strain which contributed the protein. Following replication within the host, the protein component of new virus particles is identical to that in the strain from which the nucleic acid was extracted. Evidently the nucleic acid also carries the information required for the assembly of its protein envelope.

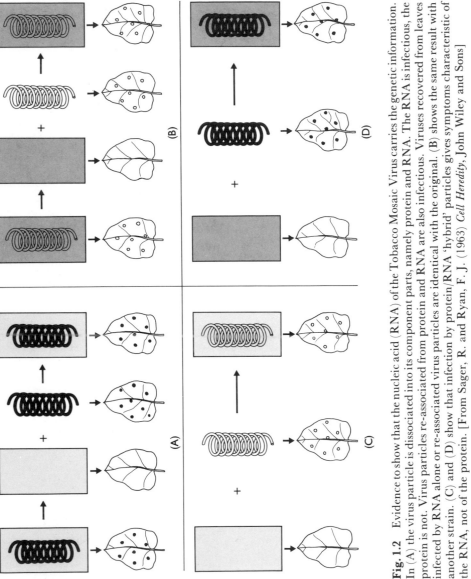

Fig. 1.2   Evidence to show that the nucleic acid (RNA) of the Tobacco Mosaic Virus carries the genetic information. In (A) the virus particle is dissociated into its component parts, namely protein and RNA. The RNA is infectious, the protein is not. Virus particles re-associated from protein and RNA are also infectious. Viruses recovered from leaves infected by RNA alone or re-associated virus particles are identical with the original. (B) shows the same result with another strain. (C) and (D) show that infection by protein/RNA 'hybrid' particles gives symptoms characteristic of the RNA, not of the protein. [From Sager, R. and Ryan, F. J. (1963) *Cell Heredity*. John Wiley and Sons]

In emphasizing that the genetic information resides in the nucleic acid within the particle this is not to say that the protein may not participate in its *expression*. The ultimate source of information is, however, unquestionably the nucleic acid, the viral RNA. The nucleic acid in TMV is not, of course, organized within a chromosome, unless we regard each individual as a chromosome in itself! Experiments in other organisms where the nucleic acid determinants are embodied within chromosomes serve equally convincingly to locate the information in the nucleic acid of these chromosomes.

Transformation

Griffith (1928) showed that when the debris of *Pneumococcus* cells killed by heat was added to a culture of living *Pneumococcus* of a different strain a small proportion of the latter underwent a permanent change such that they came to resemble in certain respects the strain from which the dead material was derived. For example, debris obtained from a strain in which the cells are highly virulent and coated with a polysaccharide capsule of immunological type III transforms cells of a non-virulent, unencapsulated strain to virulent cells encapsulated in precisely the same way as the transforming strain. The *transformation* is of a permanent, heritable nature as can be established by repeatedly infecting and re-infecting mice with the transformed strain. In 1944, Avery, McLeod and McCarthy showed that the active ingredient of transformation is the nucleic acid, in this case DNA, deoxyribonucleic acid. Purified DNA extracts, and no other single component, effects the transformation.

Bacteriophage infection

Viruses which infect and penetrate bacteria, such as the T4 bacteriophage of *Escherichia coli*, are, relative to other viruses, complex in construction. Even so they are simple in chemical composition, consisting of 40% DNA packed tightly in the head and 60% protein. Within the host the bacteriophage multiplies, utilizing the raw materials of the cell. Hershey and Chase (1952) showed that the information which enables the virus to multiply and develop within the host is vested in the DNA component of the head. They labelled the DNA of the phage with radioactive phosphorus (^{32}P) and the protein with radioactive sulphur (^{35}S). Progenies of these phages produced within the bacteria are heavily labelled with ^{32}P but have incorporated little or none of the ^{35}S. Complementary evidence showing that only the DNA of the phage enters the bacterial cells and that only the DNA is endowed with the genetic information comes from the studies of Hershey (1955). The heads of bacteriophages subjected to osmotic shocks burst to release the DNA leaving empty protein 'ghosts'. These are capable of attaching themselves to the surface of a bacterium but there is no penetration and replication, from which we infer that the DNA alone embodies the information controlling these events.

Table 1.1   Chromosome organization in prokaryotes*

Prokaryote	Nucleic acid	Chromosome configuration†	Chromosome length (μm)	Number of base pairs (or bases)
VIROIDS	1-RNA	L	< 0.1	< 200
BACTERIAL VIRUSES				
T2, T4	2-DNA	L	54.0	162 000
T5	2-DNA	L	39.0	117 000
λ	2-DNA	L	17.3	51 900
P22, P2	2-DNA	L	13.7	40 500
T7	2-DNA	L	12.5	37 500
φ6	2-RNA	L(M)	4.7	14 100
φX174	1-DNA	O	1.8	5 400
R17, F2, MS2	1-RNA	L	1.0	3 000
PLANT VIRUSES				
Wound tumour	2-RNA	L	7.0	21 000
Brome mosaic	1-RNA	L(M)	3.0	9 000
Potato virus-X	1-RNA	L	2.9	8 670
Cauliflower mosaic	2-DNA	O	2.7	8 100
Tobacco mosaic	1-RNA	L	2.1	6 300
Turnip yellow mosaic	1-RNA	L	2.0	6 000
ANIMAL VIRUSES				
Fowlpox virus	2-DNA	L?	100.0	300 000
Vaccinia	2-DNA	L	80.0	240 000
Herpes	2-DNA	L	53.0	159 000
Reovirus	2-RNA	L	11.3	34 000
Foot and mouth	1-RNA	L	3.4	10 300
Polio	1-RNA	L	2.6	7 800
Polyoma	2-DNA	O	1.5	4 500
BACTERIA				
Pseudomonas aeruginosa	2-DNA	O?	3480.0	10 440 000
Streptomyces coelicolor	2-DNA	O?	2600.0	7 800 000
Bacillus subtilis	2-DNA	O	1350.0	4 050 000
Escherichia coli	2-DNA	O	1333.0	4 000 000
Neisseria gonorrhoea	2-DNA	O	640.0	1 920 000
Acholeplasma laidawii	2-DNA	O	507.0	1 520 000
Haemophilus influenzae	2-DNA	O	505.0	1 515 000
Mycoplasma hominis	2-DNA	O	253.0	760 000

* Estimates of chromosome length and base pair number have been made using the following approximate mass-length equalities of DNA:
3000 base pairs ≡ 1 μm ≡ M.W. 2×10^6.

† L = Linear chromosome, O = Circular chromosome, (M) = Genome apparently divided into several pieces though this may be a consequence of breakage during preparation.

It is only in the smaller viruses that RNA acts as the repository of genetic information. In all other micro-organisms the role is fulfilled by DNA (Table 1.1). In many of the RNA viruses the information is embodied in a single RNA strand. A complementary strand functions only as a template during multiplication of the virus. The situation is similar in viruses which carry a single strand of DNA. Where, as is most often the case, the DNA takes the form of a double helix, only one of the two strands at any one segment (the plus strand) carries base sequences transcribed to messenger and, subsequently, to protein. One means by which this was established was to render the normally double-stranded DNA of the bacteriophage SP8 single stranded, by heating in the presence of messenger RNA (mRNA). At any one segment the mRNA forms duplexes with one only of the DNA strands. It is worth emphasizing that the same strand may be plus at one locus, minus at another.

Higher organisms
The chromosomes of higher organisms all contain DNA and it was reasonable in the light of evidence from bacteriophage and other micro-organisms to conclude that this DNA was the bearer of genetic information. In addition to DNA however the chromosomes contain protein, RNA and traces of other compounds. This complex chromosome composition and organization is characteristic of eukaryotes as distinct from the prokaryotes, in which the chromosomes are naked DNA helices (double in *E. coli*, single in bacteriophage ϕX174) or RNA helices (single in TMV, double in the animal virus, Reovirus). The view that information resides in the DNA fraction is supported by the fact that the DNA content per chromosome is constant whereas other components, the protein in particular, vary in amount (Table 1.2) and quality (Table 1.3). This observation does not, however, provide incontrovertible evidence that the chromosomal DNA carries the genetic information because one could argue that a fraction of the protein is of consistent quality and quantity. More recently, positive proof has been forthcoming in respect of many different kinds of information, i.e. of the information embodied in different genes within the chromosome. One kind of proof comes from labelling the RNA transcribed from the gene with a radioisotope, usually tritium ^{3}H, and hybridizing this RNA to the DNA *in situ* at the locus of the gene on the chromosomes. RNA/DNA hybridization is achieved under conditions which render the chromosomal DNA single stranded, as in the presence of strong alkali. In *Drosophila melanogaster*, for example, the 5S (S = Svedberg units) fraction of ribosomal RNA, labelled with tritium, hybridizes with the DNA at the 56F band in chromosome 2, the location of a cluster of genes coding for the 5S RNA fraction. There are also clusters of genes coding for 18S and 28S RNA at the nucleolus organizers in both X and Y chromosomes. In the wild type fly the 18S and 28S ribosomal cistrons are each reiterated about 130 times per haploid genome (Wimber and Steffensen, 1973). In

these instances the information along with the base sequences making up the genes which specify the information are repetitive. Other base sequences within the chromosomes of higher organisms, as we shall see later, are repeated to a far greater degree. Many sequences and the genes they represent are in contrast unique within the chromosome set or genome.

Another sensitive method for tracing the source of information within the chromosomal DNA, including that which emanates from non-repetitive, i.e. single-copy genes, has been employed by Bishop and Freeman (1973). The messenger RNA coding for haemoglobin in blood cells of the duck was extracted and from this a complementary DNA sequence, cDNA, prepared using the enzyme reverse transcriptase. The labelled cDNA was then mixed with single stranded unlabelled duck DNA. The cDNA annealed with the complementary sequences to produce double stranded segments. The results establish not only that the messenger RNA is transcribed by the DNA extract but, on the basis of the extent of hybridization, that the messenger is made up of sequences transcribed at three different loci. Since the haemoglobin includes two kinds of polypeptide chains, α and β, produced by the action of different genes, one of the loci must transcribe for α and the other two for β, or vice versa.

Table 1.2   Constancy of DNA and quantitative variation in other chromosomal components, from different tissues of the fowl. [Data of *Mirsky, A. E. and Osawa, S. (1961) *The Cell*, Vol. II, Academic Press Inc.; †Seligy, V. and Miyagi, M. (1969) *Expl. Cell Res.*, **58**, 27-34]

Component	Tissue		
	Red blood cells	Kidney	Liver
* DNA per nucleus (pg)	2.58	2.28	2.65
† Protein/DNA ratio in isolated chromatin	1.18	1.76	2.61
† RNA/DNA ratio in isolated chromatin	0.003	0.007	0.025

Table 1.3   Qualitative differences in chromosomal histone proteins during early development of Japanese newts. Lewin's nomenclature in brackets. [From Asao, T. (1969) *Expl. Cell Res.*, **58**, 243-52]

Stage of embryo	Histone fractions (ng per 10^4 nuclei)		
	FI (H1, H5)	FII (H2A, H2B)	FIII (H3, H4)
Early gastrula	0	0	58
Middle gastrula	28	47	50
Late gastrula	117	192	114
Neurula	218	409	114
Tail bud	256	526	119

1.2 Eukaryote chromosomes

The chromosomes of eukaryotes, as we have indicated earlier, are complex in structure and composition. While it is clear that the DNA fraction carries the genetic information, the way in which the DNA and other fractions are organized and arranged is by no means clear. It will be useful in the first instance therefore to consider the general morphology of the chromosomes and subsequently, to determine as far as possible how the morphology relates to the separate components and to their organization.

Morphology

Under the light microscope, details of the morphology of the chromosomes are in most species manifested only when the chromosomes are contracted during nuclear division. At interphase they are extended to such a degree that individual members of the complement are indistinguishable. At mitosis the contraction, through coiling and supercoiling, is at a maximum at metaphase. Treatment with colchicine or other spindle inhibitors, increases the contraction and also facilitates the spreading of the chromosomes. From such metaphases it is clear that the chromosomes vary enormously in number and in size among eukaryotic species but remain remarkably consistent within species. A longitudinal differentiation in the structure of individual chromosomes is equally obvious. The pattern of differentiation is important for purposes not only of identity but also of function. The 'markers' of identity and function are as follows.

(a) *Primary and secondary constrictions*

The primary constriction is called the centromere, a less densely coiled and thereby slender segment near which the chromatids are held together during metaphase. The centromere is essential for movement and *acentric* fragments, which lack a centromere, are therefore incapable of movement on the spindle because spindle fibres only attach to centromeres. The chromosomes of the vast majority of species contain one centromere. In a few species of plants, such as *Luzula campestris*, and

Fig. 1.3 Electron micrograph of one arm of a metaphase chromosome of *Fritillaria lanceolata* showing three heterochromatic (H) segments which are distinguishable by their reduced opacity to electrons. × 20 000. [Courtesy of L. F. LaCour]

animals (e.g. *Pseudococcus obscurus*) the centromeres are *diffuse* and the capacity for mobility on the spindle is spread throughout most of the chromosome. Chromosome fragments in these cases are capable of movement by attachment to spindle fibres.

Secondary constrictions mark the locations at which the nucleoli are assembled. Both primary and secondary constrictions are consistent in location. They reflect an underlying variation in the molecular organization of the chromosomal DNA. They testify also to an association between the coiling status of chromosome segments and their functional activity.

(b) *Heterochromatin*

At interphase in many species, a specific and consistent fraction of the chromosome material, the heterochromatin, is out of phase in being highly contracted when the bulk of the chromosomes, the *euchromatin*, is extended and uncoiled. These same segments are more densely coiled also at early prophase both of mitosis and meiosis. Following cold treatment in many species, the heterochromatin is also revealed at metaphase, being uncoiled and less densely stained than the euchromatin. In *Fritillaria lanceolata* even at normal temperatures the heterochromatin is sharply distinguishable from euchromatin under the electron microscope (Fig. 1.3) by virtue of its reduced opacity to electrons. However, in other species, e.g. *Scilla sibirica*, no such distinction is possible in metaphase chromosomes. The consistency in the distribution of heterochromatin among chromosomes, to which we have referred, is not invariable. Some chromosomes or segments of chromosomes are heterochromatic in some cell environments and not in others. The most familiar example of this kind of heterochromatin is found in the X-chromosomes of female mammals. One of the two X-chromosomes is heterochromatic in each cell. The heterochromatinization is a random event with respect to the two X-chromosomes such that the X from the maternal parent is heterochromatic in about half the cells, the X from the paternal parent in the other half. This class of heterochromatin is *facultative* as distinct from the *constitutive*.

We have referred above to an association between genetic activity and chromosome coiling. Nowhere is this more apparent than with the heterochromatinization of the mammalian X-chromosome (Lyon, 1963). The genes on the heterochromatic X are 'switched off'. Because one X is heterochromatic in half the cells and its homologue is heterochromatic in the other half, in a genic heterozygote one allele of any X-chromosome gene is active in half the cells, the other allele in the remainder. For example, in the female mouse heterozygous for the alleles $+$ and Mo which determine dark and light coat colour respectively, the mouse presents a mosaic of dark and pale patches (Fig. 1.4).

Constitutive heterochromatin may be distinguished from the facultative type by staining with fluorescent dyes such as quinacrine mustard and viewing under ultraviolet light. Constitutive heterochromatin

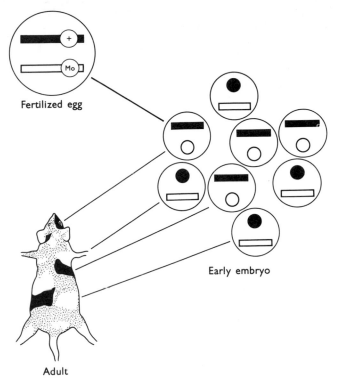

Fig. 1.4 X chromosome inactivation in the female mammal. The inactive X is shown as a circle and, the active one as a rod. [From Lyon, M. F. (1963) *Genet. Res.* **4**, 93–103]

fluoresces, the facultative heterochromatin does not. Staining with Giemsa dye, following treatment with barium hydroxide and subsequently SSC (a mixture of sodium chloride and sodium citrate), is another way of distinguishing the constitutive heterochromatin, which stains violet. Modifications of this method bring to light other segmental differences along the chromosomes: G-bands as distinct from the C-bands which mark the constitutive heterochromatin (Fig. 1.5). What the G-bands specify in terms of organization or function is not clear.

(c) *Puffs, bands and loops*

In the salivary glands of the diptera the chromosomes are multiplied side by side (up to 1000 times) such that the lengthwise variation in morphology becomes amplified to render unusually distinctly the elaborate differentiation of chromosome segments in respect of density of coiling and of staining. The heterochromatin in the salivary gland nuclei of *Drosophila* species, for example, shows up as lightly stained diffuse

(A)   **(B)**

(C)

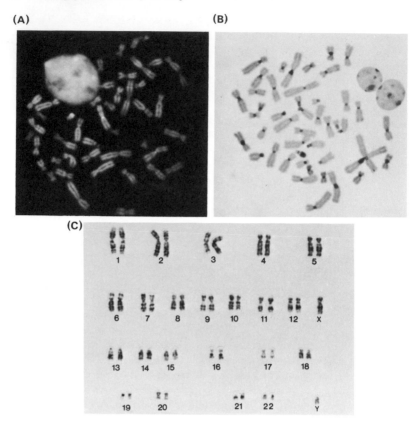

Fig. 1.5 Metaphase chromosomes of normal human males prepared from cultured blood lymphocytes: (A) Staining with quinacrine to show fluorescent Q-bands. (B) C-banding to demonstrate the location of constitutive hetero-chromatin. (C) Staining with acetic acid–saline–Giemsa to produce G-banding. [Courtesy of H. J. Evans]

matter in the vicinity of the centromeres. Densely stained bands of different size are interspersed with lightly stained interband segments. As we shall explain later the dark bands correspond with the sites of genes mapped by breeding tests. While the order of bands in these amplified (polytene) chromosomes is constant there is puffing at sites of transcription due to the RNA synthesized. Puffs develop in a regular sequence during larval development, partly at least in response to the increasing concentration of the hormone ecdysone. They mark directly the induction of transcription at specific loci (Ashburner, 1969).

Puffing in polytene nuclei is a reflection of the unravelling and spinning out of chromosome fibrils. A similar phenomenon is observable in the diplotene chromosomes of amphibian oocytes. These diplotene chromosomes are enormously long, ranging from 350 to 800 µm in

Triturus viridescens. As shown in Fig. 2.1 the varying but characteristic number of loops are spun out at specific sites (chromomeres) from the chromosomes along the axes of the bivalents. The pattern of loops which gives rise to the lampbrush appearance of the chromosomes is characteristic of the species. There are occasional mutants where the looping pattern is altered. These mutations are readily detectable in heterozygotes. More important, the fact that the structural difference with respect to loops is preserved in heterozygotes serves to emphasise that the looping pattern is a permanent structural feature of individual chromosomes. We shall consider lampbrush loops in more detail later in relation to both the molecular organization of the chromosomes and their genetic activity. In the meantime we note that they provide most detailed markers throughout the amphibian complement.

The differential coiling of segments in lampbrush and polytene nuclei finds a parallel in the chromomeres which give a beaded appearance to prophase chromosomes, especially at pachytene of meiosis. It is important to note, however, that while the chromomere pattern of prophase is specific and characteristic of a species it may vary substantially from one developmental stage to another. For example there is often a sharp reduction in the number of chromomeres at prophase of the second division of meiosis as compared with the first division, e.g. in *Agapanthus* (Lima-De-Faria, 1954). Apart from anything else the distinction serves to emphasize the changing character of the chromosome phenotype in relation to differentiation and development. It also warns against drawing conclusions from comparisons between chromomeric and banding patterns without taking account of variation due to differentiation.

Chromosome phenotype and genotype

Implicit in the discovery that the genes and other units of genetic information are located within the DNA of the chromosomes is the premise that the genotype of any individual is defined by the order of base sequences within the members of the chromosome complement. With rare exceptions (see Chapter 2), this order is preserved throughout the growth and development of each organism. As we have seen, however, the appearance and composition of chromosomes, leaving aside the composition of their DNA, is by no means constant during development. Neither is their behaviour. The chromosomes of salivary gland cells of *Drosophila* are very different from those of brain cell chromosomes. At mitosis the sequence of events and their consequences are very different from those at meiosis. Even in one tissue, such as the root meristem of seedlings in plants, the size and mass of chromosomes increase by as much as threefold from germination to three weeks (Bennett and Rees, 1969). Constancy, therefore, applies to chromosomal DNA only. In other respects the appearance and behaviour of chromosomes changes with changing cell environment. Chromosomes, in other words, have a well-defined phenotypic as well as genotypic aspect. From this we should expect that phenotypic changes in chromosome form and behaviour would be influenced by changes in both the external environment and by changes in genes which serve to control the activities of chromosomes. This, of course, is precisely what we find. There is ample evidence to show the dependence of all aspects of chromosome behaviour upon the genotype of the nucleus and, equally, upon the external environment of the organism. That the chromosome phenotype comes under the control of genes is clearly of significance from an evolutionary standpoint because it provides the means for adaptive changes affecting the transmission and distribution of genetic information (Chapter 3).

1.3 Prokaryote chromosomes

In the plant viruses and the smaller animal viruses the chromosomes comprise a single strand of RNA sometimes linear, sometimes circular (Table 1.1). In the larger animal viruses the chromosomes are of DNA, either single stranded or double stranded like the chromosomes of bacteria. As we might expect, the size of the nucleic acid molecule varies between the species but, without exception, the chromosomes consist of a naked nucleic acid thread lacking the histone and other proteins which comprise a substantial fraction of the chromosomes of eukaryotes.

One striking feature of prokaryotes is that chromosomes accept insertions of DNA segments incorporating genes derived from the same or from different species. The sex factor (F) in *E. coli* which promotes conjugation between bacteria is an example. This may exist as a particle in the cytoplasm which is independent of the chromosome or inserted to become part of the DNA helix of the normal chromosome. Such particles are called episomes. Their insertion may be facilitated by the 'naked' condition of the chromosomal DNA of prokaryotes. It may well be, on the other hand, that similar bodies are found in eukaryotes, but that they are more difficult to detect. Indeed Zhdanov (1975) provides good evidence to indicate the incorporation of the measles virus into the chromosomes of man. The measles virus is thought to interact with a latent oncornavirus, resulting in transcription of the measles virus RNA into double-stranded DNA which then integrates into the human genome.

2

Fine structure and organization

2.1 Architecture

Mononemy

Electron micrographs show that the chromosomes of eukaryotes are composed of fibres. These vary in diameter from one phase of the cell cycle to another and from one tissue to another. They consist mainly of protein and DNA. The structural organization of these fibres within the chromosome is interpretable as follows.

1 Each fibre contains one DNA helix in association with protein.
2 Variation in fibre diameter is dependent upon the density of coiling.
3 Each chromosome is made up of one fibre, and therefore one DNA helix, running continuously from one end of the chromosome to the other. Each chromosome in other words is *mononemic*, rather than *polynemic*; single-stranded rather than multi-stranded.

Compelling evidence for mononemy comes from measurements of lampbrush, diplotene chromosomes in the oocytes of the newt *Triturus viridescens* and from measurements of the DNA molecules extracted from *Drosophila* cells.

1 In *Triturus* each chromatid is represented by a single fibre (Fig. 2.1). When the protein is digested away by trypsin the diameter of the fibre is 2–3 nm. This would accommodate one, and only one, DNA helix (Miller, 1965).
2 Cells of wild-type *Drosophila* lysed with detergent and digested with pronase, yield chromosomal DNA molecules which are more or less intact, i.e. unfragmented. The mass of the largest DNA molecule in lysates corresponds closely with that of the DNA content of the largest chromosome of the complement. Furthermore, in a mutant stock where chromosome 3 is increased in length due to the attachment, by translocation, of a large piece of the *X*-chromosome, the largest DNA molecule increases in mass in direct proportion to the DNA increase in chromosome 3 (Table 2.1).

Packing and coiling

In the interphase nuclei of man the chromosome fibres have a dia-

16   *Fine structure and organization*

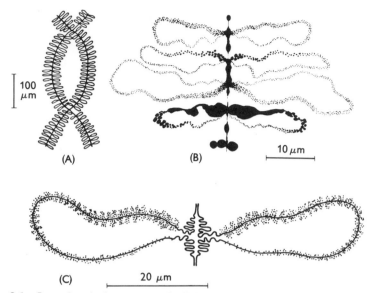

Fig. 2.1   Lampbrush chromosomes in *Triturus viridescens*. (A) A bivalent showing homologous chromosomes joined together by two chiasmata. (B) The central chromosome axis with paired lateral loops, showing differences in loop morphology and length. (C) The single stranded chromatids in the main chromosome axis are continuous with those of the lateral loops. [From Gall, J.G. (1956) *Brookhaven Symp. Biol.*, **8**, 17–32]

meter of approximately 23 nm and a dry mass of 6.1×10^{-16} g/μm. Thirty per cent or more of the mass consists of DNA. An extended DNA helix 1 μm in length weighs only 3.26×10^{-18} g. Clearly the DNA within the fibre must be tightly packed. The 'packing ratio' (extended double helix: packed double helix) has been estimated at 56:1 (DuPraw, 1970). The packing is probably achieved by compaction of the DNA within the fibre and the supercoiling of the fibre itself. Variation in the degree of compaction and of supercoiling of the fibre accounts for the variation in diameter. The indications are that coils and supercoils are maintained by histone proteins bound at specific sites to the DNA helix. The compactness of chromosomes at metaphase is achieved in two ways: (1) by

Table 2.1   Molecular weights and DNA contents of *Drosophila* chromosomes. [Data of Kavenoff, R., Klotz, L. C. and Zimm, B. H. (1973) *Cold Spring Harb. Symp. quant. Biol.*, **38**, 1–8]

Stock	M.W. of largest DNA molecule	DNA content of largest chromosome
Wild type	$41 \pm 3 \times 10^9$	43×10^9
Translocation	$58 \pm 6 \times 10^9$	59×10^9

contraction, roughly two-fold, in fibre length. This is clear from the increased DNA packing ratio of metaphase (100:1) in comparison with interphase fibres (56:1); (2) by folding and closer packing of the fibres themselves.

Recent work indicates that the compaction of the nucleoprotein (chromatin) fibre is, in the first instance, intermittent. Olins and Olins (1974) and Oudet, Gross-Bellard and Chambon (1975) have produced electron micrographs of fibres from interphase nuclei of the rat and the chicken which show clumps of chromatin separated by a fine thread. The clumps, called nucleosomes by Oudet *et al.* are 12.8 nm in diameter. The thread has a diameter of 1.5–2.5 nm. Each repeat in this string of beads contains about 200 DNA base pairs. The pattern, according to Kornberg and Thomas (1974) is determined by histone linkages with the DNA. Their evidence comes from mixing the tetramer of histone H3 and H4 along with the dimer of H2A–H2B (Lewin's nomenclature, 1975) with DNA *in vitro*. The X-ray diffraction pattern of the DNA/protein complex

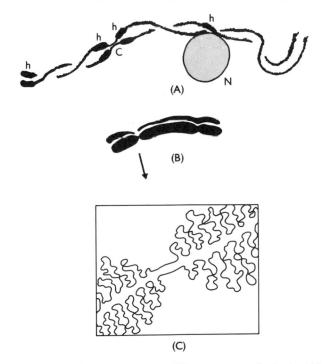

Fig. 2.2 Coiling and supercoiling at different stages of mitosis. (A) prophase showing localized densely coiled heterochromatin, h, at one end, on either side of the centromere, C, and in the region of the nucleolus organizer, N. (B) highly contracted chromosome at metaphase with chromatids unseparated in the vicinity of the centromere. (C) the fibrillar organization at the centromere, each chromatid comprising one highly convoluted and compacted fibre.

produced was identical with that of chromatin. The fifth histone, H1, and the non-histone proteins are not involved. This is not to say, however, that they play no part in compaction at a higher level, by coiling and supercoiling. The coiling and supercoiling properties of chromosomes at different stages of mitosis are illustrated in Fig. 2.2.

DNA and genes

In *E. coli* the chromosomal DNA is made up of about 4 000 000 base pairs. Approximately 1000 base pairs provide the information which specifies the protein or polypeptide chain synthesized under the control of each gene. On this basis the *E. coli* chromosome carries 4000 genes, assuming each gene to be represented by a single copy. This is almost certainly an overestimate because recent evidence suggests that the RNA messenger transcribed at each locus contains base pairs supplementary to those providing information about the amino acid sequence of the polypeptide product. Even so 4000 is an acceptable estimate and most would agree also with the generalization that the *E. coli* chromosome is made up of genes, side by side, each in single or very few copies. In eukaryotes the DNA content of chromosome complements is far greater than in *E. coli* and other prokaryotes (Table 2.2). The increase in DNA content is not surprising on grounds of increasing complexity and consequent demand for increase in the amount of information, in other words for a greater variety of genes and gene products. What is surprising, however, is the magnitude of the DNA variation between prokaryotes and eukaryotes and even within eukaryotes. For example the diploid nuclei of man, at early interphase, contain 6.4×10^{-12} g of DNA (equivalent to 5800 million base pairs as compared to 4 million in *E. coli*). If each allele is represented by a single copy the diploid complement in man is made up of almost 6 million genes, the haploid complement of half that number. Even allowing for the complexity and elaboration of eukaryote cells relative to prokaryotes the estimate is unreasonably high and human cells are by no means amongst those with the highest DNA contents (Table 2.2). The conclusion is reinforced by the wide variation in DNA amount between closely related species. *Vicia faba* has seven times as much DNA as its close relative *V. sativa*. Both are diploids and it is clearly unreasonable to assume the chromosomes of the former incorporate seven times as many kinds of genes as the latter. To account for the large amounts of DNA in eukaryote nuclei and for the magnitude of the DNA variation there are a number of possibilities:

1 A number of genes, i.e. of informative and transcribable DNA sequences, are highly repeated throughout the complement.
2 A number of sequences, either in single copies or in many copies are uninformative and not transcribable.
3 Some of the variation in DNA content reflects differences in the number and variety of genes in different species but, as we have

Table 2.2 Amounts of DNA in eukaryotes and prokaryotes (pg)

Eukaryotes

ANIMALS	DNA per 2C nucleus	$2n$
Amphiuma	168.0	24
Protopterus (Lungfish)	100.0	38
Salamandra salamandra	85.3	24
Triturus viridescens	72.0	22
Bombina bombina	28.2	24
Rana esculenta	16.8	26
Rana temporaria	10.9	26
Bos taurus (ox)	6.4	60
Man	6.4	46
Sheep	5.7	54
Mouse	5.0	40
Drosophila	0.2	8

FLOWERING PLANTS		
Fritillaria davisii	196.7	24
Lilium longiflorum	72.2	24
Tulipa gesneriana	51.5	24
Allium cepa	33.5	16
Vicia faba	28.0	12
Lathyrus latifolius	21.8	14
Ranunculus ficaria	19.2	16
Secale cereale	18.9	14
Zea mays	11.0	20
Crepis capillaris	4.2	6
Vicia sativa	4.0	12
Antirrhinum majus	3.6	16
Linum usitatissimum	1.4	30

FUNGI	DNA per haploid nucleus	n
Dictyostelium discoideum	0.384	7
Ustilago maydis	0.208	2
Aspergillus nidulans	0.048	8
Saccharomyces cerevisiae	0.026	15

Prokaryotes

BACTERIA	DNA per cell
Salmonella typhimurium	0.0143
Escherichia coli	0.0040
Diplococcus pneumoniae	0.0022
Mycoplasma hominis	0.0009

VIRUSES	
T2	0.000 220
λ	0.000 055
P4	0.000 016
ϕ X174	0.000 005

indicated, this is unlikely to account for more than a small fraction of the variation observed.

2.2 Repeated sequences

Callan and Lloyd (1960), on the basis of their observations of lampbrush chromosomes in *Triturus* oocytes, were among the first to propose that genes within chromosomes were represented by numerous copies. Their proposal is based on the premise that each chromomere in the lampbrush chromosome corresponds with the location of a gene. At each chromomere a loop is spun out and it is on this loop that the RNA messenger is transcribed and protein synthesized (see Chapter 1). The evidence for the latter comes from Gall and Callan (1962) who showed that tritiated uracil (in RNA) and tritiated phenylalanine (in protein) were incorporated by the chromomere loops and only by the loops. The average loop, however, is at least 150 000 nucleotides in length, many times longer than the 1000 nucleotides required for the synthesis of the average polypeptide. The conclusion was that each gene was represented by many copies in tandem.

A number of methods are now available to confirm the high degree of repetition within chromosomal DNA.

Cot curves

DNA extracted in its normal, 'native', double-stranded form, sheared into short fragments, e.g. by sonic disintegration, rendered single-stranded (denatured) by heating or by strong alkali can be renatured, i.e. restored to the double-stranded condition, under suitable conditions of temperature, pH and cation concentration. The rate at which the DNA renatures is dependent upon the genome size, i.e. the variety of differing base sequences and the extent to which any of the sequences are repeated. Only complementary single strands will pair and anneal and clearly the greater the repetition of sequences the greater the likelihood of collisions between complementary strands. Another factor which influences the rate of renaturing is obviously the concentration of the DNA: the greater the concentration the more frequent the collisions. For this reason the rate at which the DNA renatures is plotted as a percentage against concentration × time, expressed in moles of nucleotides × seconds per litre (*Cot*). The percentage DNA renatured at a particular *Cot* value may be established in a number of ways (Britten and Kohne, 1968). One is to pass the DNA in solution through a hydroxyapatite column which traps that fraction of the DNA which has renatured, i.e. the double-stranded DNA. Another is to measure the optical density of the DNA by spectrophotometry. Single-stranded DNA absorbs more ultraviolet than double-stranded DNA. With renaturing therefore the hyperchromicity decreases in proportion to the amount of double-stranded DNA. In

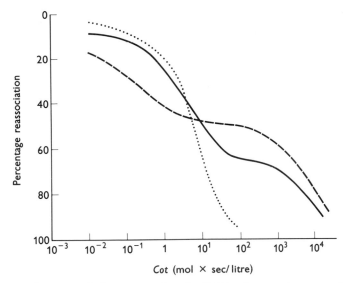

Fig. 2.3   *Cot* reassociation curves for the DNA of two eukaryotes, *Lathyrus sativus* (————) and the calf (– – – –), and of the prokaryote *E. coli* (. . . .).

Fig. 2.3 are the *Cot* curves for *E. coli*, the calf and the plant species *Lathyrus sativus*.

The distribution and dimensions of repeated sequences

Despite the fact that the variety of base sequences within the complement of eukaryotes is very much greater than in *E. coli* it will be observed in Fig. 2.2 that the percentage of the DNA of the eukaryotes (calf and *Lathyrus sativus*) renatured at low *Cot* (10^{-3} to 1) is in fact greater than for *E. coli*. This rapid renaturing at low *Cot* shows that some of the base sequences in the eukaryotes are highly repetitive. Between 1 and 30% of the DNA of the eukaryotes renatures at very low *Cot* values, up to 10^{-2}. This represents, on an arbitrary basis, the 'fast' reassociation component, the highly repetitive fraction of the DNA, sequences repeated 10 000 times or more (Southern, 1974). Beyond *Cot* 10^{-2} the moderately repeated sequences and the single copy or unique sequences are renatured. The units of repetition range in length from as few as six base pairs in the guinea pig (E. M. Southern, 1970) to several hundred.

Satellite DNA

When native DNA is centrifuged in a density gradient such as caesium chloride the buoyant density at equilibrium is dependent upon its G + C content. The higher the G + C content the higher the buoyant density. The DNA of eukaryotes has, on average, a G + C content of 30–50%. The G + C content, however, varies from one section of the DNA mole-

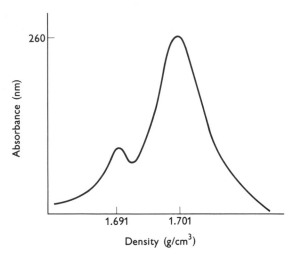

Fig. 2.4   The plot of UV light absorbance against density for mouse DNA centrifuged in a caesium chloride density gradient. The satellite DNA peak (density 1.691 g/cm^3) is on the light side of the main band DNA (density 1.701 g/cm^3). [From Evans H. J. and Sumner A. T. (1973) *Chromosomes Today*, **4**, 15–33]

cule to another by 10 % or more. For this reason the sedimentation co-efficient of the DNA from one species covers a range of buoyant densities, giving a broad band when plotted in a buoyant density curve. Exceptional DNA segments may have an unusually high or low G + C content. When plotted, these fractions appear as heavy or light satellites respectively at the tails of the 'main-band' DNA. Heavy satellites are found in the guinea pig and in human DNA. Light satellites, like the one shown in Fig. 2.4 for mouse DNA, are less common. One reported in the crab has a G + C content of only 5 %; others have been identified in *Drosophila* species.

Highly repetitive segments with distinctively high or low G + C contents, relative to the main band DNA, would clearly be detectable as heavy and light satellites respectively. It is worth emphasizing, however, that a highly repeated sequence is not necessarily detectable as a satellite. For example, a highly repeated sequence with the same base ratio as the bulk of the DNA would have the same buoyant density and 'settle' in the same density band.

The mouse satellite represented in Fig. 2.4 makes up 10 % of the total DNA. Its location within the mouse complement was established with considerable precision by annealing tritium-labelled single-stranded DNA or else radioactive RNA copied in vitro from mouse satellite DNA (complementary or cRNA) with chromosomal DNA in situ (K. W. Jones, 1970; Pardue and Gall, 1970). For this purpose chromosomal

DNA is rendered single stranded either by heating or by treatment with alkali before being brought into contact with the labelled DNA of the satellite or else labelled RNA complementary to it. Autoradiographs show that the satellite DNA and the cRNA bind to the heterochromatic segments around the centromeres and at the nucleolus organizers. Similar methods have established that the highly repetitive satellite DNA in *Drosophila* is also located mainly in the centromeric heterochromatin (Jones and Robertson, 1970).

Thomas' circles

An ingenious method developed and applied by Thomas and his colleagues shows that repetitive DNA segments are not necessarily contiguous within chromosomes. Fragments of DNA in double-stranded form are exposed to an exonuclease, e.g. exonuclease III, which digests single strands of the double helix in the 3' to 5' direction. When sufficient of the undigested single strands, at least 200 nucleotides in length, are exposed at each end, the ends with complementary base sequences will pair under conditions permitting renaturing to form stable rings (Fig. 2.5) readily distinguishable under the electron microscope (Thomas, 1970). Complementarity is, of course, evidence for repetition and the proportion of fragments forming rings is an index of the degree of repetition within the chromosomal DNA. Another factor which influences the capacity to form rings is the length of the fragment. In *Drosophila* the optimum length is 1.2 µm. With longer fragments ring formation is reduced, from which it is inferred that repetitive sequences

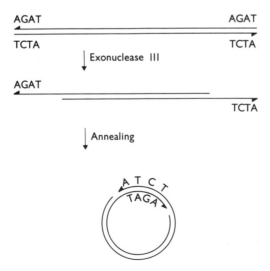

Fig. 2.5 The formation of Thomas' circles from DNA with tandemly repeating base sequences.

are in clusters, g regions, interspersed with unique sequence DNA. The average length of a g region is estimated to be 5 µm, i.e. about 22 500 nucleotides.

Twenty two per cent of the DNA in *Drosophila* is highly repetitive and located in the heterochromatic regions around the centromeres. For this reason ring formation is mainly restricted to the DNA in heterochromatin in this case (Hutton and Thomas, 1975; Peacock *et al.*, 1973; Schachat and Hogness, 1973). This is not to say there is no element of repetition elsewhere, interspersed within the euchromatic fraction of the complement.

Davidson *et al.* (1973) provide further detailed information on the distribution of repetitive sequences in animal DNAs. They incubated long strands of denatured DNA, labelled with tritium, with an excess of short, unlabelled single-stranded fragments at low Cot, i.e. in low concentration for a short period of time, such that only highly repeated sequences reassociated. The mixture was then passed over hydroxyapatite which binds to the duplexes made up of long labelled strands paired with short unlabelled fragments carrying complementary repeated sequences. Most of the bound, labelled strands included non-repeated as well as the repeated sequences with which the short fragments are paired. They concluded from this that repeated sequences throughout the genome are interspersed with unique sequences. In *Xenopus*, by using labelled strands of different length, they estimate that: (1) the majority of repetitive sequences are short, averaging 300 nucleotides; (2) each of the sequences repeated is separated by a unique DNA sequence 700–1000 nucleotides in length, some by unique sequence segments up to 4000 nucleotides; and (3) a small fraction of the total DNA, about 7 % is made up of repetitive sequences many thousands of nucleotides in length. Davidson *et al.* (1973) and Thomas *et al.* (1973) also, place particular emphasis on the alternating pattern of repeated and of unique sequences in the DNA of *Xenopus* and other animals. What this signifies in functional terms is not known.

The composition of repeated sequences

E. M. Southern (1970) and Gall, Cohen and Atherton (1973) have established in detail the dimensions and base composition of satellite DNAs in the guinea pig, and in *Drosophila*. In the guinea pig satellite the sequence is $\frac{CCCTAA}{GGGATT}$ repeated, with some variation due to mutations, millions of times. Satellites I, II and III in *Drosophila virilis* comprise thousands of repeats of the sequences $\frac{ACAACT}{TGTTGA}$, $\frac{ATAAACT}{TATTTGA}$ and $\frac{ACAAATT}{TGTTTAA}$ respectively. None of these sequences makes any sense from the standpoint of coding for protein. This is not to say, however, that repetition involves only nonsense sequences of DNA bases. One example of a transcribable base sequence repetition, to which we have already referred, is the sequence coding for ribosomal RNA. It will be recalled that in *Drosophila melanogaster* the wild type fly carries at the *bobbed* locus about 130 copies of the gene sequences which code for the 18S and 28S components of the ribosomal RNA (Ritossa and Spiegelman, 1965).

It has also been suggested that the bands in the *Drosophila* chromosomes, which correspond to the sites of single genes (Judd and Young, 1973), may represent many copies of each gene. In DNA terms each band on a chromosome includes some 25 000–50 000 nucleotide pairs, 'sufficient' for 20 or more copies of a gene. The evidence of Peacock *et al.* (1973) is against this view on the grounds that most of the repetitive material, they claim, is concentrated in the centromeric heterochromatin, rather than in the bands. One may speculate that the DNA extra to the transcribed sequences within the bands may fulfil regulatory or other functions. The truth is that one does not know.

Chromosomes and information

Until fairly recently the chromosomal DNA was considered, in the main at least, to be made up of gene sequences in tandem much like the beads on a string. For prokaryotes the concept is still largely valid. The picture which now emerges for eukaryotes is, as we have indicated, much more complex. Much of the DNA comprises a variety of untranscribable sequences, some of which are repeated many thousands of times. Many transcribable gene sequences are repeated, others are represented by single unique DNA sequences. Some of the unique sequence DNA is untranscribed, like the 'spacer' sequence between genes coding for ribosomal RNA in *Xenopus laevis* (Loening, Jones and Birnstiel, 1969). It needs to be emphasized that from a genetical as well as structural stand-point, the chromosome in heredity represents very much more than a row of genes coding for polypeptides.

2.3 Replication and division

Mitosis in eukaryotes

In a tissue of multiplying cells a proportion of nuclei are in interphase, the remainder in division. The ratio of one to the other, as we shall see, is of use in timing the sequence of events during the mitotic cycle. From the genetic standpoint the characteristic and significant consequence of mitosis (Fig. 2.6) is that the products of division are identical, embodying precisely the same genetic material as the parent nucleus. The justifica-tion for this statement comes from many sources. On cytological grounds we observe that the chromosomes in the two nuclei produced by division have the same appearance and the same DNA content as the parent nucleus. On genetic grounds we observe that the offspring of asexual reproduction, whether by budding or clonal propagation, have similar genotypes. A further compelling proof is that when the nucleus from a specialized epithelial cell of the intestine of a *Xenopus* tadpole is trans-planted into an unfertilized egg whose nucleus has been removed or inactivated, the egg can develop to produce a normal adult. We con-

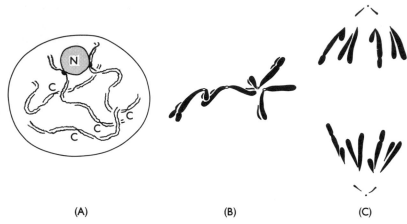

(A)   (B)   (C)

Fig. 2.6 Mitosis in an organism with two pairs of chromosomes. (A) Prophase: each chromosome comprising two chromatids. C = centromere. N = nucleolus, attached to the nucleolus organizer regions. (B) Metaphase with centromeres aligned along the equator of the spindle. (C) Anaphase showing equal separation of daughter chromosomes to opposite poles of the cell.

clude that the information embodied within the chromosomes is preserved unchanged during the many cycles of mitotic division leading up to the formation of the larval intestine (Gurdon, 1973). There are, as we shall consider below, certain exceptions to this general rule.

The mitotic cycle

Nuclei. Replication of chromosome material, including the synthesis of DNA, takes place during interphase. The onset and duration of the synthesis and of other phases of the cycle are readily established by autoradiography. The method for root meristems for example, is as follows.

1 Roots are immersed briefly in a solution containing tritiated thymidine and, after washing, returned to water or culture solution. The labelled thymidine is incorporated by the DNA synthesized in interphase nuclei.

2 Root tips are fixed at intervals of one hour over a period of 24 hours or more, then stained and squashed.

3 Coverslips are removed, the squashed cells on the slide covered with a photographic emulsion and the slides kept in darkness. After some ten days or so the emulsion is developed and fixed, after which the location of labelled DNA among nuclei and among chromosomes is established by the distribution of silver grains in the autoradiograph.

Timing the cycle and its components. The proportion of nuclei labelled at a well-defined stage, e.g. prophase, is plotted against time (Fig. 2.7). From this graph the following information is established.

The total cycle. The interval between the two peaks represents the average time for the complete mitotic cycle. The first peak in the graph represents the first batch of labelled nuclei to reach prophase, the second peak the nuclei entering prophase for the second time after labelling. The most efficient estimate of the interval between peaks is obtained from the difference between points half way up the ascending curve of the peaks (*A* and *A'* in Fig. 2.7).

The synthesis (S) phase. This is estimated from the interval, *A–S*, the midpoints on the ascending and descending curves. Another way of estimating *S* is from the proportions of nuclei labelled (say one hour after labelling) × the duration of the total cycle. The basis for this estimate is that the period spent at a particular phase of the total cycle is directly reflected by the proportion of nuclei at that phase. Thus, we expect to observe more nuclei at a particular phase if that phase is of long duration, and vice versa. The estimate holds good if the cells are unsynchronized, proceeding through mitosis independently of one another in random order. This holds true for most somatic tissues in adult organisms, including root meristems.

The post-synthesis phase (G2). This is given by the interval from the time of immersion in tritiated thymidine to *A*, the average time at which cells enter prophase.

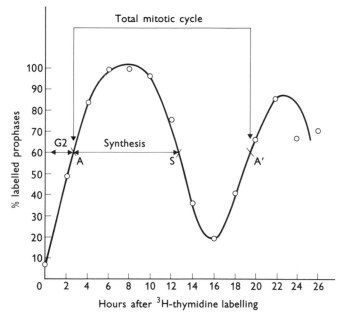

Fig. 2.7 Timing the mitotic cycle in the plant species *Nigella damascena*. [Data from Evans, G. M., Rees, H., Snell, C. L. and Sun, S. (1972) *Chromosomes Today*, **3**, 24–31, with permission]

The division phase (*D*). The proportion of cells in division, prophase to telophase, reflects directly the proportion of the total cycle time spent in division. The duration of component phases of the division may be estimated in the same way.

The pre-synthesis phase (*G*1). *G*1 is estimated by the difference, total cycle time minus the duration in division, *G*2 and *S*.

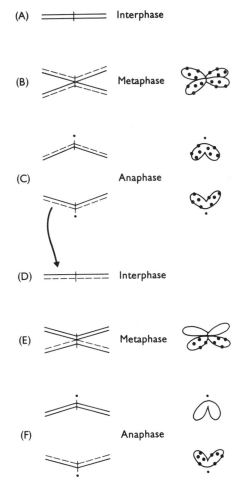

Fig. 2.8 Taylor's experiment. Following replication of the double helix (at A) in the presence of tritiated thymidine one of the DNA strands in each of the chromatids at metaphase (B) and anaphase (C) is labelled. As a result both chromatids produce silver grains in autoradiographs (right). After replication of the daughter chromosome (D) in the absence of tritiated thymidine only one of the chromatids contains a labelled DNA strand at the following metaphase (E) and anaphase (F).

The procedure for timing the mitotic cycle and its component phases is illustrated with respect to the root tip meristems of *Nigella damascena* in Fig. 2.7. The timing of the mitotic cycle is influenced by environmental factors, especially temperature, so that for purposes of comparison it is clearly important to standardize conditions of culture and experimentation. We shall be considering in more detail later some of the genetic factors which influence the duration of the mitotic cycle and its components.

Chromosome duplication

An experiment by Taylor, Woods and Hughes (1957) marked a very considerable advance in our understanding of the replication of chromosomes. After labelling the DNA during interphase with tritiated thymidine, both the chromatids of each chromosome at the subsequent metaphase carry the label. At the second metaphase following label one of the two chromatids is labelled (Fig. 2.8). The results are precisely those expected on the view that each unreplicated chromosome (chromatid) is mononemic, i.e. containing one DNA double helix which, of course, replicates semi-conservatively.

In prokaryotes such as *E. coli* labelling with tritiated thymidine also showed a semi-conservative replication of the chromosomal DNA. These experiments also demonstrated that replication proceeds bi-directionally from a single fixed point on the chromosome. In eukaryotes replication is also bidirectional but proceeding from a number of initiation points within each chromosome. In somatic cells of *Xenopus* the initiation points are spaced at intervals of 20–125 μm, in *Triturus* 125 μm or more (Callan, 1972). The duration of the DNA synthesis phase varies however between tissues. The reason is a change in the spacing between initiation points. In *Triturus* for example the S phase in spermatocytes, relative to differentiated somatic cells, is protracted and the number of initiation points reduced by wider spacing along the DNA helix. The *rate* at which replication extends from the initiation points remains constant. The same correlation between the duration of S in different tissues and of the intervals between initiation points applies in *Drosophila* (Blumenthal, Kriegstein and Hogness, 1973). Change in the rate of DNA synthesis between initiation points could also affect the duration of S between cells of different tissues but we know of no evidence of control by this means.

DNA synthesis and replication in euchromatin and heterochromatin

The synthesis of DNA is not synchronous throughout the chromosome complement. The most marked difference in the timing of DNA synthesis is that between euchromatin and heterochromatin. This was first established by Lima-De-Faria (1959). After treatment with tritiated thymidine the heterochromatin was heavily labelled late in the synthesis (S) phase when the euchromatin carried little or no label, showing that the synthesis continued in heterochromatin after cessation of DNA synthesis in the euchromatin.

The distinction between heterochromatin and euchromatin may extend to the replication and transmission of chromosome material during cell division. Normally during mitosis the products of division are genetically identical, with each other and with the parent nucleus. There are startling exceptions. In *Sciara coprophila* whole chromosomes, the L-chromosomes, are eliminated from somatic cells at about the fifth cleavage division. Three or four divisions later one of the three X-chromosomes in the female, and two of the three in the male are also eliminated from all but germ line cells. Both the L- and X-chromosomes are largely heterochromatic. Another class of 'dispensable' chromosomes, the E-chromosomes of dipterans, is shed from somatic nuclei. Like the Ls and Xs they are also heterochromatic (White, 1973; Lewis and John, 1963).

Heterochromatic segments, making up some 60 % of the chromosomes, are shed from somatic nuclei during cleavage in *Parascaris* (see White, 1973). In the polytene nuclei of the salivary glands of *Drosophila* the centromeric (α-)heterochromatin, which makes up about 30 % of most somatic nuclei, is relatively much reduced. While the euchromatin replicates normally during the differentiation of the salivary glands the α-heterochromatin fails to reproduce itself. In the oocytes of amphibia, e.g. *Xenopus laevis* (Gall, 1968), a specific chromosome segment, the nucleolus organizer coding for ribosomal RNA, is amplified relative to other segments, the amplified products in this case being detached from the chromosomes and migrating to the nuclear envelope where they eventually lead to the formation of multiple nucleoli.

The products of mitosis

These exceptional cases of non-conformity in the replication of chromosome material within nuclei are significant for a number of reasons. First, they bear testimony to the dispensability of some of the genetic material, albeit in certain tissues and organs. Second, they establish the capacity for modification of the mitotic process to generate gain or loss of specific chromosome segments. In the examples we have given the loss or gain of segments is certainly specific in distribution within the nucleus and equally specific in timing during growth and development; in short the events are well ordered and controlled. Third, the possibility is that less obvious loss or gain, involving a very small fraction of the chromosome material, may well be a not uncommon phenomenon during development. Indeed Pearson, Timmis and Ingle (1974) have confirmed a widespread and substantial variation in the satellite fraction of the DNA located in the heterochromatin of Melon chromosomes in different tissues. Exactly what this signifies in functional terms awaits a better understanding of the role and properties of satellite and other dispensable DNA fractions. While it is right therefore to emphasise the uniformity of the products of mitosis in general it is equally important to emphasise the capacity for qualitative modifications of the chromosome complement by amplification or deletion especially within differentiated, as distinct from meristematic tissues.

Variation in the quantity of genetic material, by polyploidy or endopolyploidy is, of course, common in differentiated tissues of both plants and animals which provide, in themselves, further examples of adaptive modifications in the mitotic process. Genetic quality and constancy of the products are, therefore, a characteristic but not invariable consequence of mitosis.

3

Meiosis and recombination

3.1 The meiotic cycle

At mitosis in eukaryotes the synthesis of DNA and of the chromosome constituents is succeeded by the separation of chromatids at anaphase during the cell division which follows. The two daughter cells produced contain, as a result, identical sets of chromosomes which, in turn, are identical with those of the parent cell. At meiosis, in contrast, one round of DNA replication and chromosome duplication is followed by two successive nuclear divisions. At anaphase of the first of these divisions the homologous chromosomes are separated to opposite poles. At anaphase of the second division the chromatids are separated. This results in four cells, each with a reduced haploid complement which, in heterozygotes, differ from one another due to segregation and recombination. Fusion of the haploid male and female gametes at fertilization is, of course, the mechanism which restores the complement to the diploid state in the zygote.

DNA synthesis takes much longer at meiosis than at mitosis in most dividing tissues. In *Lilium* pollen mother cells DNA synthesis at meiosis takes about three times as long as in root meristem cells undergoing mitosis. DNA synthesis at meiosis in the testis of mouse and of *Triturus* is also much protracted relative to the duration of DNA synthesis in tissues undergoing mitosis. As we have mentioned earlier the longer duration of DNA synthesis at meiosis is associated with a reduction in the number of initiation points of synthesis within the DNA molecules.

Apart from a difference in the duration of DNA synthesis there is, at meiosis, a fraction of the DNA (about 0.3% in *Lilium*) that remains unreplicated at interphase. It is replicated during early prophase, at zygotene. This fraction is specific and distinguishable from the bulk of the nuclear DNA in having a higher $G + C$ content: 50% as compared with 40% in the total DNA.

Following the S phase, $G2$ at meiosis is very short or even nonexistent. The meiotic cycle embracing both divisions, also takes much longer than the mitotic cycle (Table 3.1). In man, for example, the average duration of the mitotic cycle in cultured Hela cells is about 24 hours. Meiosis in the spermatocytes takes about 24 days. In the female, meiotic divisions commence before birth. They are arrested at early diplotene of the first division until the eggs are shed such that the completion of meiosis, to give egg cells and polar bodies, awaits puberty at the earliest and may take 30 years or more. In flowering plants, as well, the meiotic cycles are generally of much greater duration than the mitotic cycles, from 1 to 11 days. As Bennett (1971) has shown the duration is closely correlated with the amount of nuclear DNA, as is the case for mitotic cycles.

While the facts relating to the differences in duration of mitotic and meiotic cycles are, in themselves, indisputable it is fair to say that the significance of such differences remains conjectural. There is a great deal of ignorance concerning the mechanisms and processes that make up the meiotic cycle.

Table 3.1 2C DNA amounts and the mitotic and meiotic cycle times in plant species. [Cycle time data from Bennet, M. D. (1971) *Proc. R. Soc. B.*, **178**, 277–99]

Species	2n	DNA (pg)	Cycle times (hours)	
			Mitosis	Meiosis
Haplopappus gracilis	4	3.4	11.9	24.0
Secale cereale	14	18.9	12.75	51.2
Allium cepa	16	33.5	17.40	96.0
Tradescantia paludosa	12	36.6	20.0	126.0
Lilium longiflorum	24	72.2	24.0	192.0
Trillium erectum	10	74.4	29.0	274.0

3.2 The meiotic divisions

The general sequence of events during division is similar throughout most diploid eukaryote species even though, as we have mentioned, the timing and duration may vary from one species to another. The following is a brief summary of the events at division. They are also presented in Fig. 3.1. We shall deal in greater detail with certain of these events in a later section.

The first division

Division commences directly or very soon after the completion of DNA synthesis. The precocity of the onset of prophase, as compared with mitosis, may explain why the chromosomes although duplicated in respect of their DNA content at the *leptotene* stage, still appear as single threads, i.e. without evidence of replication into chromatids. At *zygotene* the intimate pairing of homologous chromosomes begins. Darlington (1937) attributes the capacity for pairing to the single-stranded organization of the chromosomes at early prophase of meiosis, which in turn is attributable to the precocity to which we have referred. Pairing can be

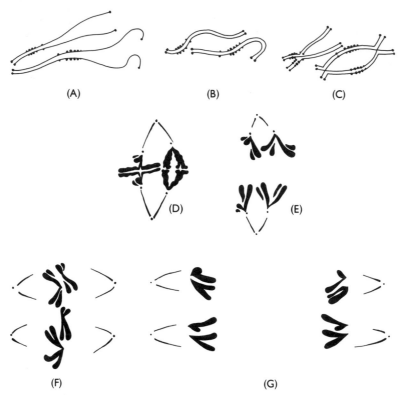

Fig. 3.1   Meiosis. Commencement of pairing between homologous chromosomes at zytogene (A); completed at pachytene (B). Chiasmata in diplotene bivalents (C) and at first metaphase (D). Separation of half bivalents at first anaphase (E), their orientation at second metaphase (F), and separation at second anaphase (G).

seen to commence at well-defined sites in many species, e.g. at the centromeres in *Fritillaria meleagris*, at the ends of the chromosomes in rye. It is probable that a similar localized initiation applies to all species. Pairing is specific in respect not only of homologous chromosomes but to homologous segments within the chromosomes. It is completed at *pachytene*. During pachytene the chiasmata are formed, marking sites of crossovers. The homologous chromosomes are held together by the chiasmata so that the complement is deployed as a set of bivalents corresponding to the haploid number. A striking feature of pachytene chromosomes is the longitudinal differentiation into a series of chromomeres. They represent specific and diagnostic sites of aggregation of material due to intensive localized coiling. At *diplotene* the attraction between homologous chromosomes lapses, indeed there is evidence of a repulsion between them. They are held together, however, by chiasmata. Both chromosomes of each bivalent are clearly resolved as double-stranded in organization, i.e. comprising separate chromatids. The shape of each bivalent at diplotene and in subsequent stages, including metaphase, is determined by the number and distribution of its chiasmata. Progressive coiling and contraction, which is continuous throughout prophase, leads to diakinesis which marks the disappearance of the nuclear membrane and the detachment and dispersion of the nucleoli. The spindle forms and at *first metaphase* (MI) the bivalents become orientated on the equator with the centromeres of homologous chromosomes directed to opposite poles. At first anaphase (A1) the half bivalents are separated and there follows some uncoiling of chromosomes at *telophase* which continues in the interphase nuclei. At the interphase there is no DNA synthesis or chromosome replication. The content of each of the pair of nuclei remains as half bivalents, equivalent to the haploid number.

The second division

The *second prophase* proceeds, much as in mitosis, with increasing contraction to the *second metaphase* (MII), with the exception, in most cases, that the chromatids lie well apart and not in close proximity as at mitosis. At *second anaphase* the centromeres of each half bivalent move to the poles so that telophase and, subsequently, interphase nuclei contain a haploid set of single-stranded chromosomes.

There are, of course, variations on this general theme (see John and Lewis, 1965). In plants a common phenomenon is the omission of cell wall formation after the first division, e.g. in *Paeonia*.

Univalents

Where, for some reason, there is a failure of chiasma formation the unpaired univalents fail to move on to the spindle equator or do so much later than the bivalents. In the former case they frequently form micronuclei and are 'lost' at the first division. In the latter they may divide at first anaphase to be included as single-stranded, chromatids in telophase nuclei. At second anaphase, however, they are incapable of movement from the equator and are often 'lost' as micronuclei at this second division. The chiasmata are, therefore, essential not only in the context of crossing over and recombination but also for the regular distribution of chromosomes to the gametic nuclei (Fig. 3.2).

An alternative, either at the first or second anaphase, is that the univalent undergoes 'misdivision', i.e. splits transversely within the centromere. The result will be two chromosomes with terminal centromeres, telocentrics. In many organisms telocentrics are unstable in the sense that they are converted to iso-chromosomes following fusion of the centromeres from identical arms (Fig. 3.3).

3.3 Chromosome pairing

While observations under the light microscope confirm the fact and the

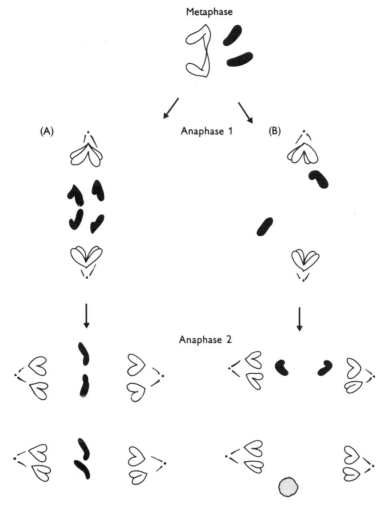

Fig. 3.2   The fate of univalents (solid black). They may divide at first anaphase (A) and be left undivided on the spindle equator at second anaphase. Alternatively (B) a univalent may be stranded on the equator at first anaphase or move to the pole and be included in the daughter nucleus. At the second division the stranded univalent is isolated in a micronucleus (shaded), the other divides normally, the daughter chromatids being included in the haploid nuclei. Another possibility is misdivision of the univalent (see Fig. 3.3).

specificity of pairing between homologous chromosomes, little is known of the forces or structures involved. More details of the structure of paired chromosomes at pachytene are revealed by the electron microscope (Moses, 1956; Westergaard and von Wettstein, 1972). Each chromosome pair is associated with a *synaptinemal complex* with the following

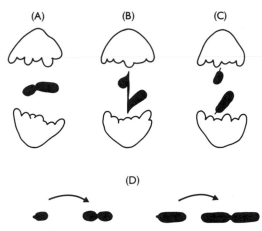

Fig. 3.3 The misdivision of univalents. The univalent (solid black) on the spindle equator at the first meiotic division (A) may undergo transverse splitting at the centromere (B) to produce two telocentric chromosomes (C). Telocentrics (D) are usually unstable and may be converted to isochromosomes.

features (see Fig. 3.4). Two dense *lateral elements* in parallel are separated by a central space which includes a central element. The lateral elements consist of protein from each of the pair of homologous chromosomes. The DNA containing chromosome fibres extend outwards in loops from these elements. The central element, like the laterals, again consists of protein emanating as loops directed inwards from the lateral elements and compressed together in the centre. The synaptinemal complex is attached at each end to the nuclear membrane by the lateral elements. The thickness of each lateral element ranges from 30–65 nm, the central space from 65–120 nm, and the central element, from 12–50 nm. The assembly of the synaptinemal complex is completed only when homologous chromosome segments are closely paired at zygotene and pachytene. During diplotene and diakinesis the synaptinemal complex disintegrates, losing first the central element and fibre loops and subsequently the lateral elements. According to Westergaard and von Wettstein (1972) the most persistent segments along the complex correspond to those at which chiasmata have formed. The implication is that the synaptinemal complex is involved in chromosome pairing and, perhaps directly, in chiasma formation. To support this view, no synaptinemal complexes are found at meiosis in male *Drosophila* where there is no crossing over or chiasma formation. The same is true of *Drosophila* females homozygous for the mutant gene $c(3)G$ which suppresses crossing over. In some species carrying X and Y sex chromosomes the synaptinemal complex forms only along the paired, homologous segments of the chromosomes, at which chiasmata are formed, but not along the unpaired, differential segments, where only lateral elements develop. More difficult to accommodate are observations in haploids of maize, tomato and barley

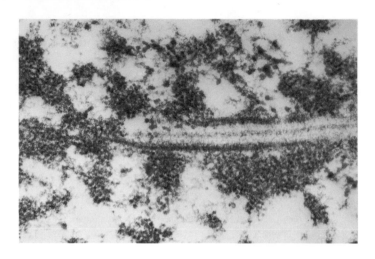

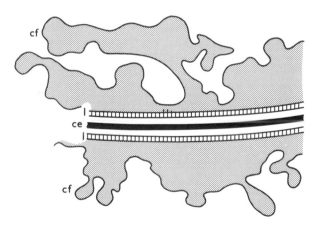

Fig. 3.4   A small section of a synaptinemal complex from a pachytene nucleus of *Lilium tigrinum*, × 45 000. ce, central element; l, lateral element; cf, chromosome fibres. [Courtesy L. F. La Cour]

which form synaptinemal complexes but where there is no chiasma formation and, apart from some degree of duplication, little homology between chromosomes. While in general therefore there are grounds for synaptinemal complexes being implicated in pairing and chiasma formation there are no clear indications of the mechanisms by which these events are accomplished.

3.4 Chiasmata

At diplotene the structure of chiasmata is readily resolved under the light microscope. The spermatocytes of grasshoppers are exceptionally favourable for this purpose. We observe that two of the four chromatids are involved in the formation of each chiasma. We infer that the chiasma results from breakage of the two chromatids followed by the rejoining of the broken ends such that there is crossing over and recombination (Fig. 3.5).

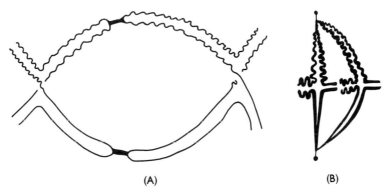

(A)   (B)

Fig. 3.5 Breakage of corresponding loci (A) followed by the re-union of broken ends (crossing over) in each arm to produce a ring bivalent (B) at first metaphase. For convenience the homologous chromosomes are distinguished by wavy and solid lines.

Co-incidence of chiasmata and crossing over

The co-incidence of chiasmata and crossing over is established from three kinds of evidence.

(a) *Cytological*

In *Lilium formosanum* Brown and Zohary (1955) found that one of the chromosomes was much shorter than its homologous partner. About two-thirds of one arm had been lost by deletion. As a result the bivalent formed is asymmetrical (Fig. 3.6). Proof of crossing over is as follows. When no chiasma forms between the unequal arms the inequality between homologous chromosomes is maintained at first anaphase. The separation is *reductional*. In contrast a chiasma in this arm leads to *equational* separation, with one arm of each of the homologous chromosomes (half-bivalents) comprising a long and short chromatid as a result of crossing over. The correspondence between the frequency with which a chiasma formed in the unequal arm and the frequency at anaphase of equational separation was extremely close, 70 and 71% respectively. Further confirmation of the correspondence of chiasma formation and of

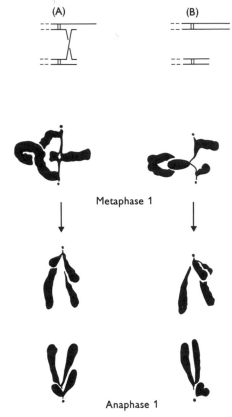

Fig. 3.6 Bivalents with one chiasma (A) and with no chiasma (B) between unequal arms of homologous chromosomes in *Lilium formosanum*. With one chiasma separation at first anaphase is equational, with no chiasma it is reductional. [After Brown, S. W. and Zohary, D. (1955) *Genetics* **40**, 850–73, with permission]

equational separation came from subjecting plants to conditions where the chiasma frequency in unequal arms was reduced to 51%. The frequency of equational separation in this case dropped, as expected, to a corresponding degree, to 50%.

(b) *Autoradiography*

In the grasshopper species *Stethophyma grossum* each bivalent has a single chiasma near to the centromere which is located terminally. By injecting tritiated thymidine into the abdomen of young males during DNA synthesis of the spermatogonial (mitotic) S-phase preceding meiosis, each chromosome at the commencement of meiosis has one labelled and one unlabelled chromatid (Fig. 3.7). At first anaphase of meiosis the

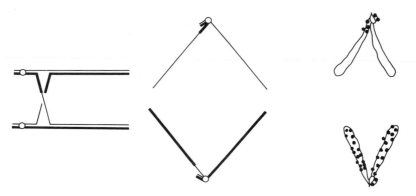

Fig. 3.7 Chiasma formation near the centromere between homologous chromosomes of *Stethophyma grossum*. One chromatid of each chromosome is labelled with tritiated thymidine (thick line), the other (thin line) is unlabelled. At first anaphase of meiosis there is a switch of label within chromatids corresponding in position to the cross-overs. [After Jones, G. H. (1971) *Chromosoma*, **34**, 367–82]

positions of exchanges between labelled and unlabelled chromatids correspond with those at which chiasmata are formed, i.e. near to the centromeres (Jones, 1971).

(c) *Linkage*

On the face of it the most convincing source of evidence that chiasmata are the results of crossing over would be to measure the amount of crossing over between marker genes within an identifiable chromosome segment and to compare this estimate of linkage with the chiasma frequency within that segment. Strange to say there is no such evidence to hand. In a more general way, however, it has been established (Table 3.2) that there is a reasonably good correlation between the chiasma frequency per bivalent at first metaphase in pollen mother cells of maize and the amount of crossing over as estimated by linkage (Whitehouse, 1969).

Chiasmata, crossing over and recombination

Crossing over, as we have indicated, is assumed to take place at the site of chiasma formation. At each chiasma two of the four chromatids will be cross-overs, the others unchanged so that two of the four gametes resulting will incorporate cross-overs, the remaining two non-cross-overs at that site. The relation between chiasma frequency and cross-over frequency is, therefore, 2:1. With suitable gene markers on each side of the chiasma and near to it, all the cross-overs will be detectable as recombinants. If, however, the markers are so far apart as to allow for more than one chiasma between them the cross-overs will, on occasion, be undetectable as recombinants. While the relation therefore between

Table 3.2 Comparisons of chiasma and cross-over frequencies in *Zea Mays*. The chiasma frequency per chromosome is derived from the mean chiasma frequency per cell, assuming that the chiasma frequency is proportional to chromosome length. The cross-over frequency is based on linkage data for the gene markers available in each chromosome. These may not embrace the entire length of each chromosome. [Data from Whitehouse, H. L. K. (1969) *Towards an Understanding of the Mechanism of Heredity*, 2nd ed., Edward Arnold, London]

Chromosome	Length in μm at mid-pachytene	Estimated mean chiasma frequency	Mean cross-over frequency
1	82	4.0	3.1
2	67	3.3	2.6
3	62	3.0	2.4
4	59	2.9	2.2
5	60	2.9	1.4
6	49	2.4	1.3
7	47	2.3	1.9
8	47	2.3	0.6
9	43	2.1	1.4
10	37	1.8	1.1
Totals	553	27.0	18.0

chiasma frequency and crossing over frequency is a constant at 2:1, the recombination value, as an index of crossing over, becomes less reliable with increasing distance between markers. However frequent the chiasma formation and crossing over within a bivalent the maximum recombination between markers will be 50%. In contrast, the frequency of chiasmata and of crossing over between markers may be, and often is, much higher. The situation with respect to two chiasmata between widely spaced markers is illustrated in Fig. 3.8.

The time of chiasma formation

In eukaryotes the evidence points firmly to the formation of chiasmata at early prophase of meiosis. That chiasmata are formed after the synthesis of DNA at the premeiotic interphase is clear from the work of Rossen and Westergaard (1966). In the Ascomycete *Neottiella rutilans*, DNA synthesis takes place in haploid nuclei before they fuse to produce the diploid nucleus which undergoes meiosis during which the chiasmata are formed.

Figure 3.9 shows the effects of heat treatment upon chiasma frequencies in the spermatocytes of *Schistocerca gregaria* (Henderson, 1970). It will be observed that treatments applied up to the zygotene/pachytene stage affect the chiasma frequencies. The conclusion is that chiasmata are formed at zygotene/pachytene but not later than this stage. There is no evidence that they are formed earlier. It is pertinent to recall that it is at

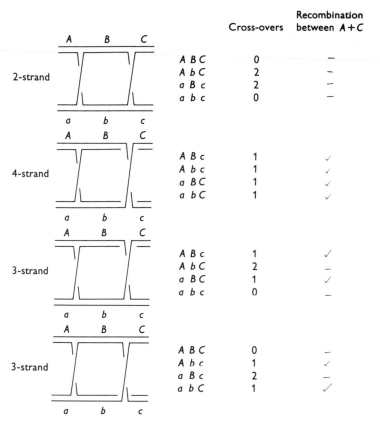

Fig. 3.8 The four possibilities when chiasmata form simultaneously between *A* and *B* and *B* and *C*. The four possibilities are realized with equal frequency showing there is no 'chromatid interference'. On the right are the recombinant gametes, the number of cross-overs between *A* and *C* and, far right, the number of recombinants for the outside markers, *A* and *C*. The ratio between the chiasma frequency and the cross-over frequency is a consistent 1:2. For any pair of markers, e.g. the outside markers A and C, the recombination overall, is 50%. In the absence of chromatid interference this is the maximum. The diagram shows that for the construction of linkage maps which are based on the cross-over frequencies it is important to choose markers as close together as possible. If, as with *A* and *C*, the distance between them allows for the formation of two or more chiasmata simultaneously the double cross-overs, A*b*C and *a*B*c* would be undetected.

this zygotene/pachytene stage that the synaptinemal complexes are fully assembled; also that they are indispensable prerequisites for chiasma formation in eukaryotes. There is little doubt therefore that chiasma formation is restricted to the zygotene/pachytene stage. This would suggest that the intimate association of homologous chromosomes

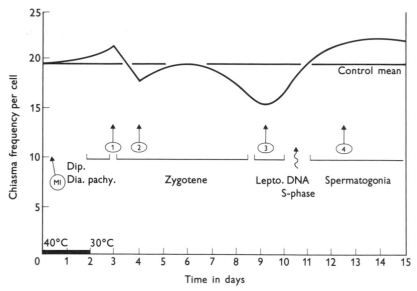

Fig. 3.9   The effects of heat 'shocks' (40°C for 2 days) upon chiasma frequencies at first metaphase of meiosis in spermatocytes of *Schistocerca gregaria*. The shocks are effective up to late zygotene (2) and early pachytene (1) but at no later stage, suggesting that chiasma formation is completed in pachytene. The decrease and increase respectively in chiasma frequencies due to shocks at leptotene (3) and the interphase preceding DNA synthesis (4) reflect disturbances of processes which affect the formation of chiasmata at the later stage. The time intervals between treatments and the manifestation of their effects at first metaphase are established by reference to the horizontal scale. [From Henderson, S. A. (1970) *Ann. Rev. Genet.*, **4**, 295–324]

initiated at zygotene is necessary for the formation of chiasmata. There are numerous observations to support this view. In maize the amount of crossing over, and therefore the frequency of chiasmata, is directly proportional to the completeness of chromosome pairing at pachytene. In the pollen mother cells of a heterozygote carrying a normal chromosome 9 and an abnormal 9 which includes a transposed segment from chromosome 3, the amount of intimate pairing at pachytene between these homologous chromosomes is reduced, and so also is the amount of crossing over (Rhoades, 1968).

Figure 3.9 shows that heat treatments applied at leptotene, and even before the synthesis of DNA, affect chiasma formation. Heat shocks (or low doses of radiation) have similar effects in *Tradescantia* (Lawrence, 1961), *Drosophila* (Grell and Chandley, 1965) and the grasshopper *Gonia australasiae* (Peacock, 1970). They are no doubt symptoms of disturbances of processes which directly or indirectly affect the formation of chiasmata at a later stage. Precisely what these processes are is not known.

The question of interlocking

At first metaphase of meiosis bivalents are, on occasions, interlocked. The cause of the interlocking is obvious enough (Fig. 3.10). What is not obvious is why interlocking is such a rare event because the chromosomes at zygotene and pachytene are still uncoiled and extended. It is certain therefore that before the intimate pairing of homologous chromosomes that is required for chiasma formation there must be some preliminary separation and alignment of homologues to prevent interlocking when the chiasmata are formed. One suggestion is that such alignment takes place at the pre-meiotic mitosis such that homologues are already contiguous at the start of the interphase of meiosis. The evidence for this is conflicting. Moreover in many haploids, such as the Ascomycetes among fungi, the homologous chromosomes are not brought together until the fusion of gametic nuclei immediately prior to meiosis itself. There are no reports of widespread interlocking in these or comparable species. We would suggest that the alignment takes place during interphase itself when, as is established from cinematographs, the chromosomes are in vigorous movement. The alignment could be achieved as follows: (1) at interphase the homologous chromosomes are associated at or near one end (telomere); (2) the paired telomeres move through the nucleus pulling the chromosomes for a distance equal to or greater than the length of the chromosomes. By this means each pair, although associated closely only at one point, is now disentangled from the remainder.

It is well established that the paired telomeres at both ends of homologous pairs of chromosomes become attached to the nuclear membrane well before the onset of pachytene (Moens, 1969a). By this means the separation of homologous pairs achieved by movement will have been consolidated and the possibility of interlocking reduced to a minimum

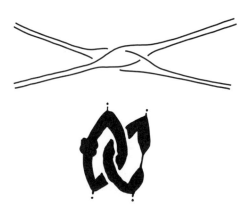

Fig. 3.10 Pairing at pachytene leading to the interlocking of bivalents at first metaphase of meiosis.

by the time the chromosomes begin their intimate association around the synaptinemal complexes.

The occasional and exceptional cases of interlocking such as may be induced by, for example, heat shock are explained by restriction of movement during interphase, or by the failure of telomere association or of their attachment to the nuclear membrane.

Chiasma distribution

(a) *The question of chromatid interference*

Where two or more chiasmata form within the one chromosome are the two chromatids involved in crossing over at any one chiasma independent of those at other chiasmata? Put in another way, is there *interference* with respect to chromatids such that a chromatid involved at one chiasma is more, or less, likely to be involved at another. In Fig. 3.8 we have, in respect of each pair of chiasmata, assumed no chromatid interference such that each of the four alternative types apply with equal frequency. The assumption is justified by evidence from linkage experiments. They show that the possibilities in Fig. 3.8 occur at random (but see also p. 52).

(b) *Chiasma localization*

In general more chiasmata are found in the longer than in the shorter chromosomes within a complement. The relation between chiasma frequency and chromosome length, however, is not linear. Mather's analysis of meiosis in pollen mother cells of numerous plant species shows that the chiasma frequency per unit length is characteristically higher for short than for long chromosomes (Fig. 3.11). The figure shows that at

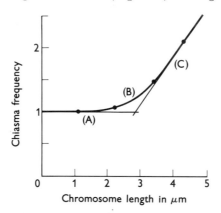

Fig. 3.11   The relationship between chiasma frequency and chromosome length. For very short chromosomes (A) there is a chiasma frequency of 1 irrespective of chromosome length. At (B) the relationship changes such that at (C), the chiasma frequency increases with increasing chromosome length. [From Mather, K. (1938) *Biol. Rev.*, **13**, 252–92, with permission]

least one chiasma is formed even in very short chromosomes. From an adaptive standpoint one readily appreciates that if these short chromosomes were unable to form a chiasma regularly at meiosis they could not persist. Chiasma failure results in univalents and, as we have explained, univalents are, more often than not, 'lost' during meiosis.

From the standpoint of mechanics the factors which determine the distribution of chiasmata are by no means clear. On a descriptive basis, however, the model proposed by Mather in 1937 is especially useful. The model envisages that chiasma formation is sequential. The first chiasma forms at a distance d, the *differential distance*, from an initiation point which coincides with the point of initial pairing at zygotene. Since even short chromosomes form at least one chiasma this distance may be short, even approaching zero (Henderson, 1963). Cytological observations show that subsequent chiasmata form at fairly regular intervals apart. The inference is that there is interference between chiasmata such that they do not cluster. The inference is verified from linkage data which show chiasma interference: i in Mather's notation. For all bivalents within a chromosome complement, therefore, the location of chiasmata follows a well-defined pattern. Moreover the frequency per bivalent will relate directly to d and i. For a short chromosome, where the first chiasma forms at a distance d from, say the end of one arm, a second chiasma is precluded if the remaining segment is shorter than i. For longer chromosomes the number of chiasmata formed subsequent to the first will be equal to the distance distal to the first chiasma, divided by i. The pattern, though well defined, is not rigid. Both d and i are subject to fluctuation. Even so the model makes abundantly clear that chiasma distribution is not random. A most important implication is that chiasmata and crossovers are more numerous in some chromosome segments than others. Put in another way, it means that some gene sequences are, to a degree, 'protected' from disruption by crossing over, others are not. The maintenance of relatively intact sequences of genes may, from an adaptive standpoint, be as important in conserving genetic variability as the disruption by recombination within other sequences to generate new variability.

There are, of course, numerous variations on the standard sequence of events which take place at meiosis including those affecting chiasma formation (see John and Lewis, 1965). It is not unusual either for the female meiosis to be substantially different from the male. A familiar example is the absence of crossing over at meiosis in *Drosophila* males in comparison with the normal chiasmate meiosis in the female. There are instances also where the *distribution* of chiasmata differs sharply between male and female. In *Stethophyma grossum* males for example the chiasmata are strictly localized near to the centromeres; in the female they are widely distributed (Perry and Jones, 1974). Extreme localization of chiasmata as in the male *Stethophyma* may stem from more than one cause. Two flowering plant species, *Fritillaria meleagris* and *Allium fistulosum*,

ıata localized mainly near to the centromeres. In *F. meleagris*
ne pairing is complete only in the vicinity of the centromeres.
ly the chiasmata are confined to the centromere regions
, 1935). The same explanation probably applies also in cases
where crossing over is clustered within very short chromosome segments
during meiosis in some fungi (Fincham and Day, 1971). In *Allium
fistulosum*, on the other hand, the explanation does not hold as pairing at
pachytene is complete.

3.5 The mechanism of recombination

Reduction of the chromosome number at meiosis is accounted for by
bivalent formation at first prophase followed by the separation of half
bivalents at first anaphase and of chromatids at the second anaphase.
Recombination between genes on different chromosomes is explained by
the random orientation of bivalents at first metáphase such that genes on
different bivalents segregate independently to yield 50% recombinant
and 50% parental type gametes (Fig. 3.12). The recombination of genes
on the same chromosome has been interpreted on the basis of breakage of
non-sister chromatids at precisely corresponding places followed by the
reunion of the broken ends resulting in crossing over (Fig. 3.5). The
mechanism depicted also accounts for the reciprocal nature of the
products of recombination and for the numerical equality of reciprocal
products characteristic of results from the vast majority of linkage
analyses. In micro-organisms however, where the precision of linkage
analysis is enhanced enormously on account of the large number of
progenies available for classification and, in the *Ascomycetes*, further
refined by the opportunity of classifying the products of individual
meiotic divisions (Fig. 3.13), the results have brought to light instances
where the products of recombination are not reciprocal. Any model
which seeks to accommodate the facts of recombination at meiosis must
take account of these non-reciprocal recombinants. The breakage–
reunion model in Fig. 3.5 does not.

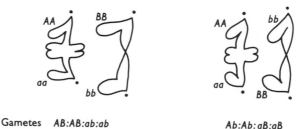

Gametes   *AB:AB:ab:ab*   *Ab:Ab:aB:aB*

Fig. 3.12 Random orientation of two bivalents at first metaphase of meiosis,
giving independent segregation of genes *A* and *B*.

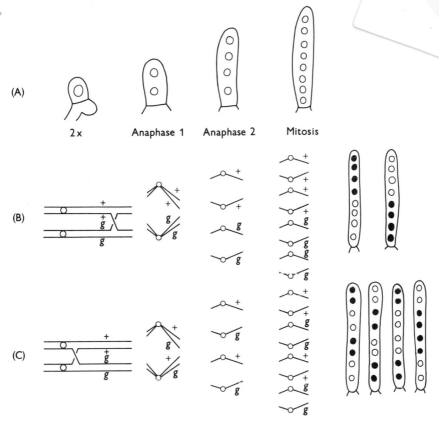

Fig. 3.13 Tetrad analysis in *Sordaria fimicola*. Each mature ascus contains the products of one meiosis and the ascospores are ordered in relation to the first and second division (A). Spore colour markers show first division segregation (B) in the absence of chiasma formation between the marker gene and the centromere and second division segregation (C) when a chiasma is formed.

Gene conversion

Crosses between a haploid strain of *Sordaria fimicola* with wild type black ($+$) ascospores and a strain with mutant grey (g) ascospores produce asci containing two basic types of spore arrangement, the *first division segregation* (reductional) type, where there is no crossing over between the spore colour locus and the centromere and the *second division segregation* (equational) type, where crossing over has occurred in this region (Fig. 3.13). In every ascus, irrespective of whether crossing over has taken place or not, we expect 4 black and 4 grey ascospores, i.e. a 1:1 segregation. This, in fact, is what we do find in the vast majority of asci examined. In others, however, instead of the exclusive occurrence of 4:4 segregations for black and grey, some 6:2 and 5:3 segregation patterns

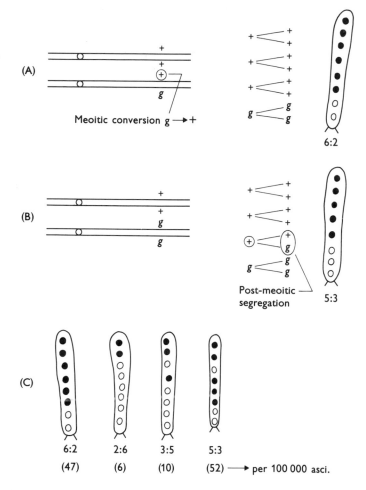

Fig. 3.14 Aberrant asci in *Sordaria fimicola*. (A) Conversion of the *g* allele at meiosis is detected as a 6:2 segregation of black to mutant ascospores. (B) 5:3 asci arise from the post-meiotic segregation of heterozygosity within a single chromatid. The frequencies of 6:2 and 5:3 conversion asci for the *g* locus of *S. fimicola* are given in (C). [Data from Kitani, Y., Olive, L. S. and El-Ani, A. S. (1962) *Am. J. Bot.*, **49**, 697–706]

turn up with low frequency (Fig. 3.14). The 5:3 segregations are particularly interesting because they represent post-meiotic segregation of differences within a single chromatid. Mutation will not account for these *aberrant* asci because, first, they are too numerous and, second, the change is always in the direction of one of the existing alleles and not at random. The alternative is *gene conversion*—'the replacement of the genetic material at a particular site in one chromatid by the genetic material from a precisely corresponding site from a non-sister chromatid' (Fincham and Day, 1971). The phenomenon accounts for the non-reciprocal recombinants to which we have referred (Fig. 3.14).

(a) *Reciprocal and non-reciprocal intragenic recombination*
Lissouba *et al.* (1962) collected several thousand pale-coloured ascospore mutants in *Ascobolus immersus*. On the basis of linkage the mutants were assigned to a number of different *series*. Within each series the mutants were tightly linked and each series was considered to represent one gene locus. Crosses between pale-spored mutants located in different genes yielded some wild type and double mutant reciprocal recombinant classes as well as the two single mutant parental types. Crosses between mutants located in the same gene (i.e. allelic mutants) also yielded some wild type recombinants but test crosses with the parental strains showed that this intragenic recombination was not reciprocal (Fig. 3.15). In *Ascobolus* there was also an exception to the general rule. Crosses between some of the mutants at one locus (in series 73) produced a high propor-

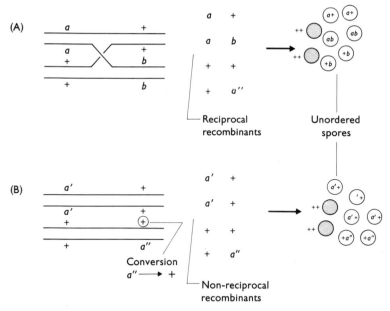

Fig. 3.15 Intergenic and intragenic recombination in *Ascobolus immersus*. (A) Crosses between mutants of different genes (*a* and *b*) yield 6:2 asci arising by reciprocal crossing over. (B) Pairwise crosses between mutants of the same gene (*a'* and *a"*) also give 6:2 asci, but these are found to contain the products of non-reciprocal recombination, arising by conversion. Note the spores in this species are not ordered as in *Sordaria*.

tion of reciprocal recombinants. Reciprocal intragenic recombination has also been detected for the *arg-4* gene of *Saccharomyces cerevisiae* (Fogel and Mortimer, 1969), so that intragenic recombinants are not exclusively the result of conversion.

(b) *Polarity*

When a number of *Ascobolus* mutants from within the one gene are crossed in all possible combinations there is a well defined hierarchy or polarity with regard to conversion such that one mutant, e.g. *a* in a series *a b c d* is converted in crosses with *b, c* or *d*; *b* when crossed with *c* or *d* and *c* when crossed with *d*. Moreover this polarity is directly correlated in *Ascobolus*, as in other but not all *Ascomytes*, with the spatial distribution of the mutant sites, i.e. their linkage relations. The polarity may, in fact, be used for mapping the mutants within the gene (Lewis and John, 1970).

(c) *Negative interference and chromatid interference*

In asci showing conversion (e.g. at the *g* locus in *Sordaria*, Kitani *et al.*, 1962) there is an unexpectedly high frequency of recombination between the gene undergoing conversion and closely linked markers on either side. Double recombinants are also unexpectedly numerous. In view of the short distances involved the double recombination is suggestive of negative interference between chiasmata and, moreover, of chromatid interference during chiasma formation, since the double recombination implicates the same pair of chromatids. As we shall explain the excess of double recombinants may be accounted for without invoking interference between chiasmata.

It is also important to emphasize that when the region is mapped on the basis of a random sample of asci, as distinct from those showing conversion, there is no detectable negative interference or chromatid interference. The reason is that conversion at the *g* locus (or any other) occurs with very low frequency, in one out of about 1000 asci. It is for the same reason that non-reciprocal recombinants are undetectable by conventional linkage analysis. Even so the three phenomena, the high recombination frequency within short chromosome segments, the excess of double recombinants and the non-reciprocal recombinants are real enough. In constructing a general model which accounts for the mechanism of recombination they must be taken into account. A number of such models have been proposed. We present below that put forward by Holliday.

The molecular basis—Holliday's model (1964)

1 Each chromatid is composed of one double helix of DNA.
2 Recombination takes place at the four strand stage after chromosome pairing and duplication.
3 The DNA from non-sister chromatids is closely paired (Fig. 3.16A).
4 Simultaneous breakage at strictly corresponding sites in two single

chains (half-helices) of the same molecular polarity is followed, first, by unwinding of the half-helices in one direction for a variable length and, second, by the union of ends in new combinations to form a 'half-chiasma' and a region of hybrid DNA (Fig. 3.16B).

5 Either the half-chiasma is resolved by iso-locus breakage in the overlapping strands, the result being no recombination for outside markers (Fig. 3.16C), or alternatively, the half-chiasma is converted into a full chiasma by iso-locus breaks in the non-overlapping strands, which leads to reciprocal recombination for the outside markers (Fig. 3.16D).

6 There is correction of 'illegitimate' pairing of bases for the mutant site spanned by the region of hybrid DNA. The result is either a 4:4 or a 6:2 ascus-type segregation, depending on the way in which the correction takes place (Fig. 3.16E 1, 2). Delay in the correction of one of the hybrid DNA molecules until the post-meiotic DNA synthesis period accounts for the 5:3 post-meiotic segregations (Fig. 3.16E 3).

The model shows that: (a) gene conversion is achieved with or without recombination for outside markers, depending upon whether a half-chiasma or a full chiasma is formed. Conversion will, of course, only take place at all when the hybrid DNA is formed within a site of heterozygosity. (b) The 'negative interference' is the outcome of conversion and of the formation of a chiasma as in Fig. 3.16C. (c) The unwinding of half-helices from a fixed point in one direction accounts for the polarity of conversion. Mutant sites nearest to the breakage point are the most likely, therefore, to undergo conversion, and vice versa. (d) Reciprocal recombinants between mutant sites within a gene are produced if both

Fig. 3.16 Holliday's 1964 model for the mechanism of recombination. The model (A) represents paired homologues at the four-strand stage and includes a single gene, with two sites of heterozygosity, two outside marker genes (*A*, *B*) and the centromeres. Each chromatid consists of a double helix of DNA drawn as a pair of straight lines. The arrows indicate the polarities of the DNA half-helices. Recombination is initiated by simultaneous breakage at strictly corresponding sites (arrowed) in each of two single chains of the same polarity in DNA helices from non-sister chromatids. The half-helices unwind in one direction for a variable length and then rejoin in a new combination to form a half-chiasma (B). The half-chiasma may either be resolved by an iso-locus break in the crossed over strands (C), when there is no recombination for the outside markers, or else become converted into a full chiasma (D) following iso-locus breaks in the non-crossed over strands, when outside marker recombination will occur (E). In either event the region of hybrid DNA spans a site of heterozygosity and thus includes the mismatched base pairs A:C/T:G. Correction of this illegitimate pairing before replication can lead to a normal 4:4 segregation of alleles at site M1①, a 6:2 segregation ②, or a 5:3 segregation ③, depending on the direction (in terms of parental types) in which the correction is made (F).

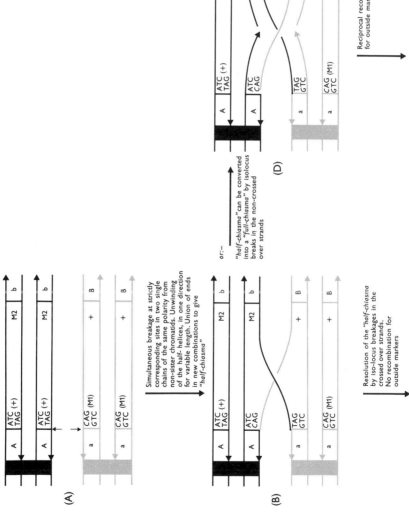

(A)

Simultaneous breakage at strictly corresponding sites in two single chains of the same polarity from non-sister chromatids. Unwinding of the half-helices, in one direction for variable length. Union of ends in new combinations to give "half-chiasma"

(B)

Resolution of the "half-chiasma" by iso-locus breakages in the crossed over strands. No recombination for outside markers

or:—

"half-chiasma" can be converted into a "full-chiasma" by isolocus breaks in the non-crossed over strands

(D)

Reciprocal recombination for outside markers

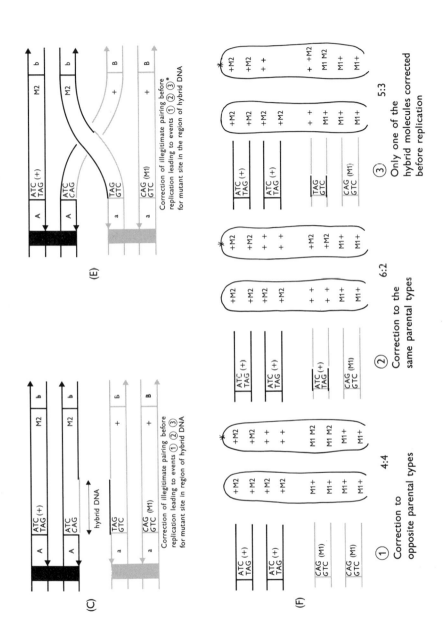

(C)

hybrid DNA

Correction of illegitimate pairing before
replication leading to events ① ② ③
for mutant site in region of hybrid DNA

(E)

Correction of illegitimate pairing before
replication leading to events ① ② ③*
for mutant site in the region of hybrid DNA

(F)

① Correction to
opposite parental types

4:4

② Correction to the
same parental types

6:2

③ Only one of the
hybrid molecules corrected
before replication

5:3

mutants are to one side of the hybrid DNA or on either side but outside the hybrid DNA region.

At this point it is worth re-emphasizing that for any one of numerous loci within a chromosome the frequency with which it undergoes conversion is low. Moreover, the consequences of conversion, such as non-reciprocal recombinants affect only a restricted chromosome segment in the immediate vicinity of the conversion. For these reasons the general relations between chiasmata crossing over and recombination correspond very closely with those which apply in models which assume single breaks in non-sister chromatids followed by reunion of the broken ends as in Fig. 3.5, even though, as we have seen, the molecular basis for recombination is substantially different from this.

Models invoking the formation of hybrid DNA such as those of Holliday and of Whitehouse (1963) are the more plausible in the light of more recent discoveries. We have already referred to the small fraction of the DNA which is synthesized at zygotene/pachytene when the 'correction' of mismatched base sequences is assumed to take place. Enzyme systems have been described also, with the capacity for inducing breakage, repair and the synthesis of DNA in meiotic cells. An endonuclease, found exclusively in meiotic cells, has the property of inducing single-strand 'nicks' in native DNA. Its activity is at a peak during late zygotene and early pachytene. A ligase 'repair' enzyme also shows a flush of activity at the same period (Howell and Stern, 1971). Stern and Hotta (1973) have also shown that DNA extracted from meiotic cells of *Lilium* and denatured contains a higher frequency of short strands during pachytene than at other stages of meiosis. They ascribe the high frequency of these short strands to the endogenous nicking by the endonuclease and their subsequent disappearance to the repair mediated by the ligase.

A full account of the events associated with gene conversion and recombination is given in a companion volume in this series (Catcheside, 1977). This also deals in detail with the various molecular models which have been proposed to account for the diverse phenomena associated with recombination.

4

Quantitative variation in genetic material

4.1 The structural basis

Evolution is the result of the selection of phenotypes displaying adaptive variation in heredity. The heritable variation stems, basically, from changes in the information which resides mainly in the chromosomal DNA. The changes are of two kinds, qualitative and quantitative. At the molecular level the qualitative changes result from base substitution within the DNA molecule, which affects the *base ratio*, or from rearrangement affecting the order of bases (Fig. 4.1). Examples of both kinds are revealed by direct comparisons between DNAs of species throughout both plant and animal orders. In the bacteria the DNA base ratio (expressed as the proportion of guanine + cytosine + 5-methylcytosine where present), ranges from 25 to 75 % (Belozersky and Spirin, 1960; Sueoka, 1961). The G + C content in higher plants ranges from 35 to 49 %.

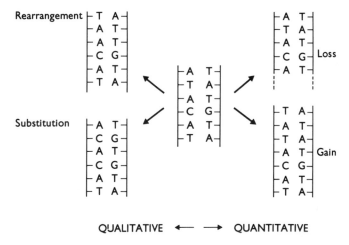

QUALITATIVE ⟵ ⟶ QUANTITATIVE

Fig. 4.1 Qualitative changes in DNA affecting the base ratio and the base sequence.

A 10% range is found even within the one family, the Liliaceae (Kirk, Rees and Evans, 1970). Changes in order are less easily established. One example, involving RNA in this case, is given by Reddi (1959) who showed differences in the order of bases within the nucleic acids of different strains of tobacco mosaic virus. Turning to *quantitative* variation we would expect to find an increase in DNA amount accompanying the evolution of more complex and sophisticated forms. The amount of chromosomal DNA in vertebrate animals and higher plants is, in fact, 1000 times greater than in bacteria (Fig. 4.2). We have commented earlier, however, that variation of this order of magnitude is difficult to explain directly in terms of increase in the variety of genes within the complement. The large variation in DNA amount between closely related species, e.g. in *Vicia* and *Lathyrus* (Rees *et al.*, 1966) is equally incomprehensible on grounds of variation in the number and variety of nuclear genes. What then is the explanation for this wide range of variation in nuclear DNA? How does it arise? What kind of DNA is involved? What does it signify in genetic terms?

In structural terms the causes of quantitative variation in nuclear DNA arise either by the addition or deletion of whole chromosomes, i.e. a change in chromosome number or, alternatively, by the addition or deletion of segments within chromosomes.

Polyploidy and aneuploidy

The addition of whole sets of chromosomes follows the failure of spindle formation at mitosis in somatic cells giving rise in a $2x$, diploid organism to $4x$, tetraploid somatic nuclei. An alternative cause of polyploidy is non-reduction during gamete formation resulting for example in the production of diploid $2x$ rather than haploid gametes at meiosis in a diploid. The fusion of $2x$ and x gametes produces $3x$, triploid zygotes, the fusion of diploid gametes, $4x$ tetraploids. With aneuploidy the gain or loss of chromosomes involves less than a whole set or genome. The gain of one chromosome of the complement gives a trisomic $(2x+1)$, of two members of the complement a tetrasomic $(2x+2)$ and so on. Monosomics $(2x-1)$ and nullisomics $(2x-2)$ are the results of the loss of one or of a pair of homologous chromosomes, respectively.

The nature of the DNA lost or gained through polyploidy and aneuploidy is clear enough. It represents exact copies of complete chromosome complements or of complete chromosomes. In the long term it is possible that small segments become redundant and are deleted from the chromosomes of tetraploids. Timmis, Sinclair and Ingle (1972) and Flavell and Smith (1974) report a loss of ribosomal cistrons in polyploid hyacinths and wheats respectively.

(a) *Autopolyploids*

These are polyploids in which the additional sets of chromosomes are structurally similar and therefore homologous to those of the diploid

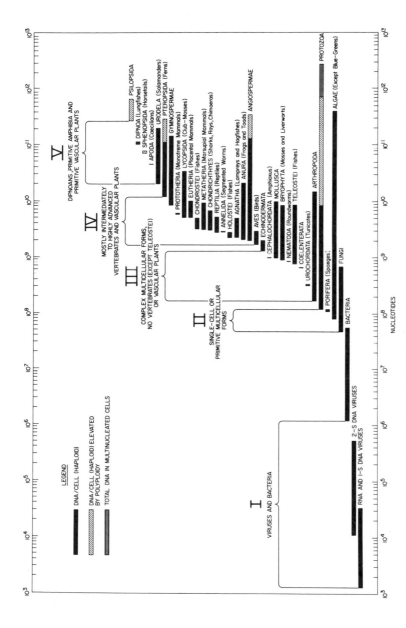

Fig. 4.2 Quantitative variation in nuclear DNA amount in a number of major groupings of plant and animal species. [From Sparrow, A. H., Price, H. J. and Underbrink, A. G. (1972) *Brookhaven Symp. Biol.*, **23**, 451–94]

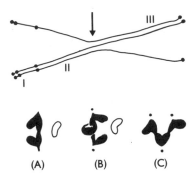

(A)   (B)   (C)

Fig. 4.3   Pachytene pairing in a triploid. Positions of centromeres are in the vicinity of the arrow. One terminal chiasma only, e.g. in region I, gives a rod bivalent and a univalent at first metaphase (A); two chiasmata, one terminal and one interstitial, e.g. at I and II, a rod bivalent with two chiasmata in one arm (B). Two chiasmata at I and III produce a trivalent (C).

population from which they derive. From the cytological standpoint the extra sets of chromosomes pose no problems at mitosis. The orientation of the chromosomes on the spindle at metaphase and their separation at anaphase are effective and efficient. At meiosis, however, the pachytene associations involve three or more homologous chromosomes. Following chiasma formation they may be transformed to multivalent rather than bivalent configurations. The formation of multivalents, in turn, affects the segregation at first anaphase of meiosis and, as a result, the distribution of genetic material to the gametes.

(i) *Meiosis in triploids.* Although the three homologues of each member are associated during pachytene the association at any one chromosome segment is restricted to a pair (Fig. 4.3). Recent electron microscope investigations, it is true, have revealed intimate contact between all three homologues (Moens, 1969b) but such contacts are restricted to very short segments and contact is between pairs for the greater part of the chromosomes. The metaphase configuration resulting from this association depends on the number and the position of chiasmata within it. Evidently with no chiasmata the result will be three univalents, with one chiasma a rod bivalent. With two chiasmata involving the same pair of chromosomes we get a bivalent and a univalent. With two chiasmata involving all three of the chromosomes the result is a trivalent. Loss of univalents during meiosis will, of course, be a cause of inviability among gametes. Even with complete trivalent formation, however, their asymmetry at metaphase results in the production of gametes with varying chromosome numbers. Fig. 4.4 shows all four types of arrangement of three trivalents in a triploid $2n = 3x = 9$. Twenty five per cent of the gametes contain either haploid or diploid (euploid) chromosome complements. The remainder are aneuploid with either one or two of the

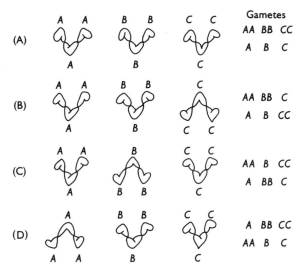

Fig. 4.4 The four alternative types of orientation of trivalents at first metaphase of meiosis in a triploid $2n = 3x = 9$. Only one (A) produces euploid gametes. The gametes from the remainder are aneuploid which, as a result, would normally be inviable or ineffective.

chromosomes in double dose. This means that the genes from the different chromosomes are unequally represented, resulting in a genetic imbalance that is often lethal (see Darlington and Mather, 1949). Assuming, in this hypothetical case, all the aneuploid gametes to be lethal only 25% of the gametes would be viable. In fact an analysis in a $3x = 9$ triploid *Crepis capillaris* by Chuksanova (1939) showed precisely this degree of gamete viability.

Basic number. With three trivalents per cell we have only 25% euploid, viable gametes, corresponding with the percentage of cells in which all three trivalents are orientated in the same way relative to the spindle poles. In a triploid with a basic number of 4, i.e. $3x = 12$ the probability of such orientation is clearly less (12.5%): and less still with higher basic numbers. Indeed triploids with basic numbers of 6 or more are almost completely sterile. Some varieties of apples, such as Blenheim Orange and Bramley's Seedling are triploids ($3x = 51$). The seed set is less than 5% and the percentage of germinating seeds even lower (Crane and Lawrence, 1934).

In almost all odd numbered polyploids, $5x$, $7x$ etc., the degree of sterility is extremely high. As in the triploid described the main cause is the production of aneuploid, genetically imbalanced gametes. There are rare exceptions. In *Hyacinthus*, for example, aneuploid gametes are relatively viable and fertile (Darlington, Hair and Hurcombe, 1951).

For this reason the species is represented by an unusually high number of aneuploid individuals.

(ii) *Meiosis in tetraploids*. Since each chromosome is represented by four homologues the pachytene associations are made up of all four, but again the contact at any one segment is restricted to a pair. As with the triploid the frequency of different kinds of configurations at metaphase: univalents, bivalents, trivalents and quadrivalents, will depend on the frequency and location of chiasmata within the pachytene associations (Fig. 4.5). Loss of chromosomes in the form of univalents, including those accompanying trivalents, are a cause of aneuploidy and consequently of inviable gametes. A high frequency of trivalents could also cause aneuploidy, as in triploids, on grounds of asymmetry of the metaphase configurations and unequal anaphase separation. In many species, however, trivalents are conspicuously rare in autotetraploids. This is especially true for species with chiasmata located mainly in distal segments of chromosomes. With the formation of 3 or 4 chiasmata near the ends of the chromosomes in pachytene associations of four, only quadrivalents and bivalents are possible. Another reason for the absence of trivalents is that the four homologous chromosomes may associate as two pairs. In most pollen mother cells of autotetraploid rye, for example, the four chromosomes of one of the seven homologous sets associate exclusively in pairs at pachytene (Timmis and Rees, 1971). Where the chiasma frequency is reasonably high (of the order of three or more per association of four) trivalents and univalents will be very rare. Equally important the quadrivalents formed, as well as the bivalents, separate in a perfectly regular, 2×2 fashion at anaphase to produce euploid, viable gametes.

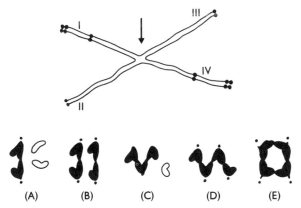

Fig. 4.5   Pachytene pairing in a tetraploid. The arrow indicates the position of the centromeres. The first metaphase configurations below result from the formation of: (A) one terminal chiasma, e.g. at I; (B) two chiasmata, one at I and IV (or at II and III); (C) two chiasmata at I and II or III (or IV and II or III); (D) three terminal chiasmata (e.g. at I, II and III) and (E) four terminal chiasmata. Univalents unshaded.

For this reason autotetraploids are far less infertile than triploids and, to generalize, even numbered autopolyploids are more fertile than odd numbered, although never as fertile as diploids.

Chromosome pairing and fertility. Univalents and trivalents are the chief causes of aneuploidy and of inviability among gametes produced by autotetraploids. It follows that a reduction in the frequency of univalents and trivalents should improve fertility. Whether this be achieved at the expense of increasing quadrivalents or bivalents should be immaterial because both disjoin regularly at first anaphase, at least in species with distally localized chiasmata. We have indicated that the frequency of trivalents and univalents is high in organisms with low chiasma frequencies. In such cases we should expect an improvement in fertility, therefore, by increasing the chiasma frequency. Investigations in rye and in *Lolium* (Hazarika and Rees, 1967; Crowley and Rees, 1968) confirm that an increase in the chiasma frequency is indeed accompanied as expected by a reduction in univalents and trivalents and also by an increase in quadrivalents. They also show an increase in fertility as measured by the seed setting.

Table 4.1 The frequencies of quadrivalents (IVs), trivalents (IIIs), bivalents (IIs) and univalents (Is) in pollen mother cells of tetraploid populations of *Lolium perenne* ($2n = 4x = 28$) and *Secale cereale* ($2n = 4x = 28$) before and after selection for improvement in fertility. Selection in *Lolium* was over five generations and over 19 generations in *Secale*. Note the increase in the frequencies of quadrivalents and the decrease in frequencies of trivalents and univalents among selected populations. [*Lolium* data from Crowley, J. G. and Rees, H. (1968) *Chromosoma*, **24**, 300–8; *Secale* data from Dr. Gul Hussain, by kind permission]

	Unselected				Selected			
	IV	III	II	I	IV	III	II	I
Lolium perenne	3.01	0.32	7.24	0.49	4.07	0.10	5.67	0.10
Secale cereale	2.97	0.68	9.04	0.75	5.12	0.34	7.74	0.19

Investigations of autotetraploid rye and *Lolium* varieties selected for improvement in fertility provide further confirmation that reducing the number of trivalents and univalents and increasing the quadrivalent frequency by increasing the chiasma frequency is effective (Table 4.1). The quadrivalents are more numerous as a direct result of an increase in chiasma frequency. From the standpoint of fertility an increase in bivalent frequency would have been equally acceptable. Even so it is worth emphasizing that multivalents, in themselves, provided they disjoin with regularity as do the quadrivalents in rye and *Lolium* are no barrier to fertility.

Finally, it is important to appreciate that infertility in polyploids is not solely due to the production of aneuploid gametes. Different tetraploid strains of rye vary considerably in seed set despite the fact that they display similar patterns of chromosome pairing at first anaphase of meiosis and each produces much the same proportion of aneuploid gametes. The variation is evidently due to *genic* as distinct from *chromosomal* causes (Hazarika and Rees, 1967).

(b) *Allopolyploids*
    At meiosis in hybrids between diploid species the pachytene pairing is often ineffective with the result that chiasma formation is severely

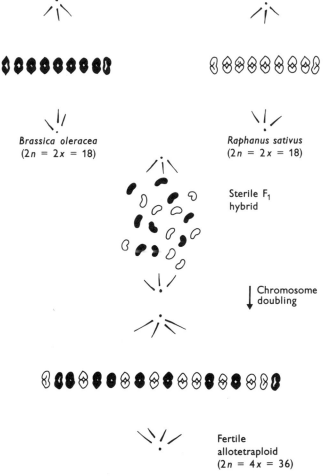

Fig. 4.6   Chromosome pairing at first metaphase of meiosis in *Brassica oleracea*, *Raphanus sativus*, their F_1 diploid hybrid and in the allotetraploid. The lack of homology between *B. oleracea* and *R. sativus* chromosomes is the cause of failure of pairing in the diploid hybrid and of bivalent formation in the allotetraploid.

Table 4.2 Polyploidy in flowering plants

Species (N = natural, C = cultivated)	Common name	Constitution	Origin
Allopolyploids			
Galeopsis tetrahit (N)	hemp nettle	$2n = 4x = 32$	*G. pubescens* × *G. speciosa* ($2x = 16$) ($2x = 16$)
Raphano-Brassica (C)		$2n = 4x = 36$	*Raphanus sativus* × *Brassica oleracea* ($2x = 18$) ($2x = 18$)
Brassica napus (N)	rape	$2n = 4x = 38$	*B. oleracea* × *B. campestris* ($2x = 18$) ($2x = 20$)
Triticale (C)		$2n = 8x = 56$	*Triticum aestivum* × *Secale cereale* ($6x = 42$) ($2x = 14$)
Triticum aestivum (C)	wheat	$2n = 6x = 42$	*T. monococcum* × *T. speltoides* × *Aegilops squarrosa* ($2x = 14$) ($2x = 14$) ($2x = 14$)
Primula kewensis (C)		$2n = 4x = 36$	*P. floribunda* × *P. verticillata* ($2x = 18$) ($2x = 18$)
Nicotiana tabacum (C)	tobacco	$2n = 4x = 48$	*N. sylvestris* × *N. tomentosiformis.* ($2x = 24$) ($2x = 24$)

Species (N = natural, C = cultivated)	Common name	Constitution	Method of propagation
Autopolyploids			
Beta vulgaris (C)	sugar beet	$2n = 3x = 27$	seed, by crossing diploid × tetraploid
Malus pumila (C)	apple	$2n = 3x = 51$	grafting
Secale cereale (C)	rye	$2n = 4x = 28$	seed
Allium tuberosum (N)		$2n = 4x = 32$	bulbs
Ranunculus ficaria (N)	celandine	$2n = 2x, 4x, 5x, 6x$ $= 24, 32, 40, 48$	bulbils
Solanum tuberosum (C)	potato	$2n = 4x = 48$	tubers
Hyacinthus orientalis (C)	hyacinth	$2n = 3x, 4x = 24, 32$	bulbs
Fritillaria camschatensis (N)		$2n = 3x = 36$	bulbs

reduced and chromosomes are found unpaired, as univalents, at first metaphase. Such hybrids whether they be mules or F_1s from crossing *Brassica* species are virtually sterile. It was in the genus *Brassica* that Karpechenko (1928) first explained how the doubling of the chromosome number in a hybrid produced a fertile tetraploid (Fig. 4.6). The chromosomes of the intergeneric hybrid between *Raphanus sativus* (radish) × *Brassica oleracea* (cabbage) show very little pairing at meiosis in the hybrid between them. They are not homologous, by which is meant that they are not sufficiently alike structurally to achieve effective pairing. This is not to say, of course, that they are structurally or genetically unrelated and for this reason the members of each 'pair' are properly described as *homoeologous* as distinct from homologous. As Fig. 4.6 shows, doubling the chromosome number permits pairing between strictly homologous chromosomes and since pairing is restricted to homologues only bivalents appear at first metaphase. In the absence of multivalents separation at first anaphase of meiosis is regular, the gametes are euploid (diploid) and viable. Such a tetraploid is an allotetraploid or amphidiploid. Other examples of allopolyploids are given in Table 4.2. Their fertility is high and comparable to that of diploids.

(c) *Segregation in polyploids*

Odd numbered polyploids are, almost invariably, highly infertile. Their propagation and survival under natural or cultivated conditions is consequently dependent upon asexual means. Examples are given in Table 4.2. Such clonal propagation means that progenies are genetically identical with one another and with their parents. In autotetraploids the segregation of the four alleles of a gene to give diploid gametes is random. The gametes formed from the alleles a^1, a^2, a^3, a^4 will be of the constitution a^1a^2, a^1a^3, a^1a^4, a^3a^4, a^2a^4, a^2a^3 and will occur with equal frequency. Gametes produced by quadruplex $(AAAA)$, triplex $(AAAa)$, duplex $(AAaa)$, simplex $(Aaaa)$ and nulliplex $(aaaa)$ tetraploids are given in Table 4.3. There is one qualification to be made. Triplex and simplex tetraploids in the Table give no double recessives (aa) or dominants (AA) respectively. Yet such gametes do arise (Mather, 1936). For example, aa gametes are produced by a triplex when: (1) a chiasma forms between the centromere and the A locus; and (2) the orientation of the chromosomes at first and second anaphase is such that a and a move to the same

Table 4.3 Segregation in autopolyploids

		Gametes	Phenotypic ratio
Quadruplex	$AAAA$	AA	1
Triplex	$AAAa$	$1AA:1Aa$	$1:1$
Duplex	$AAaa$	$1AA:4Aa:1aa$	$5:1$
Simplex	$Aaaa$	$1Aa:1aa$	$1:1$
Nulliplex	$aaaa$	aa	1

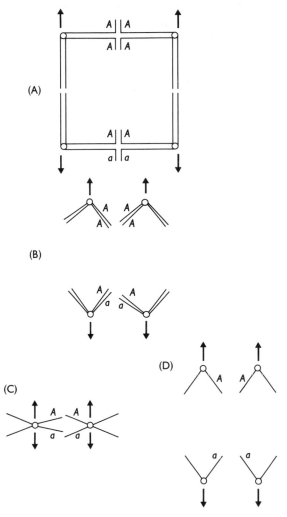

Fig. 4.7 Chromatid segregation in an autotetraploid. In a triplex *AAAa*, a chiasma between the centromere and the locus *A* (A) is followed by reductional separation at first anaphase (B). Orientation at second metaphase (C) is such that *aa* are drawn to the same pole at the second anaphase (D). The frequency with which a chiasma forms between the centromere and the locus will depend upon the distance between them. If a chiasma does form, separation at first anaphase will be reductional in half the cases and orientation at the second metaphase will, again in half the cases, give *aa* gametes. While the kinds of separation at first and second anaphase are predictable in any situation, the frequency with which the chiasma forms between the centromere and the locus can only be established by experiment. Therefore the amount of chromatid interference varies from one locus to another.

poles (Fig. 4.7). The frequency with which such gametes are produced will depend on the frequency with which chiasmata occur between A and the centromere and hence upon the distance between the centromere and the locus.

In allotetraploids, derived from chromosome doubling of a diploid hybrid, the gametes produced will be of uniform constitution. For the alleles a^1a^1 on a pair of chromosomes from the one parent and a^2a^2 on the homoeologous pair from the other, the gametes will be invariably a^1a^2. Progenies of the allotetraploid will therefore be similar to one another and to the parents. In other words the species hybrid derived by allopolyploidy is immediately true breeding.

While natural allopolyploids arise by chromosome doubling in diploid hybrids it is possible to synthesize allopolyploids by crossing autotetraploids from different species. In this case homologous chromosomes may be heterozygous, e.g. a^1a^2, a^1a^2 and such an allopolyploid would not, of course, breed true on selfing.

(d) *The classification of polyploids*

We have treated autopolyploids and allopolyploids, in terms of chromosome pairing and segregation, as sharply distinct and disparate forms. The distinction is rarely, if ever, as sharp, and homoeologous chromosomes are not always sufficiently different in structure to prevent pairing between them, so that multivalents as well as bivalents may well be formed at meiosis in polyploids produced from species hybrids. *Triticum aestivum*, the cultivated bread wheat is a hexaploid ($6x = 42$) made up from three diploid genomes, AA, BB and DD which are derived from the primitive diploid wheats *Triticum monococcum*, a species related to *Aegilops speltoides* and *Triticum squarrosa* (Fig. 4.8). At metaphase of meiosis, only bivalents are formed and separation is regular to give viable gametes, each within an A, B and D genome making up 21 chromosomes. On the face of it this is a classical example of chromosome pairing at meiosis in an allopolyploid. Riley and Chapman (1958), however, have shown that the restriction of pairing to strict homologues is not entirely accounted for by structural differences between homoeologous chromosomes. A gene, or cluster of genes, in the long arm of chromosome 5 of the B genome is an essential component of the mechanism ensuring pairing between homologues only. In the absence of this gene or gene cluster (as in a wheat plant carrying only telocentrics for the short arm of the $5B$-chromosome) there is pairing between homoeologues as well as homologues. In short there are multivalents at metaphase of meiosis as well as bivalents. It is probable that such diploidizing genes or gene complexes operate in other allopolyploids reinforcing the structural differences between chromosomes to prevent pairing and chiasma formation between other than strictly homologous chromosomes. This implies that non-homology is not, in itself, sufficient to ensure only bivalent formation at meiosis. It is clear as well that the degree of homology between chromo-

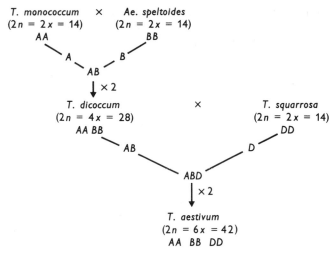

Fig. 4.8 The origin of the allohexaploid cultivated wheat, *Triticum aestivum*, by hybridization and chromosome doubling in the ancestral species and hybrids. We have given *Ae. speltoides* as the source of the *B* genome. An *Ae. speltoides*-like species would be more accurate.

somes from different species will vary from slight to substantial. Consequently, at meiosis in even numbered polyploids from hybrids between species, we may find a range of events from widespread multivalent formation typical of autopolyploids to bivalent formation only, typical of allopolyploids.

(e) *The polyploid phenotype*

In plants, polyploids, in comparison with their diploid ancestors, are generally slower in initial growth and later in maturing, more tolerant of extremes of temperature and rainfall and have larger and thicker leaves. As Stebbins (1950) makes clear there are abundant exceptions to this broad generalization. In *Tradescantia*, for example, diploids and tetraploids are indistinguishable on the basis of external appearance (Swanson, 1960). The same is true of diploid and polyploid forms of the wild potato, *Solanum tuberosum* (Dodds, 1965). The effects of doubling the chromosome number may also vary from one genotype to the other within a population. In rye, for example, the dry weights of tetraploids derived from some genotypes are twice that of the diploid, in others the dry weights of diploids and tetraploids are similar (Hutto, 1974). The generalization about the effects of polyploidy upon growth and development is, as we have stressed, subject to many qualifications. We shall return to this issue later (p. 96).

(f) *Variability*

Lewis (1967) has pointed out that a substantial portion of both root and shoot tissue in plants is composed of polyploid cells. He argues that any physiological advantages conferred by polyploidy upon individuals are readily achieved by the induction and restriction of polyploidy to differentiated, non-dividing cells. Above all this would avoid the complications at meiosis which cause infertility in autopolyploids. On this basis he considered that from the standpoint of adaptation in evolution the importance of polyploidy rests mainly on genetical rather than physiological criteria. He argues that polyploidy, above all, serves to maintain a high level of heterozygosity following hybridization. For a single gene with two alleles, *A* and *a*, the relative frequencies of homozygotes and heterozygotes in random mating populations of diploids, auto- and allo-tetraploids are given in Table 4.4. Given heterozygous advantage in the form of heterosis the generation and maintenance of heterozygosity in polyploids may well be of paramount importance. It will be observed that the combination *Aa* in the allotetraploid, following hybridization between species carrying *AA* and *aa* respectively, is permanently fixed and even with inbreeding the heterozygosity would be retained and maintained by the restriction of pairing to strictly homologous chromosomes, in this case carrying, on the one hand, *AA* and on the other, *aa* (see also Rees and Jones, 1971).

(g) *Polyploidy in animals*

   In our brief account of polyploidy and of polyploids we have confined our observations to plants. There are two good reasons for this. The first is that, in higher organisms especially, polyploidy is rare in animal species and, consequently, most of the investigations and most of the information on polyploidy refers to plants. The two reasons given to

Table 4.4   The relative frequencies of heterozygotes and homozygotes in populations of diploids, autotetraploids and allotetraploids: assuming a 1:1 distribution of alleles (*A* and *a*), random mating and no selection. [From Rees, H. and Jones, R. N. (1971) *Cellular Organelles and Membranes in Mental Retardation*, Churchill Livingstone]

	Allelic constitution	Heterozygotes
Diploid	*AA* 1 *Aa* 2 *aa* 1	50 %
Autotetraploid	*AAAA* 1 *AAAa* 4 *AAaa* 6 *Aaaa* 4 *aaaa* 1	87.5 %
Allotetraploid	*AAaa* 1	100 %

account for the rarity of polyploidy among animals are: (1) that individuals are more often unisexual rather than hermaphrodite as are members of most species of higher plants; and (2), that polyploidy may upset sex determination, in particular where this depends on a balance between X-chromosomes and autosomes as in *Drosophila*. The unisexuality of individuals is a restriction on the establishment of polyploidy in the following way. When a tetraploid arises by chromosome doubling, be it male or female, its diploid gametes can fertilize or be fertilized only by haploid gametes. The offspring will be triploid and sterile, so that in the absence of asexual means of propagation the polyploidy cannot be maintained. In hermaphrodite species of animals, among earthworms and snails for example, there is a possibility, as in plants, of self fertilization and thereby the establishment of fertile polyploid forms. As for sex determination Bridges (1922) was among the first to establish in *Drosophila*, that an X/autosome ratio of 0.5 $(X/AA$ in the normal diploid) irrespective of the presence or absence of Y-chromosomes, determined maleness; and an X/autosome ratio of 1.0, $(XX/AA$ in the normal diploid) determined femaleness. Intermediate ratios resulted in the development of intersexes. On this basis tetraploids derived from chromosome doubling to give $XX/AAAA$ and $XXXX/AAAA$ from diploid males and females respectively would develop as normal males $(X/A = 0.5)$ and females $(X/A = 1.0)$. With regular segregation to give X/AA and XX/AA sperm and eggs, however, fertilization would result in $XXX/AAAA$ $(X/A = 0.75)$ sterile intersexes. In other groups, however, the determination of sex should not be upset by polyploidy. In the mammals, for example, the sex determining mechanism is not based on an X/autosome balance. In the main it depends on the presence of Y to give males, on the absence of a Y to give females. Thus an XY mouse is a male, an XO mouse (missing a Y-chromosome) is a female and a fertile female at that (Morris, 1968). The X/A ratio is, of course, the same in both. Yet polyploid adult individuals, let alone populations, are virtually unknown in mammals. It is clear, therefore, that factors other than those we have considered above restrict the establishment of polyploidy in mammals and other groups. In humans for example only a very small number of polyploid fetuses survive anywhere near a full term pregnancy whereas 0.6–0.9% of fertilizations give rise to polyploid embryos (Hamerton, 1971). Of four triploid births recorded two were stillborn and the other two died soon after birth (Schindler and Mikamo, 1970). Evidently there is a physiological restriction upon the growth and the development of polyploids among higher animals. The causes are not known. Incompatibility between the polyploid embryo and the diploid mother may be one.

(c) *Aneuploidy*

Aneuploids, like polyploids, are again commoner among higher plants than higher animals. The higher incidence of polyploid higher plants is one cause. Most aneuploids are the direct result of the irregular disjunction of chromosomes at meiosis in polyploids which give rise to aneuploid gametes. Among the progenies of autotetraploid *Lolium perenne* from 6 to 23% are aneuploids (Ahloowalia, 1971). Even under experimental conditions, however, only a limited fraction of aneuploids are effective as gametes or viable as zygotes. Of these the 'fittest' have chromosome numbers close to the euploid values, e.g. $4x + 1$ and $4x - 1$, where the genetic imbalance is minimal. For the same reason the addition of one chromosome to the diploid complement, to give $2x + 1$ trisomics, is likely to cause less imbalance than the addition of two or three chromosomes. This is not to say, of course, that trisomics are free of the symptoms of imbalance. On the contrary they are almost invariably abnormal in development and frequently highly infertile. As one might expect the imbalance generated by trisomy is partly dependent on the particular chromosome involved. This is reflected by surveys within the cultivated tomato (Table 4.5). The table shows that the frequency of different kinds of trisomics among the progenies of triploids varies considerably, indicating that the viability of trisomics, as gametes or zygotes, is closely related to the particular chromosome addition. Trisomics for chromosomes 4 and 5 are relatively numerous in comparison with trisomics for 1 and 11. This fact also makes clear that the degree of imbalance is not related directly to chromosome length (see Table 4.5) and therefore the amount of genetic material which it carries. This is just as true for humans as for tomatoes. Among the most common aneuploids of man are those involving the X-chromosome which is by no means the smallest of the complement.

Table 4.5   The relative frequencies of the 12 trisomics in the progenies of triploid tomatoes. [From Rick, C. M. and Barton, D. W. (1954) *Genetics*, **39**, 640–66]

Chromosome number	Frequency of trisomics as percentage of total progeny	Chromosome length in microns
1	0.5	52.0
2	3.3	42.1
3	1.6	40.3
4	9.9	35.0
5	6.3	31.4
6	0.7	31.3
7	3.5	27.5
8	4.6	27.5
9	2.7	26.8
10	5.5	25.7
11	0.1	24.8
12	4.0	22.5

A few among higher plant species are exceptional in displaying tolerance to aneuploidy. The best studied is the common hyacinth, *Hyacinthus orientalis*. Garden varieties include diploids ($2x = 16$), triploids ($3x = 24$) and tetraploids ($4x = 32$). In addition, however, there is a whole range of aneuploids with chromosome numbers of 17, 19, 20, 21, 23 etc. Even here, however, some aneuploid combinations are more common, more tolerated than others. One supposes that in *Hyacinthus* efficient gene action is less dependent upon the interaction of genes on *different* chromosomes than is the case in other species (Darlington, Hair and Hurcombe, 1951). One must emphasize, however, that this situation in hyacinth is quite uncharacteristic. In the vast majority of species the consequences of aneuploidy are so deleterious on grounds of imbalance as to render aneuploidy of little consequence from the adaptive or evolutionary standpoint.

B-chromosomes

The chromosome complements of individuals within a species exhibit a remarkable degree of constancy in terms both of the number and form of their chromosomes. A notable exception to this general rule is provided by *B*-chromosomes. These chromosomes are supernumerary to the normal basic complement and are found in numerous species of both plants and animals. They are calls *B*s to distinguish them from the normal (*A*) chromosomes and they are characteristically smaller than the *A*-chromosomes, not homologous with any of the *A*s in that they never pair with *A*s at meiosis, variable in number both between populations within species and among individuals within populations. The *B*-chromosomes, on grounds of their size and pairing behaviour at meiosis, have diverged substantially in structure from the *A*-chromosomes from which presumably they derive. The DNA which they contribute to the nucleus, therefore, does not represent a precise copy of the nuclear DNA fraction as is the case with aneuploidy. Their evolution involves a qualitative change superimposed upon a change in amount. Even in those populations with a high frequency of *B*s, however, there are some individuals without *B*s. Evidently they are dispensable, and not essential for normal growth, development and reproduction. For this reason it is not surprising that they were often considered to be genetically inert and of little adaptive or functional significance. This view was reinforced by lack of evidence of any major gene effects attributable to *B*s. More recently, however, it has become increasingly clear that, although dispensable, the effects of *B*s upon the phenotype are manifold and in some instances quite startling. The indications are that they are of substantial adaptive significance.

(a) *Occurrence*

B-chromosomes were first distinguished and classified by Longley (1927) and Randolph (1928) in maize. The numbers of higher plants and

Table 4.6   *B*-chromosome distribution among groups of plant and animal species

Group	No. of species
PLANTS	
Gymnosperms	7
Angiosperms	
Dicotyledons	318
Monocotyledons	319
Total	644
ANIMALS	
Platyhelminths	2
Molluscs	2
Insects	158
Amphibians	4
Reptiles	1
Mammals	11
Total	178

of animal species now known to carry *B*-chromosomes is considerable, as Table 4.6 shows. Contrary to previous assertions, *B*-chromosomes are found in about the same proportion of polyploid as of diploid species. In fact in some cases, such as in *Leucanthemum* populations in Yugoslavia, *B*s are found only within the polyploid populations. In other species, e.g. *Ranunculus ficaria* in Britain, the opposite situation prevails and the *B*s are restricted to diploids. *B*-chromosomes have not been reported in inbreeding species.

Table 4.7 gives the range of variation in numbers of *B*s found among natural populations. Under experimental conditions even higher numbers, up to 34 in *Zea mays* and 22 in *Centaurea scabiosa*, are reported. These two examples serve to illustrate the very considerable amount of additional DNA that may be accumulated within the nucleus due to *B*-chromosomes. In maize the addition of 34 *B*s to the diploid complement of 20 *A*-chromosomes is equivalent to the amount of DNA that would be added by having the *A*-chromosome set multiplied five times, equivalent that is to pentaploidy.

(b) *Inheritance*

Somatic cell division. In many species the transmission of *B*-chromosomes during somatic cell division is normal and in consequence their frequency is constant throughout the somatic cells within the organism. Among plants this is true, for example, in *Secale cereale*, *Anthoxanthum odoratum*, and *Clarkia elegans*; among animals, for *Acrida lata*, *Myrmeleotettix maculatus*

Table 4.7 The range of *B*-chromosome numbers in some plant and animal species

Species	Number of *B*-chromosomes																	
	0	1	2	3	4	5	6	7	8	9	10	11	12	13	14	15	16	Σ
Centaurea scabiosa Scandinavia and Finland	5688	611	564	333	218	128	63	64	39	8	15	1	1	1	2	1	1	7738
Festuca pratensis Sweden	854	97	68	8	11	3	3	2	—	—	—	—	—	—	—	—	—	1046
Crepis conyzaefolia Italy	133	31	11	6	1	—	—	—	—	—	—	—	—	—	—	—	—	182
Tainia laxiflora Japan	5	10	49	17	24	5	2	1	4	2	—	—	—	—	—	—	—	119
Ranunculus ficaria Britain	45	8	12	15	5	2	9	3	—	—	—	—	—	—	—	—	—	99
Reithrodontomys megalotis U.S.A.	39	89	94	45	15	2	0	1	—	—	—	—	—	—	—	—	—	285
Myrmeleotettix maculatus Wales	90	41	9	—	—	—	—	—	—	—	—	—	—	—	—	—	—	140

and *Pseudococcus obscurus*. In other cases the *B*s are unstable at mitosis so that their frequency varies between different tissues and organs. The instability takes the form of lagging at anaphase or of non-disjunction. At the time of flower initiation in *Crepis capillaris* the *B*-chromosomes regularly undergo non-disjunction in shoot meristems. The non-disjunction is polarized such that the *B*-chromosome frequency is increased in the germ line cells. In *Xanthisma texanum, Aegilops speltoides* and *Haplopappus gracilis*, *B*-chromosomes are excluded from the roots but are present in the shoots and inflorescences. Among animals species mitotic instability is common during the development of the testis, leading to variation in *B* frequency among cells within and between the follicles, e.g. *Locusta migratoria, Pantanga japonica* (Kayano, 1971; Sannomiya, 1962).

Germ line cells. When a rye plant carrying one *B*-chromosome is crossed to a rye plant without *B*s, the progeny resulting from this cross consist mainly of 0*B* and 2*B* individuals. If two rye plants each carrying 2*B* chromosomes are crossed together the bulk of the offspring each possess 4*B*s. Clearly the mode of inheritance of these additional *B*-chromosomes is non-Mendelian and leads to a numerical accumulation of the *B*s over a succession of generations with a preponderance of the even numbered combinations. The mechanism which accounts for this unorthodox mode of inheritance is now firmly established for rye. Meiosis itself is fairly regular (see Fig. 4.23). With 1*B*, segregation in pollen mother cell is such that two of the four products carry 1*B* and the other two none. If 2*B*s are present they form a bivalent at the first division and segregate one *B* to each of the four microspores of the tetrad. There is never pairing between *A*s and *B*s. It is during the first mitotic division of the microspore, i.e. first

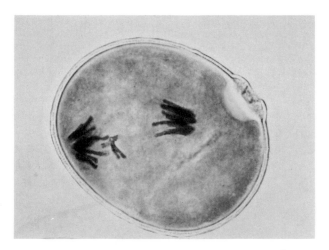

Fig. 4.9   Anaphase of the first pollen grain mitosis in rye showing a single *B*-chromosome undergoing non-disjunction.

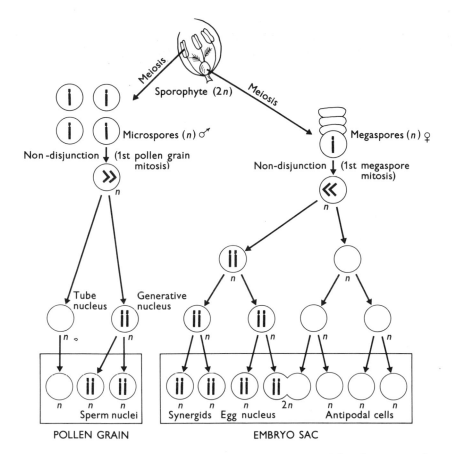

Fig. 4.10 The transmission of B-chromosomes into male and female gametes of a rye plant with two B-chromosomes. The non-disjunction at the first pollen grain and at the first megaspore cell mitosis foster the accumulation of Bs among progenies. Note that sperm and egg nuclei both contain two B-chromosomes; hence a cross between rye parents each with two B-chromosomes yields progeny with 4 B-chromosomes. Only the B-chromosomes are drawn.

pollen grain mitosis, that the Bs behave abnormally. The haploid complement of A-chromosomes divides mitotically with one set passing into the generative nucleus and the other to the tube nucleus of the developing male gametophyte. At the anaphase of this first pollen grain mitosis, however, the daughter chromatids of the B-chromosomes fail to disjoin even though the centromeres have divided, and they pass together preferentially into the generative nucleus (Fig. 4.9). They divide quite normally at the second pollen grain mitosis. In the embryo sac a similar sequence of events takes place with the Bs undergoing a polarized non-disjunction at the first mitosis of the functional megaspore, again in such a direction that they ensure their perpetuation in the germ line. The scheme is outlined in Fig. 4.10.

Some element of non-disjunction and accumulation is found in most of the plant species investigated, although the details of the mechanism vary somewhat from one species to another. In *Zea mays*, for example, polarized non-disjunction occurs at the *second* pollen grain mitosis, while transmission through the female gametophyte is quite normal. Preferential fertilization by the *B*-carrying male gametes, is a further cause of distortion from the normal Mendelian pattern in maize (Roman, 1948).

In rye, non-disjunction is brought about by failure of chromatid separation at two sites, one each side of the centromere. While the 'sticking' of chromatids is near the centromeres, the capacity for 'sticking'

Table 4.8   Mechanisms of *B*-chromosome accumulation in plants

Mechanism	Species
Preferential meiotic segregation in egg mother cells	*Lilium callosum* *Tradescantia virginiana* ' *Phleum nodosum*
Preferential fertilization by *B*-carrying male gametes	*Zea mays*
Directed non-disjunction: First pollen grain mitosis	*Aegilops speltoides* *Alopecurus pratensis* *Anthoxanthum aristatum* *Briza media* *Dactylis glomerata* *Deschampia caespitosa* *Festuca pratensis* *Haplopappus gracilis* *Phleum phleoides*
Second pollen grain mitosis	*Zea mays*
First pollen grain mitosis and first egg cell mitosis	*Secale cereale*
First pollen grain mitosis of extra divisions	*Sorghum purpureo-sericeum*
Somatic non-disjunction coincident with flower initiation	*Crepis capillaris*
No apparent mechanism	*Centaurea scabiosa* *Poa alpina* *Ranunculus ficaria* *Xanthisma texanum*

is under the control of a 'gene' located in the terminal block of hetero-chromatin in the long arm. Following deletion of the controlling site the B-chromosome loses its capability of undergoing non-disjunction. Ward (1973) has established that the controlling site for non-disjunction in maize lies in a short euchromatic segment distal to the major hetero-chromatic region of the long arm, at which the 'sticking' takes place. In maize the B-chromosomes are also the cause of sticking at heterochro-matic 'knobs' in the A-chromosomes at the second anaphase of meiosis with the result that segments of A-chromosomes carrying knobs are eliminated (Rhoades and Dempsey, 1973).

The accumulation mechanism in *Lilium callosum* differs from those in rye and maize in that it operates at meiosis and not in the post-meiotic divisions in the gametophyte. Transmission through the pollen is normal, but at female meiosis there is a preferential distribution of *unpaired Bs* in egg mother cells towards that half of the spindle nearest the micropylar end of the embryo sac. In *Lilium* the functional megaspore is proximal to the micropyle so that the Bs are accumulated in the egg cells. A summary of some of the systems of B-chromosome inheritance in plants is given in Table 4.8. By far the commonest is that involving polarized non-disjunction at the first pollen grain mitosis with normal transmission through the female.

B-chromosomes of course cannot accumulate indefinitely. The tendency to increase in number over successive generations is countered, as we shall see, by a severe impairment of reproductive fitness when high numbers of Bs are present, as well as by some loss through mechanical inefficiency at meiosis and some infidelity in the non-disjunction process.

In *Locusta* and *Pantanga* species, as we have already indicated, there is frequently variation in the frequency of the B-chromosomes among and within the follicles of the testis. It arises from non-disjunction at mitosis. There is some evidence that, as a result of non-disjunction, cells without B-chromosomes contribute disproportionately fewer descen-dants to the germ line spermatocytes than cells which carry Bs. There is, in other words, an element of somatic selection for Bs (Kayano, 1971).

(c) *Structure and organization*

B-chromosomes are generally smaller than the A-chromosomes. Their behaviour at meiosis shows they are not homologous with A-chromo-somes. This suggests a structural divergence from the As which has been confirmed by comparisons of the chromomere patterns of As and Bs at pachytene of meiosis in rye and maize. B-chromosomes from the same population are generally similar in structure and their pairing behaviour at meiosis confirms their homology. Structural variants are by no means infrequent, however. An extreme case is found in *Aster ageratoides*. A standard form of B is accompanied by as many as twenty-four structural variants within a single population (Matsuda, 1970). The standard and morphological variants of the B-chromosomes in rye are shown in

Fig. 4.11. Polymorphism in respect of *B*s is common also in animal species. In *Melanoplus femur-rubrum*, for example, there is a large metacentric *B*, called B^m and two telocentrics, B^l and B^t, derived from it by breakage of the centromere.

The genetic organization of *B*-chromosomes is not clear. No Mendelian genes with major effects have been located on them other than those concerned with the control of non-disjunction, which we have described, and over the pairing of chromosomes in species hybrids (see p. 83). Neucleolus organizer regions are also conspicuously absent in *B*-chromosomes, although there is evidence of genes coding for ribosomal RNA in the *B*-chromosomes of rye (Flavell and Rimpau, 1975). The effects of *B*-chromosomes are much like those attributed to polygenic systems, producing a continuous rather than a discontinuous variation in the phenotype that is not readily distinguished from variation due to the environment. It was for this reason that they were initially considered to be inert, a view reinforced by the claim that *B*-chromosomes are frequently, but by no means always, heterochromatic (Jones, 1975).

The DNA composition of *B*-chromosomes, as revealed by biochemical analysis, does not differ in any significant way from that of the *A*-chromosomes in the three plant species so far investigated. In rye, maize and *Aegilops speltoides* the composition of the *B*-chromosome DNA is indistinguishable from that of the *A*-chromosomes, in respect of base ratio and the proportion of repetitive and non-repetitive sequences. The same applies to the grasshopper species *Myrmeleotettix maculatus* (Dover and Henderson, 1975). If the non-repetitive DNA in *B*s, like that of *A*s, embodies structural genes of major effect one must assume that the information they embody either is not transcribed or else that it merely duplicates to a large degree that contained elsewhere in the complement.

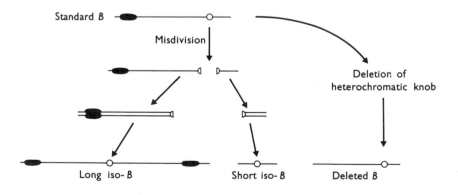

Fig. 4.11   *B*-chromosome polymorphism in rye, due to centromere misdivision and deletion of the standard fragment.

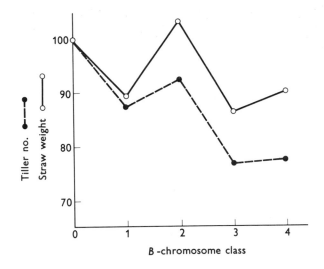

Fig. 4.12 The influence of *B*-chromosomes on vegetative characters in rye expressed as percentage of control (i.e. 0*B* plants). [Data from Müntzing, A. (1963) *Hereditas*, **49**, 371–426]

(d) *Effects*

Although most of the effects of *B*s upon the phenotype are of a quantitative rather than a qualitative nature, comparisons between large samples of individuals carrying different numbers of *B*s reveal that their effects are manifold and impinge on all stages of development. Fig. 4.12 and Table 4.9 show the effects of *B*s on morphological characters and fertility in rye plants. The general reduction in fertility and in vigour, as reflected by the morphological changes, with increase in *B* frequency is clear. We shall return later to consider the striking zig-zag pattern of variation in relation to odd and even numbers of *B*s in this species. The evidence from numberous other species conforms in demonstrating reduced growth and fertility associated with high numbers of *B*s, e.g. in the mealy bug, *Pseudococcus obscurus*. In passing it is worth noting that enhanced vigour is attributable to *B*s in low frequency in *Centaurea scabiosa* (Fröst, 1958). In

Table 4.9 Fertility, expressed as percentage seed set in rye plants with 0 to 8 *B*-chromosomes

B class	0	1	2	3	4	5	6	7	8
% seed set	49.5	31.4	34.2	21.5	5.1	7.1	1.7	0.1	0

view of the widespread distribution of *B*s among natural populations we might indeed expect that under certain circumstances they may enhance fitness.

Effects of *B*-chromosomes are equally clearly manifested at the cell level. The duration of the mitotic cycle in maize, *Lolium perenne* and rye is increased with the addition of *B*-chromosomes (see Fig. 4.23). Cell size in rye is also increased by the addition of *B*-chromosomes. In addition there is a substantial variation in the amount of RNA, and of protein due to *B*s (Fig. 4.13). It will be observed that total protein and RNA decrease overall in cells with high numbers of *B*s. The figure shows the same characteristic zig-zag relationship between phenotypic change and *B* frequency as applied to the morphological characters plotted in Fig. 4.12. The correspondence is not surprising because the morphological varia-

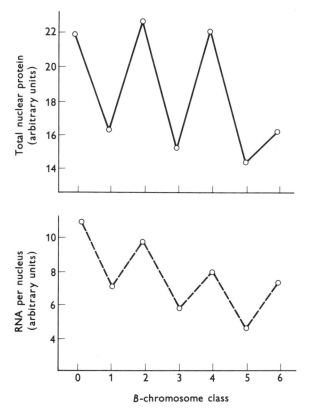

Fig. 4.13   The influence of *B*-chromosomes on the nuclear phenotype in rye. Quantitative assessments of total nuclear protein and RNA were made by micro-densitometry after staining isolated root meristem nuclei with dinitrofluoro-benzene (for total protein) and methyl green-pyronin *Y* (for RNA). [Data from Kirk, D. and Jones, R. N. (1970) *Chromosoma*, **31**, 241–54]

tion must, ultimately, be a reflection of variation in cell metabolism of the kind we have described.

(e) *Meiosis and recombination*

In 1956 Darlington suggested that *B*-chromosomes may serve to regulate the amount of variability among diploid populations. Support for this suggestion comes from work on rye by Moss (1966) who showed that the range of heritable variation for quantitative characters was greater among the progenies of plants with *B*s than those without *B*s. The increase in variability was over and above that due to the presence of *B*s among the progenies themselves. The control exerted by *B*s over the variability can be accounted for by their influence upon chiasma formation at meiosis.

In *Myrmeleotettix maculatus* the chiasma frequency of the *A*-chromosomes is increased in the presence of *B*-chromosomes. In spermatocytes without *B*s the chiasma frequency in one population averaged 14.13 per cell. With *B*s the frequency was 15.60 (John and Hewitt, 1965). It was shown also that the variation in chiasma frequency between cells and between bivalents within cells increased in the presence of *B*s, such that some cells and some bivalents had unduly high chiasma frequencies, others low. Both the effects on the mean and the variation in chiasma frequency could generate additional variability among gametes and progenies, the first by increasing the amount of recombination in general, the second by increasing the recombination in particular cells or particular chromosomes. It is the variation in chiasma frequency, both between and within pollen mother cells that is mainly affected by *B*-chromosomes in rye. They increase with increasing *B* frequency. It is worth emphasizing that in species like rye an increase in the chiasma frequency within a bivalent is accompanied also by a change in chiasma distribution. In rye bivalents with one or two chiasmata the chiasmata are restricted to the ends of the chromosomes. Additional chiasmata are formed interstitially. The effect therefore is not only to increase recombination generally but to generate recombination and novel recombinants from different regions of the chromosomes.

Figure 4.14 shows the effects of *B*-chromosomes on mean chiasma frequencies in maize, *Listera ovata* and in *Lolium perenne*. The *Lolium* results are of interest in showing that *B*s, while unquestionably controlling the chiasma frequency do so in this instance by depressing the number. This applies also in *Aegilops speltoides* (Zarchi *et al.*, 1972). While the effects of *B*s on chiasma formation and recombination vary somewhat from species to species there is no doubting their influence upon recombination and thereby upon the variability of gametes and of offspring.

(f) *Chromosome pairing in hybrids*

The chromosomes of *Lolium perenne*, a perennial outbreeding grass, are about 30% smaller and contain about 30% less DNA than the chromo-

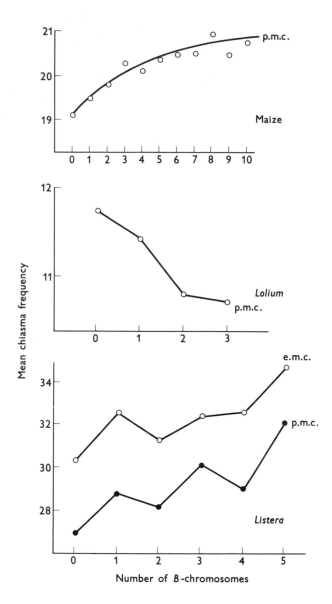

Fig. 4.14 The influence of *B*-chromosomes on the mean chiasma frequency of the *A*-chromosomes in pollen mother cells (p.m.c.) and in embryo sac mother cells (e.m.c.) in three species of flowering plants. [The maize data is taken from Ayonoadu, U. and Rees, H. (1968) *Genetica*, **39**, 75–81; the *Lolium* data from Cameron, F. M. and Rees H. (1967) *Heredity*, **22**, 446–450; and the *Listera* data from Vosa, C. G. and Barlow, P. W. (1972) *Caryologia*, **25**, 1–8]

somes of *L. temulentum*, an annual, inbreeding species of the same genus. Both are diploids with seven pairs of chromosomes. At meiosis in the diploid hybrid an average of 6, out of the possible 7, bivalents are formed in pollen mother cells. This is surprising because each bivalent comprises structurally different chromosomes, homoeologues rather than strict homologues. When the *B*-chromosomes of *L. perenne* are introduced into this F_1 diploid hybrid, however, there is a startling reduction in the bivalent frequency from six to four or fewer per cell. In tetraploids synthesized from the diploid hybrids the effect of the *B*s is even more dramatic (Fig. 4.15). Without *B*s there is pairing between both homoeologous and homologous chromosomes resulting in multivalent formation. The chromosome behaviour in fact is typical of that in an autotetraploid. When *B*s are present pairing is restricted to homologous pairs only, and all metaphase I configurations are bivalents as in an allotetraploid. That the bivalents result from the pairing of homologous chromosomes is established by their symmetry (Evans and Macefield, 1972). A similar 'diploidizing' effect of *B*-chromosomes has been reported in polyploid hybrids between the outbreeding diploids *Aegilops speltoides*, *Aegilops mutica* and the cultivated wheat *Triticum aestivum*. Hexaploid wheat normally behaves as an allopolyploid due to the presence of a gene on chromosome 5B which suppresses homoeologous pairing between chromosomes from the three constituent genomes (see p. 68). The *B*-chromosomes can substitute for the activity of this 5B gene in experimental situations where the 5B gene is absent (Dover and Riley, 1972).

(g) Odds and evens

One of the more puzzling aspects of the activity of *B*-chromosomes is the contrast between their effects upon the phenotype in odd and even numbers in certain plants. This is illustrated in respect of morphological and cytological characters in Figs. 4.12, 4.13, 4.14. The basis for such effects is not understood. A possibility put forward by Jones and Rees (1969) is that the *B*s may associate in pairs at interphase and that their activity in contiguous pairs differs from that as single chromosomes.

It will be recalled that the transmission of *B*s in many species is such as to generate an excess of gametes and progenies with even numbers. It is tempting to speculate therefore that the non-disjunction mechanism by which this is achieved is no mere fortuitous aberration but is an adaptive device ensuring a preponderance of individuals with even numbers of *B*s. The argument is strengthened by the fact that the effects of *B*s upon growth, in even numbers, are the least 'deleterious' (see Fig. 4.12).

(h) Selection and adaptation

Many of the phenotypic effects attributable to *B*-chromosomes are deleterious to growth and fertility. This is especially true for organisms carrying high numbers of *B*s. For this reason it has been argued that *B*s are nuclear 'parasites' and that their maintenance within populations is

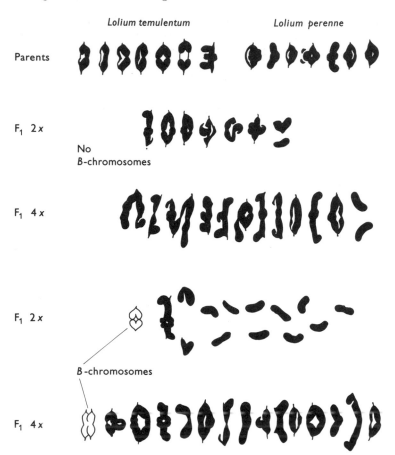

Fig. 4.15 B-chromosome control over chromosome pairing in *Lolium* species hybrids. In the absence of Bs pairing takes place between *homoeologous* chromosomes in the F_1 diploid and derived tetraploid hybrids. In the presence of Bs there is almost complete suppression of homoeologous pairing leading to bivalents only in the tetraploid. Pairing relationships in the hybrids are readily determined because of the size differences between the two parental complements.

solely dependent upon the mechanism of their transmission in heredity which tends to accumulate their numbers relative to the normal chromosomes of the complement. Yet B-chromosomes are widespread within many species and there are grounds for concluding that they confer superior fitness upon individuals and populations under certain experimental conditions.

(i) *Individuals.* B frequencies vary among populations within species. The variation, furthermore, is related to environment. Among populations of *Myrmeleotettix*, *Phleum* and *Centaurea* the B frequencies vary in relation to soil type or climate. In *Myrmeleotettix*, for example, populations in drier warmer areas have more Bs than those in colder, wetter regions of Mid Wales (Hewitt, 1973). Such observations leave no doubt that the B frequency variation is dependent upon the environment. At the same time they do not establish that B-chromosomes enhance the fitness of individuals because one could argue that the frequency of B-chromosomes may increase within a population simply because the selection pressure is not sufficient to counter their accumulation during reproduction, and vice versa. There is, however, experimental evidence which shows that individuals carrying B-chromosomes display superior fitness in competition with individuals without Bs under conditions of severe selection pressure.

   (ii) *Selection in Lolium.* Seedlings from a population of *Lolium perenne* carrying Bs were planted at different sites, one at sea level, the other at 900 feet (270 m); at different spacing, 6 inches (15 cm) and 2 feet (60 cm) apart at two seasons (spring and autumn). The mortality among seedlings varied, as would be expected, with the different 'treatments'. Of particular interest is the fact that the proportion of plants carrying Bs among those surviving to maturity increased directly in relation to mortality. The conclusion is that under these conditions of increasing stress the fitness of individuals carrying B-chromosomes was superior to plants without Bs (Rees and Hutchinson, 1973).

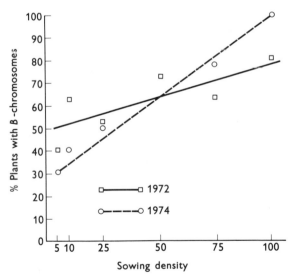

Fig. 4.16 The proportion of plants with B-chromosomes amongst survivors of *Lolium perenne* populations grown at different densities. The data from both years show the increase in the proportion of Bs with increasing density. In 1974 only B plants survived in the high density populations in which the mortality is most severe. [Data from S. B. Teoh]

Further experiments in *Lolium* support this conclusion. Seeds from a population with *B*s were sown in small, 2 inch (5 cm) pots at densities varying from 5 to 100 per pot. Competition for nutrients etc. leads to increasing mortality with increasing sowing density. Fig. 4.16 shows the proportion of plants with *B*-chromosomes among survivors at the end of the first and third year. The proportion increases with increasing density of sowing. After three years as the graph shows, only plants with *B*-chromosomes survive the most severe selection pressures under conditions of high density. There is no doubt, therefore, that under certain conditions the *B*-chromosomes confer adaptive advantage. Fig. 4.16 is instructive, also, in showing that plants without *B*s in this experiment survive better than *B* plants in conditions of low sowing density. It will be observed that the proportion of *B* plants among survivors *drops* between 1972 and 1974 at a sowing density of five per pot. It is clearly important therefore to stress that the fitness in relation to *B*s is very much influenced by the particular environmental conditions which pertain.

(iii) *Populations.* *B*-chromosomes in most of the species investigated affect the frequency and distribution of chiasmata at meiosis. They must inevitably influence the amount and distribution of recombination and thereby the variability of offspring among populations. By these means they presumably influence the variability of the population as a whole and its capacity to adapt to a particular environment or to a changing environment. From the standpoint of selection and adaptation, therefore, the *B*-chromosome systems present a dual aspect. On the one hand they affect the fitness of the individual in terms of growth and development. On the other they affect the fitness of the population through their influence upon the variability made available for selection and adaptation.

(i) *Origin*

The way in which *B*-chromosomes originate is not known. In plants it is assumed they are derived from *A*-chromosome fragments which, subsequently, become structurally modified to a degree that conceals any residual homology with the chromosome ancestor. In many animal species the *B*-chromosomes resemble and show affinities with the *X*-chromosome (Hewitt and John, 1972). They often associate with the single *X* at prophase of meiosis although there is no case known of a chiasmate association. Even so the affinity between the *B* and the *X* is indicative of some degree of homology between the two with the implication that the *B* is a derivative of the *X*.

Amplification and deletion within chromosomes

The chromosomes of different species vary not only in number but also in size. Fig. 4.17 shows the complements of four diploid species of *Lathyrus*, each with 14 chromosomes. While the number is constant there is a 3-fold difference in total chromosome size between *L. hirsutus*, with

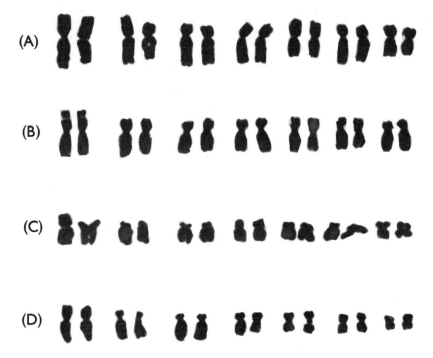

Fig. 4.17   The chromosome complements of four *Lathyrus* species. (A) *L. hirsutus*;
A(B) *L. ringitanus*; (C) *L. articulatus*; and (D) *L. angulatus*.

the largest, and *L. angulatus* with the smallest chromosomes. The variation in size is very closely correlated with DNA content (Fig. 4.18). This
survey in *Lathyrus* is typical of many which have established that substantial variation in nuclear DNA often accompanies the divergence and
evolution of species, both in plant and animal genera. The magnitude of
the DNA variation between species, aside from that due to polyploidy,
is of a particularly high order in higher plants, in algae and in the
amphibia (Fig. 4.2). In the higher plants for example, the diploid *Lilium
longiflorum* has 100 times more DNA than *Linum usitatissimum* (see Rees,
1972). Even within the one family, the Ranunculaceae, there is an 80-fold
variation. The variation among other groups such as the mammals and
fishes is much less. Why this should be so is not known.

Two explanations have been put forward to explain the evolutionary
changes in chromosomal DNA. The first is a *differential polynemy*, namely
a change in the number of DNA strands within the chromosomes of
different species (Fig. 4.19). It is now generally accepted, however, that
the chromosomes, other than in specialized cells, are *mononemic* in all
species and this proposal is therefore untenable. The second explanation
is *lengthwise amplification and deletion* of segments to increase or decrease the

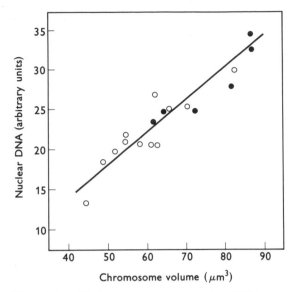

Fig. 4.18   The relationship between the mean nuclear DNA amount and total chromosome volume in 18 *Lathyrus* species. ●, perennial outbreeders; O, annual inbreeders. [From Rees, H. and Hazarika, M. H. (1969) *Chromosomes Today*, 2, 158–65]

amount of DNA within chromosomes (Fig. 4.19). The repetition of base sequences within the DNA of many species, to which we have referred earlier, is testimony enough to the capacity for amplification of chromosome segments. A substantial amplification and, for that matter, deletion of repeated sequences may be achieved during a short period of time. The 'bobbed' mutation in *Drosophila* is associated with loss of cistrons which code for ribosomal RNA in the X and Y chromosomes. The chromosome carrying the bobbed mutation may be restored to the wild-type condition by amplification of the rDNA cistrons to the number characteristic of the normal chromosome (Ritossa and Scala, 1969). In

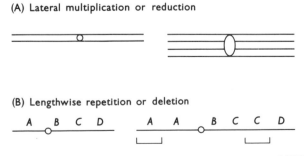

Fig. 4.19   Alternatives for quantitative change in chromosomal DNA.

Linum usitatissimum the number of rDNA cistrons and, indeed, of a large fraction of the DNA elsewhere in the chromosome complement is induced by specific environmental treatments (Timmis and Ingle, 1973; Evans, Durrant and Rees, 1966). Again, in a small proportion of cells in the hybrid *Nicotiana otophora* × *N. tabacum* a heterochromatic segment in one of the chromosomes is amplified to a prodigious extent to produce a chromosome 20 or 30 times the normal size and this during the span of one or a few interphases (Gerstel and Burns, 1966). Other evidence to confirm a lengthwise amplification as a cause for DNA increase within chromosomes comes from comparisons between related species with differing amounts of nuclear DNA.

(a) *Chironomus*

Chironomus thummi thummi has 27 % more nuclear DNA than *Chironomus thummi piger*. The DNA content of certain of the bands in the polytene salivary gland chromosomes of *C. thummi thummi* is greater than in corresponding bands of *C. thummi piger*. The DNA difference between bands ranges from a factor of 2, to 4, 8, 12 and 16. The conclusion is that the higher DNA content of *C. thummi thummi* chromosomes results from localized lengthwise DNA repetition within bands. The series 2, 4, 8 etc. indicates a repeated duplication of the original bands or their copies (Keyl, 1965).

(b) *Allium*

The DNA content and the total chromosome volume in nuclei of *Allium cepa* are each 27 % greater than in *A. fistulosum*. Despite the disparity in DNA amount the species hybridize readily. At meiosis in the F_1 hybrid all eight bivalents are asymmetrical, showing that the DNA changes associated with the divergence of the two species affected all chromosomes of the complement. Some bivalents, however, are more asymmetrical than others (Fig. 4.20). In some the difference in length between 'homologous' chromosomes at metaphase is no more than 10%. At the other extreme there is a 70 % difference between the lengths of 'homologues'. At pachytene unpaired loops and overlaps range, as would be expected, with lengthwise repetition, from 10 to 70% of the length of the paired chromosome segments (Fig. 4.20 and Fig. 4.21). Similar loops and overlaps have been described in a *Lolium* species hybrid (Rees and Jones, 1967).

The number of loops or overlaps in each pachytene configuration is restricted to one or two. This suggests that amplification within chromosomes is highly localized. In the paired regions, furthermore, the pairing appears to be regular and complete which indicates that the chromosomes of *A. cepa* and *A. fistulosum* are structurally similar except for the localized regions of amplification in *A. cepa*. Restriction of amplification to certain regions of the chromosomes means either than such regions

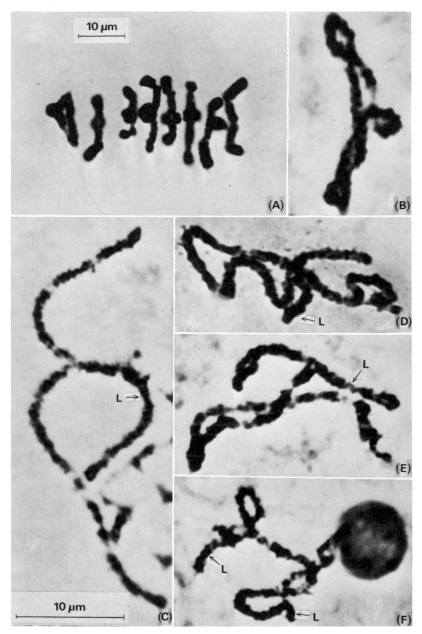

Fig. 4.20 Chromosome pairing in the F₁ hybrid between *Allium cepa* × *A. fistulosum*. (A) Metaphase 1 of meiosis showing 8 asymmetrical bivalents. (B) Diplotene with a pairing loop. (C)–(F) Pachytene bivalents with pairing loops (L). [From Jones, R. N. and Rees, H. (1968) *Heredity*, **23**, 591–605]

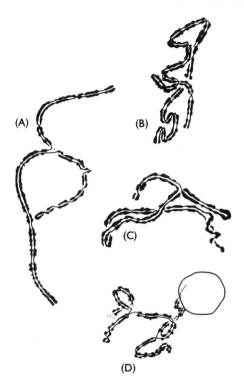

Fig. 4.21 The loops at pachytene in *Allium cepa* × *A. fistulosum*. (A), (B), (C) and (D) correspond with (C), (D), (E) and (F) in Fig. 4.20.

are prone to repetition or else that the consequences of repetition of these segments are tolerated whereas in other segments they are not. In either case the indications are that a special fraction of the chromosomal DNA is more likely to be amplified.

(c) *The location of quantitative DNA change*

Among *Lathyrus* species (Narayan and Rees, 1976) the amount of repetitive DNA is closely correlated with the total nuclear DNA amount (Fig. 4.22). The non-repetitive fraction does not vary significantly among species. We conclude that quantitative DNA variation during the evolution of these species is achieved through change in the amount of the repetitive component. It is generally assumed that the structural genes, i.e. those which are transcribed and translated into protein, are located mainly in the non-repetitive fraction. If this is so the variety of structural genes in *Lathyrus*, despite the massive alteration in DNA, remains relatively constant. The most obvious implication is that the change in total DNA concerns regulatory material but precisely what

94

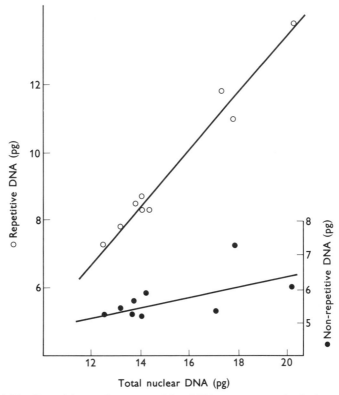

Fig. 4.22 Repetitive and non-repetitive DNA components in *Lathyrus* species plotted against the total nuclear DNA amount. [From Narayan, R. K. J. and Rees, H. (1976) *Chromosoma*, **54**, 141–54]

kind of regulation is involved is unknown. In *Lathyrus*, also, increase in DNA is associated with increase in the proportion of heterochromatin within the nucleus. We infer that the repetitive fraction is therefore located mainly in the heterochromatin. It will be recalled that pro-digious DNA repetition is associated with centromeric heterochromatin in mammals and in *Drosophila* species. It will also be recalled that a heterochromatic segment in the *Nicotiana otophora* × *N. tabacum* hybrid is the site of rapid and extensive proliferation. This is not to say, of course, that DNA repetition is confined to heterochromatin. Indeed one of the earliest examples of repetition is that of the euchromatic *Bar* locus in *Drosophila* (Bridges, 1936). At the same time it is worth noting that a high degree of repetition is not necessarily reflected by a high heterochromatic content within nuclei. In wheat, for example, 60 to 80 % of the DNA is repetitive, yet the amount of heterochromatin is very small indeed (Britten and Kohne, 1968; Smith and Flavell, 1975).

Fox (1971) finds in *Dermestes* beetles, as in *Lathyrus*, that DNA variation among species implicates mainly the repetitive fraction and, on a cytological basis, it is the heterochromatin which is increased disproportionately to euchromatin. Heterochromatin, it is frequently asserted, fulfils a regulatory role in gene activity but, as we have mentioned above, the mechanism of such regulation remains conjectural. One concludes, therefore, that the DNA variation among species appears, to a large extent, to be restricted to particular nuclear components. What this signifies, in physiological and in evolutionary terms, is by no means clear. It is worth, perhaps, suggesting that the answer may not be found in terms of one particular function but of many, embracing not only the activities of particular genes but the mechanics of chromosome organization and behaviour in general.

(d) *DNA loss*

Taking the broad view, the evolution of living organisms from microorganisms to higher plants and animals is accompanied by DNA increase. Yet within groups, both plant and animal, the evolution of (taxonomically) more advanced and specialized species is often accompanied by DNA loss. This is true for example of the bony fishes and of reptiles (Bachmann, Goin and Goin, 1972). Within plant groups also, the more advanced species often have less DNA than the more primitive, as in the genus *Crepis* and in *Lathyrus* (see Rees and Jones, 1972). In *Lathyrus* the loss is of the order of 60 % of the total nuclear DNA (Rees and Hazarika, 1969). In one sense this is surprising because the loss even of minute chromosome segments, induced by ionizing radiation for example, is more often than not lethal to the cell. An alternative explanation for the decreasing DNA amounts with specialization would be that DNA amounts had *increased* in older primitive forms while they remained more or less unchanged in the more advanced forms. Equally surprising are the spectacular increases in DNA associated with the evolution of groups such as the lungfishes, which, along with the urodele amphibia, have the highest DNA contents among eukaryotes (Morescalchi, 1973).

Ohno (1970) argues that sudden substantial increase in DNA marks the initiation of many evolutionary pathways; indeed that the duplication of genes, whether by polyploidy or repetition within chromosomes, is essential for the development of new, more complex forms. He maintains, for example, that DNA increase through polyploidy, among fish or amphibian ancestors was of crucial importance to the evolution of reptiles, birds and mammals. The basis of Ohno's compelling argument is that the evolution of a new gene from an existing one is achieved only by numerous alterations in the base composition at that locus. For this reason the function of one gene is lost before that of the new form is acquired. With duplication, one copy retains its functional efficiency while its copy accumulates those DNA base changes that ultimately specify the new product of what is now a new gene. This view, that major advances in evolution are dependent on a massive increase in DNA, is accompanied, paradoxically, by the argument that the 'redundant' component of the increased nuclear DNA is lost as species become more specialized. We have cited examples above of such apparent DNA loss in both plant and animal groups. Whatever the general pattern, however, the situation is by no means predictable or consistent. While in *Lathyrus* the more specialized, inbreeding species have more DNA than the more primitive outbreeders, in *Lolium* the more advanced inbreeders have more DNA than the outbreeding species (Rees and Jones, 1967). Inbreeding species in *Pinus* also have more DNA than outbreeders (Dhir and Miksche, 1974). In *Allium* the advanced groups of species with 14 and 18 chromosomes have more and less nuclear DNA respectively than the primitive group of *Allium* species with 16 chromosomes (Jones and Rees, 1968). Since we do not know precisely what kind of DNA is involved by these changes or how it functions there is no means by which we can evaluate the significance of DNA loss as opposed to gain during the divergence and evolution of new species.

4.2 DNA amount and phenotype

It will be recalled that polyploidy is frequently, although not invariably, identified with well-defined morphological and physiological modifications in the phenotype. In one sense this is surprising because the additional genetic material resulting from polyploidy varies in quality according to the species and, indeed, the genotype of the individual. The question arises, therefore, as to whether the addition of genetic material, of nuclear DNA *per se*, has certain inevitable effects upon the growth and development of the phenotype. If so we might well expect to find an effect on the duration of cell division on the naive assumption that the greater the nuclear DNA amount the longer the time required for the replication and division of the chromosomes.

DNA and duration of the mitotic cycle

In Fig. 4.23 the duration of the mitotic cycle in the root meristem of flowering plant species is plotted against the nuclear DNA amount. It is immediately apparent that the duration of the cycle increases with increasing DNA amount (by about 0.3 hours per picogram of DNA). This is despite the fact that the increase in DNA is achieved in a variety of ways, namely polyploidy and aneuploidy, the amplification of chromosome segments within diploid complements or the addition of *B*-chromosomes; in other words by the addition of DNA of widely differing quality and composition. This is not to say that the duration of the cycle is completely independent of DNA quality. Fig. 4.23 shows the mitotic cycles to be about 4 hours longer in dicotyledons than in monocotyledons with the same nuclear DNA amounts. Also, the rate of increase in the

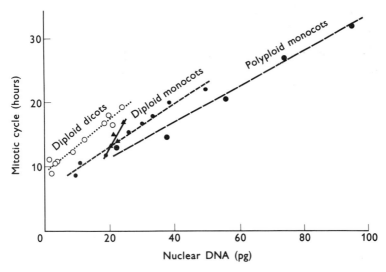

Fig. 4.23 Mitotic cycle times in root meristems at 20°C plotted against nuclear DNA amounts in flowering plants species. The black triangles are for rye plants with 1, 2, 3 and 4 *B*-chromosomes, in increasing order of cycle time. [From Rees, H. (1972) *Brookhaven Symp. in Biol.*, **23**, 394–418]

duration of the cycle is greater when due to the addition of *B*-chromosomes than of other chromosomes in the complement. These fluctuations, however, do not mask the overriding and highly predictable increase in cycle time with increasing DNA. As we have observed earlier there are equally well-defined correlations between the duration of meiosis and the nuclear DNA amount (Bennett, 1971).

DNA and cell size

Cell size in root meristems of flowering plants is also closely correlated with nuclear DNA amount. The increase in size follows DNA increase by both segmental amplification of chromosomes among diploids (Martin, 1966) and the addition of *B*-chromosomes.

Cell size and the duration of mitotic cycle together determine the rate of growth at the cell level. In tissues such as the root meristem both, as we have shown, are to a surprising extent dependent on the amount of DNA in the nucleus. There are indications that the relationship is of adaptive significance. Stebbins (1966) has pointed out that species of plants in temperate lands have larger chromosomes, and higher DNA content, than species in tropical regions. Bennett (1972) has demonstrated that short-lived species have a lower than average nuclear DNA content. The inference is that the restrictions upon such characters as the duration of mitotic cycles imposed by the nuclear DNA amount have a profound influence upon the distribution and evolution of species.

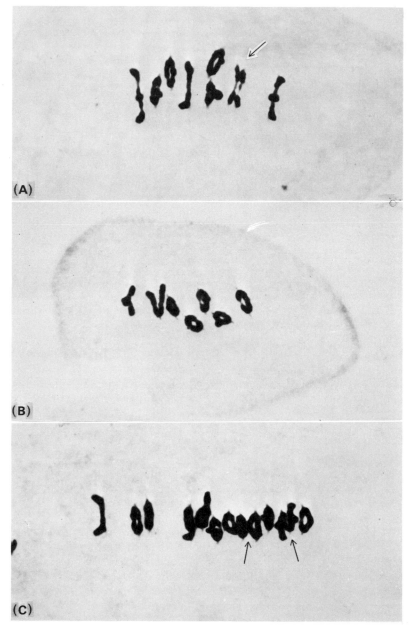

Fig. 4.24 Numerical chromosome change in rye. (A) First metaphase of meiosis with 4 *B*-chromosomes, as two bivalents (arrowed). (B) First metaphase in a trisomic, with trivalent. (C) First metaphase in an autotetraploid, 10 bivalents and 2 quadrivalents ($2n = 28$).

It is important in this context to bear in mind that the duration of mitotic cycles and the cell size differ very considerably as a normal consequence of growth and development between organs and tissues within the one individual. There is no doubt about the existence and operation of genetic mechanisms which control and regulate cell cycles and cell growth quite independently of nuclear DNA amount. In view of this it is perhaps even more surprising that the correlation between DNA amount, mitotic cycles and cell size among species is so consistent and predictable. It implies a crude and rigid constraint upon growth and development which appears to exert a profound influence upon the distribution and adaptation of species.

5

Qualitative change

Chromosome breakage, followed by the reunion of broken ends, is a normal and essential feature of crossing over at meiosis. Breakages of various kinds may also occur spontaneously during development. Their effects, like those of gene mutations, are unpredictable although usually lethal. They are lethal because most of the products of breakage are mechanically inefficient at mitosis. Acentric fragments, i.e. fragments lacking a centromere, are incapable of movement on to the spindle. Ring chromosomes and dicentrics resulting from breakage, with reunion of broken ends, are incapable of regular disjunction at anaphase. In each case the result is loss of genetic material. Mechanically efficient structural changes make up a small proportion of the products of spontaneous chromosome breakages. Even so they have important genetic consequences. The two most important are inversions and interchanges.

5.1 Inversions

Inversions are the result of two breaks within a chromosome followed by the reunion of broken ends in reverse order. When both breaks are in one arm the result is a *paracentric* inversion, a break in each of two arms gives a *pericentric* inversion. In either case a chromosome polymorphism may be generated with the inversion 'floating' within the population, comprised of inversion heterozygotes and inversion homozygotes alongside the structurally normal homozygotes. The frequency of each class will depend upon the relative fitness of the inversion in the structural heterozygotes and homozygotes relative to the normal genotype.

Because the two strands of the DNA helix are of different polarity the base sequences in the helix within the inverted segment are switched from one strand of the chromosome helix to the other. Such transfer does not in itself impair transcription. This was demonstrated, for example, within inverted chromosome segments in the *lac* operon region of *E. coli* (Hayes, 1968). It follows that transcription is selective in respect of the two strands within the DNA helix and also that sequences within the chromosome are transcribed from different directions. At the same time it must be supposed that where the inversion breaks occur within a unit of transcription the consequence would be to impair or prevent transcription. To some degree this may account for the lethal nature of many inversions in homozygous or hemizygous states.

Meiosis

At mitosis inversions do not impair the efficiency of transmission during cell division. Meiosis, also, is normal in the inversion homozygotes. In the heterozygotes, however, pairing between the normal and inversion chromosome is complicated and from the standpoint of fertility and recombination the consequences are profound.

(a) *Paracentric inversion heterozygotes*

At pachytene complete and effective pairing in the segment spanned by the inversion is achieved by the formation of a pairing loop (Fig. 5.1). The size of the loop, and hence the likelihood of chiasma formation within it, will depend upon the length of the inversion. If no chiasma forms within the loop, segregation is normal at both meiotic divisions. When a single chiasma forms within the loop the two chromatids involved form a dicentric bridge and an acentric fragment (Fig. 5.1 and Fig. 5.10). Loss of the fragment and random breakage of the bridge at first anaphase are the causes of deficiency in two of the gametes which in turn are inviable (in plants, generally) or incapable of producing viable offspring (in animals). The two chromatids not affected by chiasma formation in

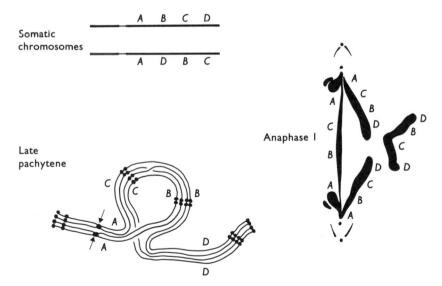

Fig. 5.1 Chiasma formation within the loop of a paracentric inversion heterozygote leading to bridge formation at anaphase, with acentric fragment. The chromatids not involved in chiasma formation are intact. Arrows indicate the centromeres.

the loop produce effective gametes. In each the gene sequence in the inversion segment will be of the non-recombinant, parental type. The result is that no recombinants for genes within the inverted segment appear among the offspring. The inversion thereby effectively suppresses

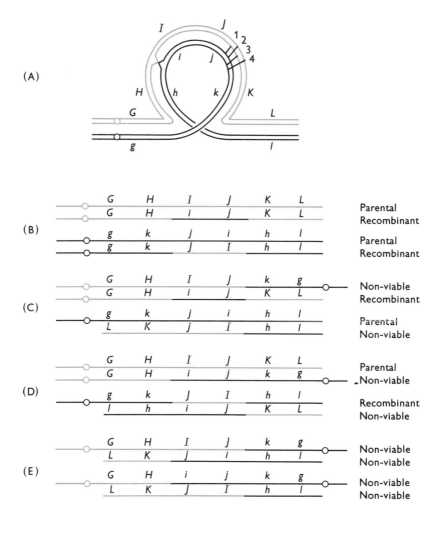

Fig. 5.2   The consequences of double cross-overs within the loop of a paracentric inversion (A). Assume the first cross-over to be constant, the second produces a 2-strand double cross-over (B), a 3-strand double (C) and (D) and a 4-strand double (E). Some viable recombinant gametes arise from the 2-strand and 3-strand doubles.

recombination throughout its length. The exception is where two-strand or three-strand double cross-overs take place within the loop. They are formed only when the inversion segment is a long one. Four-strand double cross-overs produce a double bridge and fragment and yield no effective gametes (Fig. 5.2). Leaving aside the other possible permutations of simultaneous chiasma formation within the loop and between the centromere and the loop it is sufficient to summarize the consequences of loop formation in the inversion heterozygote which are inviable or ineffective gametes and the suppression of recombination among genes in the inversion segment. Species of *Drosophila* and of related genera are exceptional in that inversion heterozygosity causes little or no reduction in fertility. There are two reasons for this: (1) there is no chiasma formation at meiosis in the male and (2) in the female the four nuclear products of meiosis are in linear order. The middle two nuclei are those involved in bridge formation at first anaphase and the functional egg nucleus is derived from one of the peripherals which carries no deficiency and is therefore viable and effective (Fig. 5.3). Thus there is effective suppression of recombination without inviability. This is one reason why paracentric inversions are particularly common in natural populations of *Drosophila*.

(b) *Pericentric inversions*

Crossing over within the loop of a pericentric inversion heterozygote does not result in bridge and fragment formation at anaphase I. Even so the gametes carrying recombinant chromatids are deficient for some

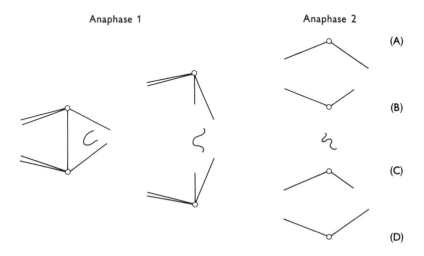

Anaphase 1   Anaphase 2

(A)

(B)

(C)

(D)

Fig. 5.3 Meiosis in the oocyte of *Drosophila* heterozygous for a paracentric inversion. The functional egg nucleus contains an intact, non-recombinant chromosome from either (A) or (D).

gene sequences while other sequences are duplicated (Fig. 5.4). As with paracentric inversions the consequences are inviable or ineffective gametes and the suppression of recombinants among the progenies for genes within the segment spanned by the inversion. However, pericentrics are particularly common among grasshoppers. The explanation is that no loops are formed in heterozygotes and, because pairing in the inverted segment is between non-homologous chromosome segments, no chiasmata are formed within it (White, 1973). It follows that recombination is suppressed but at no cost to fertility.

Distribution and selection

The effective suppression of recombination due to inversion implies that gene sequences within a segment spanned by inversion are maintained intact and segregate as one unit, a *supergene*, within a population. The frequency of an inversion within the population will therefore depend upon the relative fitness of individuals carrying these supergenes in homozygous or heterozygous combinations, bearing in mind that any advantage due to inversion must more than compensate for the infertility which, in most species, results from inversion heterozygosity. The widespread distribution of inversions in natural populations of both plants and animals testifies to their importance in adaptation and evolution. Among flowering plants inversions are reported in numerous species. In *Lolium perenne* 21 out of 50 populations investigated by Simonsen (1973) carried inversions. He also found (Simonsen, 1975) that 42 % of plants

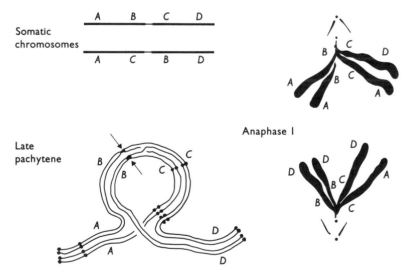

Fig. 5.4 Chiasma formation within the loop of a pericentric inversion heterozygote giving duplication and deficiency of segments in chromatids involved in chiasma formation. Arrows indicate centromeres.

within a population of a related grass, *Festuca pratensis*, were inversion heterozygotes. There is no information about the effects of selection upon the inversions in these or, indeed, in any plant species. In animals, in *Drosophila* and in grasshoppers particularly, there is a wealth of information.

The polytene nuclei of *Drosophila* larvae make the detection and identification of inversions relatively easy. Stone (1955) listed 592 different paracentric inversions in 42 species. He calculates that at least 6000 and possibly more than 30 000 inversions were associated with the evolution of the 650 species in this genus. Dobzhansky (1943) was the first to demonstrate the adaptive significance of inversions among natural populations. Chromosome III of *Drosophila pseudoobscura* in populations at Mount San Jacinto, California, is polymorphic for three inversion sequences, the Arrowhead (AR), the Chiracahua (CH) and the standard (ST) which is probably the 'wild type'. From 1939 to 1946 samples showed a consistent pattern of change within years. From March to June AR and CH increased at the expense of ST. From July to October ST increased while AR and CH declined in number. The conclusion was that selection favoured AR and CH gene sequences in the warm spring, ST in the hotter months of the year. The conclusion was confirmed by keeping flies in population cages. A sample from the wild population made up of ST/ST and CH/CH homozygotes and ST/CH heterozygotes were kept in cages at 25°C. The frequency of the ST chromosome, initially 11%, increased to 70%. At this point, with 70% ST and 30% AR chromosomes, the population attained equilibrium indicating that the CH chromosomes were maintained within the population due to the superiority of the heterozygote ST/CH over other genotypes (Dobzhansky, 1948). In cages kept at 16°C the relative frequencies of ST and CH remained unchanged to confirm that ST, as expected from the surveys of wild populations, is at an advantage only at high temperatures.

An inversion may of course be 'fixed' in a population in the homozygous form. In the *repleta* section of *Drosophila* 92 of 144 inversions in 46 species are fixed in this way (Wasserman, 1963). The remaining 52 'float' within populations which are polymorphic for inversion homozygotes and heterozygotes as in the *Drosophila pseudoobscura* stocks investigated by Dobzhansky.

In *Keyacris scurra* (formerly *Moraba scurra*), Australian populations are polymorphic for pericentric inversions located in two chromosomes (White, 1956; 1957a; 1957b; 1957c). The CD chromosome is found in one of four forms, *Standard, Blundell, Molonglo* and *Snowy*. The EF chromosome is found as *Standard* and *Tidbinbilla*. In both CD and EF forms the *Standard* chromosome is metacentric, the others are acrocentric or nearly so. Within populations, which produce one generation annually, the relative frequencies of the chromosome types are constant from year to year. Between populations, however, there is substantial variation indicating selection for particular sequences. In this species the effects of the inversion sequences upon the phenotype are directly recognizable. Adults homozygous for the *Standard* sequences are larger than *Blundell* and *Tidbinbilla* homozygotes. The gene sequences are additive in effect so that the heterozygotes are of intermediate size. In terms of viability, however, there is interaction. The inversion heterozygotes exhibit heterosis. Moreover, there is interaction between gene sequences on the CD and EF chromosomes. The least viable genotype is homozygous *Standard* for CD and homozygous *Tidbinbilla* for EF. Apart from demonstrating an adaptive polymorphism for pericentric inversions White's results are especially instructive in emphasizing that the effects of the gene sequences maintained by the inversions are manifold and complex. Changes in the frequencies of inversion sequences brought about by selection are therefore of a correspondingly complex nature both from the standpoint of gene action and interaction and the growth and development of the phenotype.

5.2 Interchanges

Interchange is the result of one break in each of two non-homologous chromosomes followed by re-union of the broken ends as in Fig. 5.5. The reciprocal exchange of segments involves no loss of genetic material, nor does it affect the efficiency of transmission of chromosomes at mitosis. In the interchange homozygotes meiosis is also normal, with regular bivalent formation and disjunction at anaphase of both divisions. There is consequently no cytological impediment to the production of viable gametes. In interchange heterozygotes, however, the exchange of segments between non-homologous chromosomes complicates the pairing at pachytene which, in turn, affects the subsequent progress of meiosis and of gamete formation.

Meiosis in interchange heterozygotes

Complete and effective pairing of all the homologous parts is possible only by association between the four chromosomes implicated in interchange, the two interchange chromosomes and the 'wild type' chromosomes from which they derive (Fig. 5.5). The fate of the interchange configuration and the nature of the gametes produced depend on the frequency and distribution of chiasma.

Consider the formation of four chiasmata, one at the distal end of each of the four arms of the association of four chromosomes, resulting in a ring. At first metaphase the ring may be orientated either as an 'open' or 'zig-zag' configuration. As an open ring the separation at first anaphase, although numerically and symmetrically disjunctional with two chromosomes moving to each pole, is non-disjunctional from the genic standpoint because the chromosomes at each pole are deficient for some chromosome segments and carry duplicates of others (Fig. 5.5 and Fig.

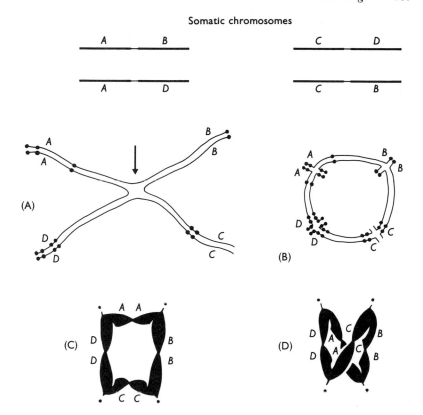

Fig. 5.5 Pachytene pairing (A) in an interchange heterozygote, followed by the formation of four terminal chiasmata (shown at diplotene, B). The open ring (C) at first metaphase leads to genetic non-disjunction, the zig-zag (D) to disjunctional separation. The arrow indicates the position of the centromere.

5.10). The gametes will be inviable or ineffective. If the ring orientation is zig-zag, disjunction is perfect and the gametes are viable and effective. The proportion of inviable gametes produced by the interchange heterozygote will depend on the proportion of cells in which the ring is orientated non-disjunctionally, as an 'open' configuration at first metaphase as opposed to a 'zig-zag', disjunctional orientation. While the theoretical expectation is 50% disjunctional orientation and 50% non-disjunctional, in many species, e.g. rye the orientation is disjunctional in about 65% of cells. It is worth noting that this proportion may be increased by selection with consequent improvement in the fertility (Thompson, 1956).

With three chiasmata in distal segments rather than four, a chain is produced. Again only a zig-zag arrangement gives disjunction and viable gametes. There is evidence to show that the proportion of chains which orientate disjunctionally is higher than of rings (Rees and Sun, 1965). It follows that a relatively low chiasma frequency improves fertility. If the distribution of chiasmata is such as to form two bivalents, there will be a 50% chance that their orientation is disjunctional.

If a chiasma forms between the centromere and the break point, i.e. in the *interstitial* segment and not distally as above in the *pairing* segment, half the gametes produced will be inviable even with a zig-zag orientation, i.e. with alternate centromeres moving to opposite poles. The additional infertility due to chiasmata in interstitial segments probably explains why interchanges are more common in populations of species in which chiasma formation is mainly distal in distribution.

Recombination

The chromosomes which segregate to opposite poles from zig-zag disjunctional arrangements at first metaphase are in two combinations, the original wild type pair *AB*, *CD* in Fig. 5.5 and the interchange pair *AD*, *CB*. Only gametes with these 'parental' combinations are effective. Any recombinants, *AB/AD*, *AB/CB*, *AD/CD* or *CD/CB* are inviable or else produce inviable zygotes. These are, of course, the products of non-disjunctional orientation. It follows that genes on the separate chromosomes, *AB* and *CD*, segregate together. The same is true for genes on *AD* and *CB*. Interchange therefore effectively prevents recombination between genes on different chromosomes by creating 'linkage' between them. The exceptions are genes near to the ends of the chromosomes where chiasmata may form. The suppression of recombination by interchange makes possible the maintenance of particular combinations of genes located on different chromosomes which, in theory, could be of adapative advantage in heterozygous or homozygous combinations.

Adaptation and selection

Paeonia californica is distributed along the California coastal strip from San Diego northwards to Monterey. Its centre of origin is at Ventura, about midway between San Diego and Monterey. In the Ventura region the populations comprise structurally normal diploids with 10 chromosomes which form 5 bivalents at meiosis. Towards both the southern and northern limits populations become increasingly heterozygous for interchanges (Walters, 1942). The distribution suggests an increasing adaptive advantage of interchange among the colonizing populations. Some contain individuals with four interchanges such that rings or chains are formed which involve all ten chromosomes at meiosis. A similar situation applies in *Oenothera*. The *hookeri* group in California form seven bivalents at meiosis. The *strigosa* and *biennis* groups, in inland eastern regions, are structurally heterozygous for interchanges forming rings or chains which, like some *Paeonia* populations, embrace all members of the complement. A common and important factor in both genera is that the interchanges are found in inbreeding, i.e. self-pollinating popula-

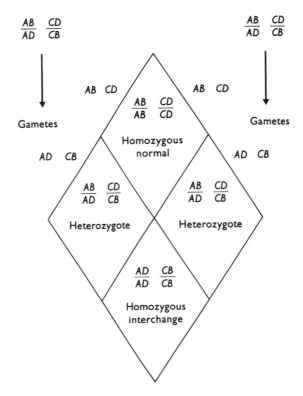

Fig. 5.6   The distribution of normal, interchange heterozygotes and homozygotes expected on selfing or crossing a single heterozygote.

tions as distinct from the bivalent forming ancestral populations which outbreed. The conclusion is that heterozygous gene sequences maintained by interchange are at an advantage under the conditions of enforced inbreeding due to isolation associated with migration and colonization. John and Lewis (1957) have described a comparable case in populations of the American Cockroach, *Periplaneta americana*. In coal mines in South Wales, where the isolation enforces inbreeding, interchange heterozygotes are common.

That the interchanges are favoured as conservers of particular heterozygous gene combinations with inbreeding is confirmed by experiment. The distribution of normal homozygotes, interchange heterozygotes and interchange homozygotes expected on selfing or intercrossing a single interchange heterozygote is in the ratio 1:2:1 (Fig. 5.6). On selfing there is an excess of interchange heterozygotes among the progenies in *Campanula persicifolia*, but not on intercrossing (Darlington and La Cour, 1950). In a rye stock with two identifiable rings of four due to two interchanges, A and B, the proportion of interchange heterozygotes among progenies increased with increasing generations of inbreeding. They outnumbered both homozygous classes by 2 : 1 (61 to 33). It is instructive that the B interchange heterozygotes showed no advantage over the homozygotes. Only some interchanges therefore—those which maintain particular gene sequences—are favoured. The extreme is found in the genus *Oenothera* where the interchanges are 'fixed' as heterozygotes and the homozygotes are lethal. Bearing in mind the infertility accompanying interchange heterozygosity the selective advantage of given gene sequences held together by interchange has to be substantial. We have indicated infertility may be reduced by heritable change in the frequency of disjunctional orientation of interchange configurations. In *Oenothera* species, for example, orientation of large rings or chains is disjunctional in 90% of cells.

Finally, although the experimental results show superiority only in respect of interchange heterozygotes, it may well be that interchange homozygotes may confer adaptive advantage over the normal wild type. In rye for example the different species are homozygous for different interchanges. *Secale montanum* and *S. cereale* differ by three interchanges. Each is homozygous for the interchange sequences. They are distinguishable by the ring and chain configurations at meiosis in the species hybrids.

Chromosome evolution and speciation

So far we have been discussing reciprocal interchanges between chromosomes which are relatively symmetrical and which lead to strictly qualitative changes in the organization of genetic material. Not all interchanges are of this kind. Some, particularly those involving acrocentric chromosomes which are common in the chromosome complements of animals, are extremely unequal and play a major role in chromosome and species evolution. They provide a mechanism for change in basic chromosome number and are the cause of quantitative as well as qualitative variation in genetic material.

One of the better known cases in plants is found in the genus *Crepis* (Tobgy, 1943). Tobgy hybridized *Crepis neglecta* $(2n = 2x = 8)$ and *C. fuliginosa* $(2n = 2x = 6)$ and found that at meiosis in the F_1 hybrid there was almost complete homology between the two parental sets of chromosomes. In many of the cells examined at first metaphase of meiosis he found two bivalents and an association of three chromosomes. Tobgy's interpretation of the chromosome changes associated with the evolution of *C. fuliginosa* from *C. neglecta* is presented in Fig. 5.7. As a result of an unequal interchange between the B and C chromosomes of *C. neglecta* one of the products, a small, 'inert' fragment is assumed to be dispensable. Its loss reduces the basic chromosome number from 4 to 3, as in *C. fuliginosa*. This is not to say that interchange is the only chromosome 'mutation' associated with the divergence of *C. fuliginosa* and *C. neglecta*. The chromosomes have diverged considerably in terms of size and DNA

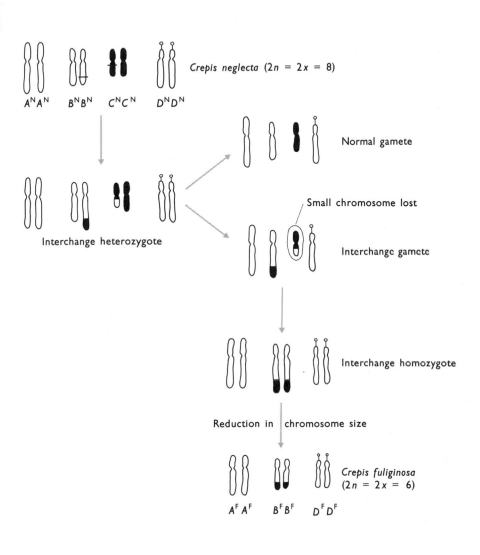

Metaphase 1 in the F_1 hybrid
C. neglecta × *C. fuliginosa*

Fig. 5.7 The origin of *Crepis fuliginosa* from *Crepis neglecta* by unequal interchange and a reduction in the basic chromosome number. [After Tobgy, 1943]

content due to segmental amplification or deletion within the chromosomes (Jones and Brown, 1976).

In the *Bovoidea* the haploid number of chromosomes among the 50 or so species investigated ranges from $n = 15$ to $n = 30$. In the former the chromosomes are mainly metacentric, in the latter acrocentric, so that the number of chromosome *arms* is remarkably constant, from 29 to 30 in each haploid complement. For this reason it is postulated that the variation in chromosome number accompanying divergence and evolution of species within the group is, in the main, due to 'fusion' of the acrocentrics to produce metacentrics, with an accompanying reduction in the chromosome number. For example, in a species with $n = 30$, made up of acrocentrics, reduction to $n = 29$ is achieved by fusion of two acrocentrics to give one metacentric and 28 acrocentrics (Wurster and Benirschke, 1968). The phenomenon is called Robertsonian fusion (after Robertson, 1916). The mechanism could well be through unequal interchange of the kind described in *Crepis* above. On the other hand some workers claim that there is direct fusion of centromeres, in which case the chromosomes described as acrocentrics would in reality be telocentrics, i.e. with strictly terminal centromeres. Which of the two mechanisms applies is conjectural. White (1973) favours the view that reduction is by interchange and that true telocentrics are unstable and not, therefore, found as persistent members of chromosome complements. John and Hewitt (1968), on the other hand, claim that in some *Orthoptera* species at least there are stable telocentrics and instances of true centric fusion.

An instructive example of 'centric fusion' is described by Jones (1974) in the genus *Gibasis* (Commelinaceae). Two 'species' have, respectively,

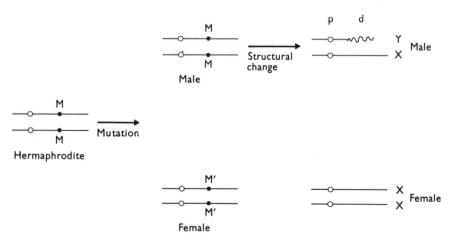

Fig. 5.8 Mutation, followed by structural divergence to account for sex chromosome polymorphism. The male is depicted as the heterogametic sex, i.e. producing gametes with different sex chromosomes. In many species it is the female that is heterogametic. p = pairing segment; d = differential segment.

$2n = 10$ and $2n = 16$ chromosomes, yet the latter is a tetraploid derivative of the former. The explanation is that the basic chromosome set of the diploid comprises 2 metacentrics and 3 acrocentrics. In the tetraploid, due to fusion, the basic set comprises 3 metacentrics and one acrocentric. Another example where polyploidy and centric fusion may, on the face of it, be confused, is in the *Salmonidae*. The chromosome numbers of species within this family correspond closely to the series $2n = 60, 80,100$ and 120, strongly suggestive of polyploidy. Yet the number of chromosome arms in each species is relatively constant (c. 104) indicating Robertsonian fusion. The latter is confirmed by nuclear DNA measurements. The nuclear DNA amount in the sea trout (*Salmo trutta*, $2n = 80$) is similar to that of the Atlantic salmon (*Salmo salar*, $2n = 58$). Polyploidy is ruled out (Rees, 1964). Change in the basic chromosome number, as in the few of many examples reported, means change in the number of linkage groups. This in turn means change in the amount of recombination by independent segregation and the rate of release of genetic variability to gametes and progenies (see Chapter 6).

Sex chromosomes

The most widespread kind of chromosome polymorphism is that concerned with sex determination. Most animal species are *dioecious* (consist of unisexual individuals) and most species of higher plants are *monoecius* (hermaphrodite). In lower plants, such as mosses and liverworts sexual differentiation is more common (Mittwoch, 1967). It is widely accepted that mechanisms of sex determination evolved within monoecius species by gene mutation in such a way that the alleles segregate to specify different sexes (Fig. 5.8). Subsequent structural divergence, often involving diminution of the *differential* as distinct from the *pairing* segments of the chromosomes carrying the different alleles is the cause of the sex chromosome polymorphism characteristic of most dioecious species. The smaller of the sex chromosomes in mammals is the Y, in the red and white campions the X. In many species such as the grasshoppers the Y is absent.

Interchange between one of the sex chromosomes and an autosome often reinforces the chromosome divergence between the sexes. One such case is presented in Fig. 5.9. It accompanied the evolution of a number of species of both marsupial and placental mammals where males are invariably the heterogametic sex. Interchange between the X and an autosome is succeeded by loss of a small, presumably inert chromosome (Fig. 5.9). The final result is that the female is homozygous for a pair of interchange X/A chromosomes. The male is heterozygous for the interchange $X/A:A$ and also carries the Y. Consequently the male has one more chromosome than the female. In the marsupial *Potorus tridactylus*, for example, the male is $2n = 13$, the female $2n = 12$ (Sharman and Barber, 1952). Such a system has been described as XY_1Y_2/XX on the basis of trivalent formation in the male (Fig. 5.9), but as Ohno (1969) makes

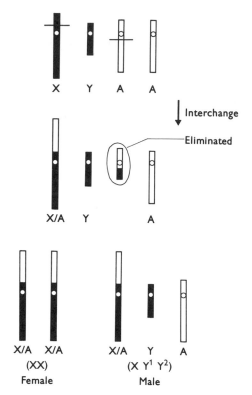

Fig. 5.9   Interchange between an X chromosome and an autosome (A), followed by loss of a small centric fragment. The consequence is that the chromosome number in the female is reduced by 1. Trivalent formation involving the X/A chromosome, the Y and the A in the male is the reason why the system has been described as an XX/XY^1Y^2 mechanism. [After Ohno, 1969]

clear such a classification is unnecessarily confusing in that it implies the development of a new sex determination mechanism whereas the mechanism determining sex is unchanged. It is simply that one of the sex chromosomes happens to be implicated in interchange. Comparable but often more elaborate interchanges involving sex chromosomes are by no means uncommon (Mittwoch, 1967; Ohno, 1969). The significance of such changes is readily explained. We must suppose that the complex of genes embraced by the sex chromosome and a particular member of an autosomal pair confers superior fitness and that the complex is maintained intact by interchange. Evidence to support this view comes from Barlow and Wiens (1975). In an East African mistletoe (*Viscum hildebrandtii*) all the males are heterozygous for one or more interchange. The number of interchanges associated with the sex chromosome, moreover, increases among populations from south to north, precisely as would be

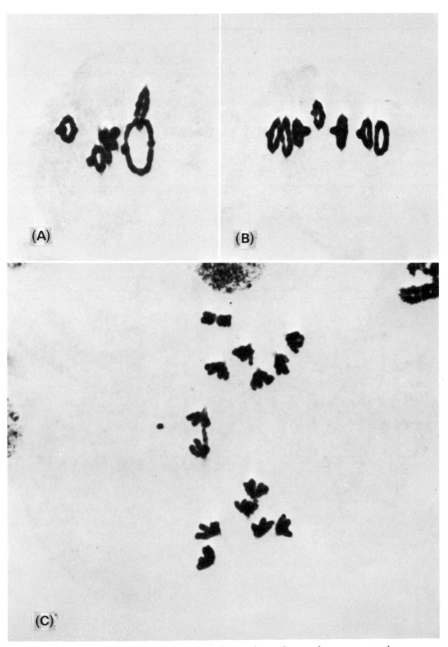

Fig. 5.10 First metaphase of meiosis in an interchange heterozygote in rye; (A) with an 'open' ring, (B) a zig-zag ring of 4. (C) First anaphase in a rye plant heterozygous for a paracentric inversion.

expected with selection for particular interchange combinations adapted to different environmental conditions.

Finally, we wish to draw attention to one other phenomenon in many, but not all species showing sex chromosome polymorphism. It is that of 'compensation'. In the mammals the male has one active X and a largely heterochromatic, relatively inert Y. As we describe elsewhere one of the X-chromosomes in the female is 'switched off' such that transcription of sex chromosome genes is similar in the two sexes. In *Drosophila* 'compensation' is not achieved by switching off one of the two Xs in the female but is imposed, in the main at least, by 'compensation' genes (regulators), located within the X itself, which depress the activity of other X-chromosome genes in the double X females or in any individual with more than one X (Muller and Kaplan, 1966). Polymorphism in relation to sex chromosomes is often elaborate not only in terms of structure but also of metabolism.

6

The regulation of recombination

6.1 Genetic systems and the states of variability

Heritable variation derives in the first instance from mutation of the genetic material. Mutation, however, is but one of many factors contributing to the heritable variation within a population. Consider, for example, two genotypes $AAbb$ and $aaBB$. Let us assume that the alleles A and B have similar effects upon the phenotype and, likewise, that the effects of a and b are similar to one another. The situation is typical of genes which make up a polygenic complex. The result would be that $AAbb$ and $aaBB$ would be of similar phenotype. By intercrossing we get F_1 heterozygotes $AaBb$ and by further intercrossing or selfing a range of genotypes and phenotypes in the F_2. In the parental forms $AAbb$ and $aaBB$ the variability is of a potential kind (a *homozygotic potential* state) (Fig. 6.1). It is converted to the *heterozygotic potential* by intercrossing and, subsequently, to *free variability*, as in $AABB$ and $aabb$ genotypes, by recombination at meiosis in the $AaBb$ heterozygotes (see Mather, 1973). The example suffices to illustrate that three ingredients: (1) the initial mutations at A and B; (2) the breeding system in relation to intercrossing; and (3) recombination at meiosis, contribute to the release of heritable variation among the F_2 phenotypes. The three together constitute what Darlington (1958) has called the *genetic systems* of populations. It was Darlington who perceived that adjustments in any one of these components serve to regulate the amount and kind of heritable variation within populations and, moreover, that the adjustments in themselves could be adaptive in controlling the variability made available for selection.

6.2 The flow of variability

In organisms which reproduce exclusively by asexual means the only source of variability is mutation. The same is true for sexual inbreeders. Because of their homozygosity, chiasma formation in these organisms yields no recombinants. Among the homozygotes, as in $AAbb$ and $aaBB$ above potential variability persists in the absence of hybridization. In an outbreeding population, in contrast, there is a continual flow of variability due to intercrossing and recombination, from the free to the

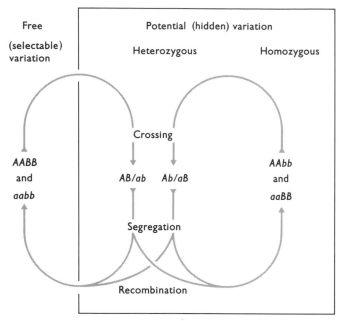

Fig. 6.1   The states of variability. The model assumes the effects of *A* and *B* upon the phenotype to be similar. *a* and *b* are also assumed to have the same effect on the phenotype. Free variability is expressed as phenotypic differences among genotypes (e.g. *AABB* and *aabb*) and may be fixed by selection. The homozygous potential variability is concealed (c.f. *AAbb* and *aaBB* which have similar phenotypes) and is therefore not discriminated by selection. The homozygous potential is converted to the heterozygous potential by crossing and, subsequently, to free variability following recombination. The amount of recombination determines the rate of flow from the heterozygous to the free state. The arrows show the direction and causes governing the flow of variability from the one state to the other. [From Mather, K. (1973) *Genetical Structure of Populations*. Chapman & Hall, London]

potential state, from the potential to the free; from the homozygotic to the heterozygotic potential and vice versa (Fig. 6.1). The *rate* of flow of variability even within the outbreeding populations is, however, subject to regulation, imposed by the amount and distribution of recombination at meiosis. The aim of this chapter is to indicate briefly how such regulation is achieved and to illustrate some of the consequences.

6.3 The regulation of recombination by structural and numerical changes

Structural changes

For genes located on different chromosomes segregation is independent such that gametes from the heterozygote *AaBb* are *AB*, *Ab*, *aB*, *ab* in equal numbers with the recombinants making up half the total. *Interchange* is one means whereby the recombination may be effectively suppressed. Likewise, *inversion* suppresses recombination between genes on the same chromosome (Chapter 5). A consequence of both kinds of structural change is to conserve particular gene combinations which may confer superior fitness in particular environments.

Numerical changes

Numerical changes of two kinds serve, like inversion and interchange, to restrict the flow of variability by reducing recombination. The one is allopolyploidy which, on account of a restriction of pairing to homologous and genetically identical chromosomes at meiosis, produces gametes similar to the two parental genotypes (Chapter 4). The second is a reduction in the basic chromosome number. The consequence is a smaller number of genes which segregate independently. The extreme is found in *Ascaris megalocephala* where all the genes are linked to one another on the single chromosome in the complement. Grant (1959) reports that short-lived species of higher plants have, on average, lower basic chromosome numbers than the longer-lived species. He argues that the difference is adaptive, on the following grounds. For short-lived species the demand is for uniformity so that the offspring, which have to re-establish the population in each generation, are similar to the parental forms adapted to that environment. Reducing the basic number and thereby the number of genes which segregate freely and independently is one way of promoting uniformity. Conversely, an increase in the basic number fosters a freer flow of variability through greater recombination.

Autopolyploidy, simply by increasing the number of genes and of chromosomes, increases the number of possible gene combinations and thereby the range of variation. There is one other kind of numerical change which affects recombination and the variability of offspring and populations, namely *B*-chromosomes (Chapter 4). Almost invariably they increase the amount of recombination among the chromosome complement as a whole or in particular chromosomes or chromosome segments. Either way *B*-chromosomes promote a rapid rate of flow of variability.

It will be observed that the regulation of recombination by structural and numerical chromosome changes is, for the most part, an 'all or nothing' situation. With interchange or inversion the recombination is completely suppressed among the chromosomes or chromosome segments affected. There is little or no element of fine adjustment. The regulation imposed by the action of genes upon recombination, on the other hand, allows for such adjustment.

6.4 Genotypic control

We have emphasized earlier that the chromosomes have a dual aspect, that of genotype and that of phenotype and that the latter is influenced not only by changes in cell environment imposed by differentiation and by the external environment, but also by the action of genes which are borne by the chromosomes themselves. The control imposed by genes over the form and behaviour of the chromosomes is aptly named *genotypic control*. We shall consider here those aspects of genotypic control which impinge upon recombination.

In *Neurospora crassa* a number of *recombination genes* regulate the frequency of recombination within other genes or between genes within short segments of the chromosomes (Catcheside, 1977). One of these, a recessive gene *rec-1*, controls the frequency of recombination at the *his-1* locus which is located on the same chromosome but at some distance from *rec-1*. In the homozygote *rec-1*, *rec-1* recombination within the *his-1* locus is increased ten-fold above normal. It is characteristic of these recombination genes in *Neurospora* that they influence recombination in specific chromosome segments at some distance away.

A single recessive gene, *as*, on chromosome 1 of maize causes almost complete failure of chiasma formation throughout the whole chromosome complement (Beadle, 1930). A gene in *Hypocheris radicata* causes failure of chiasma formation in one particular chromosome, chromosome IV (Parker, 1975). Heterochromatic segments which segregate as Mendelian genes increase chiasma frequencies in a number of grasshopper species (John, 1973). In others they reduce the frequency (D. I. Southern, 1970). These are but a few of numerous examples of genes at a single locus causing variation in chiasma frequency and in chiasma distribution. There are instances where the control is exerted by many genes, i.e. by polygenic systems, as shown for example by investigations of inbred lines.

Chiasma frequency in inbred rye

Secale cereale $(n = 14)$ is an outbreeder. Inbred lines may be produced by enforced self-fertilization. The lines are homozygous, but as a result of segregation during inbreeding, homozygous for different gene combinations. Table 6.1 shows the mean chiasma frequencies in four such lines. The differences between means are significant and heritable showing that the segregation during inbreeding involved genes determining the frequency of chiasmata in the pollen mother cells.

The chiasma frequencies in all lines are substantially lower than in normal, heterozygous rye plants. Homozygosity evidently is a cause of inbreeding depression in respect of chiasma formation as for other more familiar aspects of the phenotype such as plant height and grain yield. There is a parallel also in the *heterosis* displayed in F_1 heterozygotes derived from crosses between the inbred lines. The F_1 chiasma frequencies are substantially higher (Table 6.1).

That the control over the chiasma frequency is polygenic is evident from the continuous nature of the variation among inbred lines and F_1s. The consequences upon the regulation of recombination are, therefore,

Table 6.1 The mean chiasma frequency per pollen mother cell in 4 inbred lines of rye and their F_1 hybrids. [From Rees, H. and Thompson, J. B. (1956) *Heredity*, **10**, 409–24]

Lines and hybrids	Chiasma frequency
P3	12.95
P6	11.97
P12	11.90
P13	12.11
P3 × P6	14.21
P3 × P12	14.49
P3 × P13	14.63
P6 × P12	14.31
P6 × P13	14.25
P12 × P13	14.48

very different from those imposed by structural change which allow for no fine adjustments such as may be achieved through genotypic control.

Chiasma distribution in inbred rye

In bivalents with one or two chiasmata, the chiasmata are located distally, near the chromosome ends. In bivalents with three or more chiasmata the extra ones are formed interstitially (Fig. 6.2). An increase in the bivalent chiasma frequency in rye, as in many other species, is therefore accompanied by a redistribution of chiasmata within chromosomes. In terms of recombination it means that in bivalents with two or fewer chiasmata the gene sequences at all but the ends of the chromosomes are unaffected. With three or more chiasmata gene sequences in the middle regions of the chromosomes are vulnerable to disruption by crossing over. Bivalents with three or more chiasmata are of course more numerous in plants with high chiasma frequencies. In these not only is there more recombination but recombination in different parts of the chromosome complement.

A redistribution of chiasmata both between bivalents and within bivalents may also be achieved without change in frequency. Jones (1967) has described a rye line with a relatively low chiasma frequency, about 13 per cell, but which has an unusually high frequency of bivalents with interstitial chiasmata (Fig. 6.3).

The examples suffice to establish the capacity for heritable adjustments in the frequency and distribution of chiasmata and of recombination. That control over recombination affects the variability of progenies and of populations is also established in both experimental and natural populations.

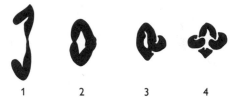

Fig. 6.2   Rye bivalents at first metaphase of meiosis with 1, 2, 3 and 4 chiasmata. Note the change in chiasma distribution with changing frequency. All chiasmata are distal in bivalents with a frequency of 1 and 2. Interstitial chismata are found in bivalents with 3 or more.

6.5 Recombination and variability

McPhee and Robertson (1970) crossed two strains of *Drosophila melanogaster* differing in the average number of sternopleural bristles, a character under polygenic control. The F_1 hybrids were of two kinds, one being heterozygous for inversions in chromosomes II and III, the other 'normal', *i.e.* without inversions. The range of variation in bristle number among the F_2 progenies of F_1 inversion heterozygotes was very much less than among F_2s from hybrids without inversions. This is precisely as expected. In the former there is effective suppression of recombinants and of the release of variability among progenies. The consequences of suppressing recombination were further manifested by the results of selection for high and low bristle numbers among F_2s and their derivatives. Measured as the difference in bristle number between

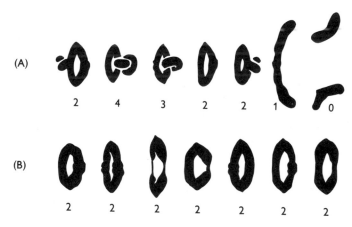

Fig. 6.3   First metaphase of meiosis in pollen mother cells of two rye genotypes with a similar chiasma frequency (14) but differing in chiasma distribution. In (A) there is wide variation in the number of chiasmata among bivalents within the cell (from 0 to 4). (B) is more characteristic of the species with a uniform distribution of chiasmata amongst the bivalents. [After Jones, G. H. (1967) *Chromosoma*, **22**, 69–90]

high and low lines the selection was only 25% as effective among derivatives of inversion heterozygotes as compared with lines derived from F_1 hybrids without inversions.

Moss (1966) has shown how the redistribution of chiasmata, brought about by *B*-chromosome action, influences recombination and variability in experimental populations of rye. In this species, as already mentioned, the *B*-chromosomes increase the frequency of chiasmata in the interstitial segments of the chromosomes. The localized increase in recombination is reflected by an increased variability among the progenies of parents with *B*s as compared with those without *B*s (Chapter 4).

6.6 Chiasmata and variability in natural populations

That the regulation of recombination, exercised by genotypic control over chiasma frequency and distribution, influences the variability within natural populations in an adaptive way is established from surveys among populations of *Lolium* and *Festuca* grasses (Rees and Dale, 1974).

Variation

Within *Lolium perenne*, *L. multiflorum* and *Festuca pratensis*, all diploids with 14 chromosomes, there is a significant variation in the mean chiasma frequencies among natural populations. The variation is heritable. As in other species of Gramineae the chiasmata are localized in terminal regions of the chromosomes when the chiasma frequency is low, spreading to interstitial regions as the frequency increases.

Variability and chiasma frequency

The variability between plants within populations for a number of characters such as flowering time is lower in populations with high chiasma frequencies and vice versa (Fig. 6.5). This relationship, which is consistent throughout all three species, is at first glance at variance with the experimental results of McPhee and Robertson and of Moss to which we refer above. The contradiction is only apparent. As shown in Fig. 6.4 the consequence of high chiasma frequencies is, initially, to release potential variability and thereby to expose the free variability to selection. The effect of selection in turn, is the fixation of some genotypes at the expense of the dissipation of the remainder and, as a result, an overall reduction in the variability, both free and potential within the population. By contrast, in populations with low chiasma frequencies much of the variability is preserved, protected from disruption and release by the absence or rarity of interstitial chiasmata. Gene sequences within each chromosome segregate virtually as single units in heredity, as supergenes. Since each chromosome type within the complement may be allelic for a number of such supergenes it follows, also, that the population is genically highly polymorphic and, in comparison with populations with high chiasma frequencies, highly variable.

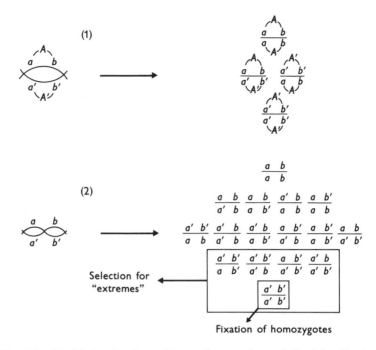

Fig. 6.4   Model showing how chiasma frequencies and distal localization (1) hold together supergene sequences, A and A'. Components of the supergene are ab and $a'b'$. In (2) interstitial chiasmata and high chiasma frequencies have fragmented the supergenes and released the variability. Part of the variability is fixed by selection, the remainder is dissipated. [From Rees, H. and Dale, P. J. (1974) *Chromosoma*, **47**, 335–51]

Chiasmata and selection

Selection for early and later flowering among eight populations of *Festuca pratensis* showed selection to be more effective among populations with low chiasma frequencies. This we expect on the grounds that, in the short term, the response to selection would be rapid through reassortment of highly diverse chromosomes; diverse because they segregate as supergenes protected from disruption in the absence or rarity of chiasmata and, in the long term, by the release of potential variation following the occasional formation of chiasmata in interstitial segments. In populations with high chiasma frequencies the variability is already largely dissipated with the result that the response to selection is restricted. In *Lolium perenne*, as in *F. pratensis*, selection is more effective in populations with low as compared with high chiasma frequencies.

Chiasmata and specialization

Short-lived populations of *Lolium perenne* and *L. multiflorum* have higher chiasma frequencies than more perennial populations from which they

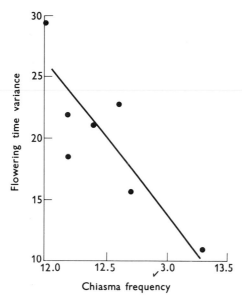

Fig. 6.5 The between-plant variance for flowering time plotted against the mean chiasma frequency in populations of *Lolium multiflorum* (From Rees, H. and Dale, P. J. (1974) *Chromosoma*, **47**, 335–51).

have evolved (Fig. 6.6). The same applies in rye (Sun and Rees, 1964). Intensively bred strains of *Lolium perenne* have higher chiasma frequencies than populations in 'natural pastures'. The model in Fig. 6.4 accounts for the high chiasma frequencies in these specialized forms. It is through recombination, particularly in interstitial segments, that new variability may be generated. In these populations the high chiasma frequency is a relic of the recombination essential for their evolution. To confirm this interpretation it would be expected that intensive selection for specialized forms would be accompanied by, albeit unconscious, selection for high chiasma frequencies. This precisely was the finding of Harinarayana and Murty (1971). Selection for late and early flowering within populations of *Brassica campestris* was accompanied by an increase in chiasma frequency.

Chiasmata, fitness and flexibility

The situation in the *Loliums* and in *Festuca* illustrates how increasing the chiasma frequency, especially within interstitial regions of the chromosomes where chiasmata are normally infrequent, generates new forms by recombination that may be fixed by selection imposed by nature or by the breeder. In the short term this facilitates rapid adaptation and high fitness. The very means which provides for this rapid adjustment,

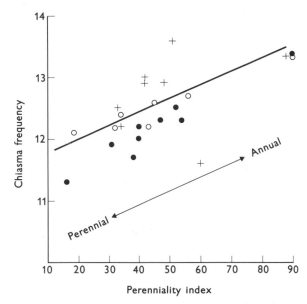

Fig. 6.6 Longevity (Perenniality index) plotted against the mean pollen mother cell chiasma frequency in populations of *Lolium perenne* (●), *L. multiflorum* (○), and *Festuca pratensis* (+). The perenniality index is the percentage of plants flowering during the first season after sowing. [From Rees, H. and Dale P. J. (1974) *Chromosoma*, **47**, 335–51]

the increase in recombination, is, however, instrumental in dissipating the variability such that the population has a lower capacity to adapt to further demands imposed by selection in a changing environment. Fitness is therefore at the expense of flexibility. As Darlington and Mather (1949) have emphasized the requirements for fitness in the short term and flexibility in the long term are often in conflict. In terms of chiasmata the conflict may be resolved by compromise. In most natural populations of the *Lolium* and *Festuca* species, for example, the chiasmata are mainly restricted to distal segments such that a variety of gene sequences, relatively well adapted to the environment, are maintained intact within the population. At the same time, occasional chiasmata in interstitial segments release a limited amount of variability such that there is the capacity for some flexibility in the face of changing selection pressures.

We have restricted our comments to the evidence for the regulation of variability imposed by genotypic control over recombination. In a wider context we should take account of the distribution and the mode of action and interaction of the genes which determine the variability. Moreover, we have focused attention on populations within a few species of grasses. The reason is that little information is available from other sources. Even though it be limited, the information from *Lolium* and *Festuca* nevertheless provides strong evidence of the adaptive consequences of adjustments in chiasmata upon the variability of populations and their capacity to respond to selection.

7

Recombination at mitosis

Recombination is a characteristic of meiosis. The extent of recombination between particular genes allows us to 'map' the chromosomes, i.e. to locate the genes within the chromosome complement. The methods are well established and familiar. In diploid organisms the mapping is achieved by scoring the proportion of phenotypes arising by recombination among F_2 or backcross progenies which, in turn, reflects the distribution of recombination among chromosomes at meiosis in the F_1 hybrids. In haploids, such as the fungi, the genotype of the gametes can be established directly and yields the same information. That recombination may also take place during mitosis was first established by Stern (1936) in *Drosophila*. The phenomenon has since been investigated in detail in *Aspergillus* by Pontecorvo (1953) and Pontecorvo and Kafer (1958), and is fully discussed in a companion volume of this series (Catcheside, 1976). More recently it has been used to provide information about the location and distribution of genes in the human chromosome complement.

7.1 Drosophila

Females of *D. melanogaster* of the genotype $\dfrac{+\,y}{sn\,+}$, heterozygous for the two recessive genes y (yellow body) and sn (singed bristles) on the X-chromosome are normally wild type in appearance. Stern observed in some cases 'twin spots' of yellow and singed on the body and also spots of yellow or singed alone. He ruled out mutation on the grounds that twin spots (which would require two simultaneous mutations) were more common than single spots. His explanation was that crossing over took place during interphase of mitosis, following chromosome duplication, to give twin or single spots as shown in Fig. 7.1. The size of any one spot would depend on the amount of cell multiplication following the crossing over. That twin spots were more numerous than yellow and singed is precisely as expected from the distribution of y and sn on the X-chromosome. Twin spots are due to a cross-over between the centromere and sn, yellows to a cross-over between sn and y and singed spots to a

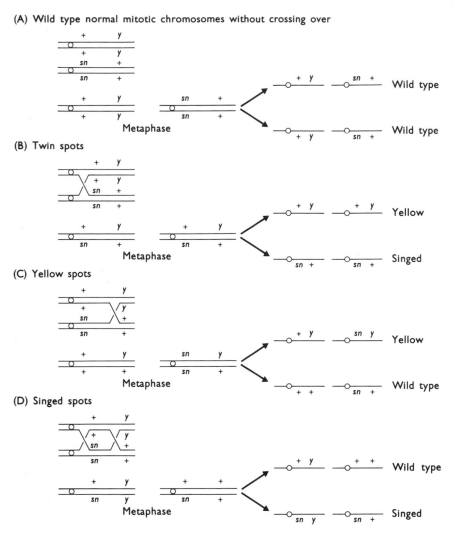

Fig. 7.1 Somatic crossing over in *Drosophila*. (A) a normal mitotic division; (B) crossing over between the centromere and the *sn* locus can give rise, assuming the metaphase orientation shown, to adjacent daughter cells homozygous for *y* and *sn*. Subsequent cell multiplication gives the twin spots of *yellow* and *singed* tissue. (C) Crossing over between *sn* and *y* gives spots of tissue with the yellow phenotype and in (D) a double cross-over gives spots with the singed phenotype.

double cross-over. There was evidence that crossing over at mitosis was genetically controlled. It was enhanced, for example, in the presence of the gene *minute*. In 1970 Stern demonstrated crossing over at mitosis between mutants w and w^{65}, within the white eye locus, also located in the X-chromosome of *Drosophila*. In 4 out of 6137 female heterozygotes, $\dfrac{w+}{+w^{65}}$, the intragenic crossing over was detectable as red, wild type, patches within the white eyes. Mutation was ruled out because no trace of pigmentation was found among 27 557 white eyed homozygotes and hemizygotes (males) used as controls. While Stern's work in *Drosophila* serves to demonstrate the occurrence of mitotic crossing over, the detailed analysis of the process comes from the investigations of Pontecorvo and his colleagues in the ascomycete *Aspergillus nidulans*.

7.2 Parasexuality in aspergillus

Wild type *Aspergillus* grows on minimal agar and produces dark green asexual spores, conidia. Nutritional (auxotrophic) mutants, which are unable to grow on minimal agar, are available to mark all eight chromosomes of the haploid complement. Strains which carry a spore colour mutation as well as a nutritional mutation are particularly useful in analysis, as is explained below.

Heterokaryon formation

When yellow and green conidia from two auxotrophic strains are mixed together, some of the hyphae within the mycelium will contain nuclei from both strains, as a result of fusion. These are heterokaryons as distinct from the homokaryons which contain one kind of nucleus only. The heterokaryon is readily isolated by growing the mycelium on minimal agar. In the heterokaryon one nucleus complements for the nutritional deficiency in the other and vice versa. The result is that the heterokaryon grows successfully, the homokaryons do not and are thereby screened out. During asexual reproduction the heterokaryon produces chains of conidia. All conidia in each chain contain one or other of the two nuclei. For this reason the homokaryons are readily recovered.

Aspergillus nidulans also reproduces sexually. In this case there is nuclear fusion as well as hyphal fusion prior to meiosis and ascus formation. Pontecorvo showed, however, that nuclear fusion in heterokaryons may occur independently of this sexual process. Such fusions are rare (1 in 10^6). The diploid products may be recovered as hyphae from sectors with exclusively green conidia within the mycelium of the heterokaryon; green because green is dominant to yellow in the diploid nucleus. Alternatively the diploids may be obtained by germinating conidia on minimal agar. Only the diploid conidia will thrive by virtue of complementation (Fig. 7.2).

Segregation and recombination

Diploid colonies of *Aspergillus* are unstable. When grown on a complete

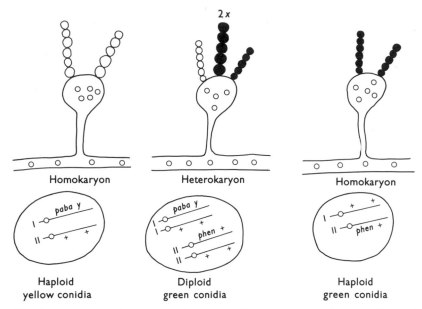

Fig. 7.2   One homokaryon, on chromosome I, has genes *y* (yellow conidia) and *paba* (a nutritional mutant which causes failure of growth in the absence of para-aminobenzoic acid). The other homokaryon carries the mutant *phen* on chromosome II and requires phenylalanine for growth.

or supplemented medium the green mycelium of the diploid breaks up to give yellow sectors. Sub-culturing shows that some of the yellow sectors are haploid but recombinants with respect to the nutritional mutants. Others are diploid and, again, may be recombinants. The haploids are thought to derive from progressive elimination of chromosomes by non-disjunction during mitosis. There is no preference for elimination of the chromosomes of either parent (Pontecorvo and Kafer, 1958).

The diploid yellow sectors (which ultimately tend to become haploid by chromosome elimination) arise by mitotic crossing over (Fig. 7.3). The principle of mapping is much the same as applied in *Drosophila*. With three genes in the one chromosome arm in the order, centromere:

$\dfrac{a\ \ b\ \ c}{+\ +\ +}$, only *c* will become homozygous on its own following a cross-over,

b and *c* the only pair to become homozygous together (double cross-overs are rare). The relative frequencies of the three classes of homozygotes give the map distances between the genes and between the genes and the centromere. These are not strictly comparable with map distances obtained from analysis of sexual ascospores because the total number of diploid nuclei from which the recombinants derive cannot be ascertained. The order of genes as adduced from analysis of mitotic recombination

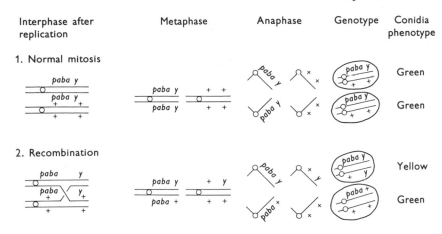

Fig. 7.3   Somatic crossing over in heterozygous diploid *Aspergillus*. The genetic consequences of a normal mitosis are shown in (1). In (2) a homozygous yellow sector results from a single cross-over between the *paba* and the *y* locus.

does, however, correspond with that found from scoring recombinants at meiosis. The relative distances do not correspond, from which it is inferred that the distribution of cross-overs is not the same at mitosis and meiosis.

Together, the formation of the heterokaryon and the nuclear fusion to give diploids, followed by recombination through mitotic crossing over or, indeed, by random chromosome elimination to produce haploid recombinants, constitute the parasexual cycle. The variation generated is probably of particular significance in organisms like the fungi imperfecti which have no sexual cycle. The principles of mapping on the basis of mitotic recombination which were developed in the fungi have proved important when applied to mammalian complements, the human chromosome complement in particular.

7.3 Mouse × man hybrids

The location and mapping of genes by analyzing recombination at meiosis following the standard test crosses are clearly not feasible in *Homo sapiens*. For this reason chromosome maps were until recently based upon painstaking analysis of family pedigrees. The situation was changed dramatically by the development of methods for making intraspecific and interspecific heterokaryons and diploid cell hybrids in mammals.

Heterokaryons and cell hybrids

Harris and Watkins (1965) were the first to produce interspecific heterokaryons in mammalian cells. They contained nuclei from the mouse and man. The hybrid cells arose within a mixture of human *Hela* cells, cultured from cervical carcinoma tissue, and Ehrlich mouse tumour cells which grow as a suspension in the peritoneal cavity of mice. The cell fusions were induced in the presence of a Sendai virus, one of the para-influenza group of myxoviruses. That the multinucleate cells in the mixture were heterokaryotic was verified by labelling the nuclei of the *Hela* cells with tritiated thymidine and showing both labelled and unlabelled nuclei within single cells. Proliferation of the hybrid cells depends on nuclear fusion in binucleate cells (Fig. 7.4). The two nuclei enter mitosis together, a single spindle forms and the daughter cells contain a single nucleus composed of the chromosome complements of both mouse and man. More recently cell hybrids have been produced using normal human cells from a variety of different organs, liver and kidney, also blood cells and nerve cells. In the mouse × man hybrids the fused nuclei initially contain the full diploid chromosome complements of both the parental species. They are, however, unstable much as the diploid nuclei in *Aspergillus* heterokaryons.

The mouse × man cells with hybrid nuclei may be screened out by a method closely allied to that used for screening heterokaryons in fungi where each nucleus compensates for a deficiency in the other. For example, a mixture of human and mouse cells is grown on a medium containing hypoxanthine, aminopterin and thymine. The human cells

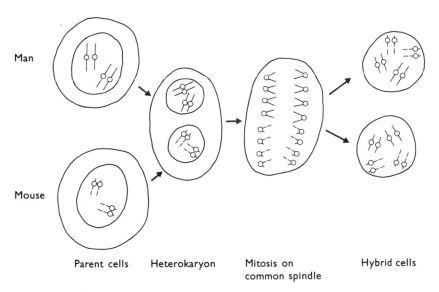

Man

Mouse

Parent cells   Heterokaryon   Mitosis on   Hybrid cells
                                common spindle

Fig. 7.4 Mouse × man hybrid cell formation. Only two pairs of chromosomes are shown for each parental cell line. Cell fusion leads initially to some binucleate heterokaryons. When both the nuclei enter mitosis synchronously the chromosomes are all attached to a common spindle. Following mitosis both the chromosome sets are contained within the one hybrid cell nucleus.

are blood cells or cells from other non-dividing tissues. The mouse cells are from a strain lacking the enzymes HGPRT (hypoxanthine guanine phosphoribosyltransferase) and TK (thymine kinase). The aminopterin in the medium blocks the synthesis of DNA nucleotides from sugars and amino acids by the regular pathway. The human nuclei which can produce the enzymes HGRRT and TK can synthesize DNA nucleotides by an alternative 'salvage' pathway. Even so the human cells are unable to proliferate because they are from a non-dividing tissue. Neither can the mouse cells multiply because their nuclei, lacking HGPRT and TK, cannot synthesize DNA nucleotides. The fusion nucleus, on the other hand, has the capacity for synthesizing DNA nucleotides, deriving from the human chromosomes, and the capacity for nuclear division, deriving from the mouse chromosomes. For these reasons clones of cells with hybrid nuclei are readily isolated.

Chromosome segregation

In mouse × man cell hybrids the chromosomes of the human complement (the segregant set) are unstable and are eliminated from the cell line. The mouse complement (dominant set) is retained complete. In other situations (Table 7.1), as in the Syrian hamster × mouse hybrids, the dominance relationship may be reversed and it is the mouse set which undergoes elimination.

Table 7.1 Patterns of chromosome segregation in some interspecific cell hybrids

Dominant set		Segregant set
Mouse	×	Man
Chinese hamster	×	Man
Syrian hamster	×	Mouse
Chinese hamster	×	Kangaroo rat
Mouse	×	Rat
Mouse	×	Drosophila

Chromosome elimination of a comparable kind is also found following normal interspecific hybridization between plant species. For example Davies (1958) and Kasha and Kao (1970) found that when they made reciprocal hybrids between a 28 chromosome autotetraploid cultivated barley (*Hordeum vulgare*) and a tetraploid *H. bulbosum*, nearly all the progeny were diploid (i.e. containing gametic chromosome numbers) and resembled the cultivated *H. vulgare*. When crosses were made using the diploid *H. vulgare* the cultured embryos were haploids with chromosomes derived exclusively from the *H. vulgare* maternal parent. The haploids arise by loss of *H. bulbosum* chromosomes during the early mitotic divisions of the embryo. In another cross, using the Chinese

Spring variety of hexaploid cultivated wheat $(2n = 6x = 42)$ and both diploid and tetraploid forms of *H. bulbosum* as pollen parents, Barclay (1975) has obtained haploid wheat with 21 chromosomes, again after embryo culture. Chromosome counts at mitotic metaphase in the zygote confirm that fertilization was quite normal and the segregation and loss of *H. bulbosum* chromosomes occurred at the subsequent divisions of the embryo. What determines that the chromosomes of a particular species be eliminated is not known. The same applies to the mammalian cell hybrids (Table 7.1).

Chromosome segregation forms the basis of the mapping procedure for the assignment of genes to chromosomes in the mouse × man hybrids. There is some suggestion that chromosome segregation takes place very soon after cell fusion, possibly at the first mitosis following heterokaryon formation. One of the consequences of this is that the numbers and kinds of chromosomes are fairly constant within young clones, but there is variation between the different clones (Ruddle, 1972). Each of the human chromosomes $(2n = 2x = 46)$ is identifiable by its banding pattern (Fig. 1.5, p. 12) and readily distinguished from the chromosomes of the mouse $(2n = 2x = 40)$.

Mapping

There is transcription and translation of information from many genes within both man and mouse chromosomes. Where the end product (enzyme) of a gene in the human complement is identifiable, the way to find out which chromosome carries the gene is to examine the chromosomes in clones which do not produce the enzyme and determine which particular chromosome lost by segregation is common to each. The practical aspects have been presented by Ruddle and Kucherlapati (1974) and are summarized in Fig. 7.5.

This kind of genetic analysis is limited to those genes which are expressed in the artificial environment of cell culture and it is clearly not possible to study character differences involving organs let alone the form and functioning of the organisms as a whole. The genes most amenable to analysis are those coding for constitutive enzymes continuously produced throughout the life of most or all cell types. Not surprisingly many of these genes are concerned with basic cellular functions like respiration. Genes specifying inducible enzymes which are more likely to be expressed in specialized differentiated cells can often be investigated if the appropriate cell types concerned, such as liver, kidney or blood cells, are used in the initial cell fusions. Alternatively some genes with highly specialized activities in human cells can, even in cells where they are not normally active, be derepressed by the presence of homologous genes in the mouse complement to permit a linkage analysis.

It is not yet possible to map the positions of genes relative to one another within human chromosomes by somatic crossing over, as it is in some fungi. The approximate positions of some genes, however, in relation to

| | | Hybrid clones | | | | |
		A	B	C	D	E
Human enzymes	I	+	−	−	+	−
	II	−	+	−	+	−
	III	+	−	−	+	−
	IV	+	+	+	−	−
Human chromosomes	1	−	+	−	+	−
	2	+	−	−	+	−
	3	−	−	−	+	+

Fig. 7.5 Experimental procedure for linkage analysis. The analysis is based on the correspondence between the presence or absence of particular human enzymes and chromosomes over a number of hybrid clones. In this hypothetical scheme enzymes I and III are both seen to be linked and assignable to chromosome 2. Enzyme II is assigned to chromosome I. [From Ruddle, F. H. and Kucherlapati, R. S. (1974) *Scientific American*, **231**, 36–44]

chromosome arms or parts of arms can sometimes be ascertained cytologically by the use of translocations, virus-induced lesions or deletions.

8

Conclusion

We have alluded to the variety in the form and behaviour of chromosomes which accompanied the divergence and evolution of species and to the widespread chromosome polymorphism which contributes to adaption within species. Recent evidence for very large scale quantitative variation in the chromosomal DNA, by segmental amplification or deletion, adds a new dimension to the pattern of adaptive chromosome change. The genetic significance of such variation is unresolved. One of the features of surveys describing the quantitative DNA variation among groups of plants and animals is worth emphasizing. It is that quantitative DNA variation *among* species, even closely related species, is commonplace and often on a very large scale, whereas variation *within* species, is uncommon and even then of a very restricted degree. We must conclude that within that group we call a species there are rigid limits to certain kinds of change which in turn argues for a finely adjusted balance among the various chromosome components. It follows that components such as those made up of very highly repetitive DNA segments, seemingly untranscribed and untranscribable, are redundant only in the context of transcription, not in the sense of fitness. Taking a more familiar point, we have known for years that the chromosome number within species is usually constant and so also, is the relative distribution of genetic material and of genes among the chromosome complement. Yet it is by no means clear why this should be so, at least in species comprising widely scattered populations. The point is perhaps best made by dwelling on an exception. In the dog whelk, *Nucella lapillus* (Bantock and Cockayne, 1975), the basic chromosome number and thereby the number of linkage groups varies from one locality to the other, from $2n = 26$ to $2n = 36$. Another example comes from a gerbil, *Gerbillus pyramidum* (Wahrman and Gourevitz, 1972). The creation or disruption of linkages is on the face of it a powerful adaptive option. Yet the option is rarely taken up. We must assume that in general the architecture of each chromosome has evolved to accommodate its particular role and activity. We have little understanding of how this is achieved but the rigidity with regard to chromosome number and DNA content may well be related. The amplified DNA segments could well serve as the 'bricks and mortar' which determine the construction and thereby the functioning of the chromosome. While large scale changes in chromosome architecture are restricted within a species it is manifestly clear that speciation within many groups of plants and animals is accompanied or succeeded by changes of cataclysmic proportions. To say the least there is a surprising *discontinuity* of chromosome organization *among* species.

Finally, it is worth considering the implications of recent information about the molecular organization of the chromosomes to the formal concepts of gene action and interaction. If, as is suggested above, the untranscribed fraction of the chromosomal DNA is nevertheless instrumental in regulating by one means or another the activity of 'structural' genes of major effect or of genes with minor effects which make up a polygenic system, the untranscribed fractions must be considered no less 'factors' of inheritance than Mendel's genes in peas. The distribution of such material among the chromosomes, in terms of linkage, its mode of action and of interaction within the nucleus, will ultimately need to be incorporated into any model which seeks to describe the transmission and expression of genetic information.

References

AHLOOWALIA, B. S. (1971). Frequency, origin and survival of aneuploids in tetraploid ryegrass. *Genetica*, **42**, 129–38.

ASHBURNER, M. (1969). Genetic control of puffing in polytene chromosomes. *Chromosomes Today*, **2**, 99–106.

ASAO, T. (1969). Behaviour of histones and cytoplasmic basic proteins during embryogenesis of the Japanese newt, *Triturus pyrrhogaster*. *Exp. Cell Res.*, **58**, 243–52.

AVERY, O. T., MACLEOD, C. M. and MCCARTHY, M. (1944). Studies on the chemical nature of the substance inducing transformation of pneumococcal types. *J. exp. Med.*, **79**, 137–58.

AYONOADU U. W. U. and REES H. (1968). The influence of *B* chromosomes on chiasma frequencies in Black Mexican Sweet Corn. *Genetica*, **39**, 75–81.

BACHMANN K., GOIN, O. B. and GOIN, C. J. (1972). Nuclear DNA amounts in vertebrates. *Brookhaven Symp. Biol.*, **23**, 419–50.

BANTOCK, C. R. and COCKAYNE, W. C. (1975). Chromosome polymorphism in *Nucella lapillus*. *Heredity*, **34**, 231–46.

BARCLAY, I. R. (1975). High frequencies of haploid production in wheat (*Triticum aestivum*) by chromosome elimination. *Nature*, **256**, 410–11.

BARLOW, B. A. and WIENS, D. (1975). Permanent translocation heterozygosity in *Viscum hildebrandtii* Engl. and *V. engleri* Tiegh. (*Viscaceae*) in East Africa. *Chromosoma*, **53**, 265–72.

BEADLE, G. W. (1930). Genetical and cytological studies of mendelian asynapsis in *Zea mays*. *Cornell Univ. Agric. Exp. Stat. Mem.*, **129**, 1–23.

BELOZERSKY, A. N. and SPIRIN, A. S. (1960). In *The Nucleic Acids*, Vol. 3, ed. CHARGAFF, E. and DAVIDSON, J. N., Academic Press, Inc., New York and London.

BENNETT, M. D. (1971). The duration of meiosis. *Proc. R. Soc. B.*, **178**, 277–99.

BENNETT, M. D. (1972). Nuclear DNA content and minimum generation time in herbaceous plants. *Proc. R. Soc. B.*, **181**, 109–35.

BENNETT, M. D. and REES, H. (1969). Induced and developmental variation in chromosomes of meristematic cells. *Chromosoma*, **27**, 226–44.

BISHOP, J. O. and FREEMAN, K. B. (1973). DNA sequences neighboring the duck hemoglobin genes. *Cold Spring Harb. Symp. quant. Biol.*, **38**, 707–16.

BLUMENTHAL, A. B., KRIEGSTEIN, H. J. and HOGNESS, D. S. (1973). The units of DNA replication in *Drosophila melanogaster* chromosomes. *Cold Spring Harb. Symp. quant. Biol.*, **38**, 205–23.

BRIDGES, C. B. (1922). The origin of variations in sexual and sex-limited characters. *Am. Nat.*, **56**, 51–63.

BRIDGES, C. B. (1936). The bar 'gene' a duplication. *Science, N.Y.*, **83**, 210–11.

BRITTEN, R. J. and KOHNE, D. E. (1968). Repeated sequences in DNA. *Science, N.Y.*, **161**, 529–40.

BROWN, S. W. and ZOHARY, D. (1955). The relationship of chiasmata and crossing over in *Lilium formosanum*. *Genetics, Princeton*, **40**, 850–73.

CALLAN, H. G. (1972). Replication of DNA in the chromosomes of eukaryotes. *Proc. R. Soc. B.*, **181**, 19–41.

CALLAN, H. G. and LLOYD, L. (1960). Lampbrush chromosomes of crested newts *Triturus cristatus* (Laurenti). *Phil. Trans. R. Soc. B.*, **243**, 135–219.

CAMERON, F. M. and REES, H. (1967). The influence of *B* chromosomes on meiosis in *Lolium*. *Heredity*, **22**, 446–50.

CATCHESIDE, D. G. (1977). *The Genetics of Recombination*. Edward Arnold, London.

CHUKSANOVA, N. (1939). Karyotypes of pollen grains in triploid *Crepis capillaris*. *Comp. Rends. (Doklady) Acad. Sci. U.S.S.R.*, **25**, 232–5.

CRANE, M. B. and LAWRENCE, W. J. C. (1934). *The Genetics of Garden Plants*. Macmillan, London.

CROWLEY, J. G. and REES, H. (1968). Fertility and selection in tetraploid *Lolium*. *Chromosoma*, **24**, 300–8.

DANIELLI, J. F. (1958). Studies of inheritance in amoebae by the technique of nuclear transfer. *Proc. R. Soc., B.*, **148**, 321–31.

DARLINGTON, C. D. (1935). The internal mechanics of the chromosomes II. Prophase pairing at meiosis in *Fritillaria*. *Proc. R. Soc. B.*, **118**, 59–73.

DARLINGTON, C. D. (1937). *Recent Advances in Cytology*. J. & A. Churchill, London.

DARLINGTON, C. D. (1956). *Chromosome Botany*. George Allen and Unwin, London.

DARLINGTON, C. D. (1958). *Evolution of Genetic Systems*, 2nd ed. Oliver & Boyd, Edinburgh.

DARLINGTON, C. D., HAIR, J. B. and HURCOMBE, R. (1951). The history of the garden Hyacinths. *Heredity*, **5**, 233–52.

DARLINGTON, C. D. and LA COUR, L. F. (1950). Hybridity selection in *Campanula*. *Heredity*, **4**, 217–48.

DARLINGTON, C. D. and MATHER, K. (1949). *The Elements of Genetics*. Allen and Unwin, London.

DAVIDSON, E. H., GRAHAM, D. E., NEUFIELD, B. R., CHAMBERLIN, M. E., AMENSON, C. S., HOUGH, B. R. and BRITTEN, R. J. (1973). Arrangements and characterisation of repetitive sequence elements in animal DNAs. *Cold Spring Harb. Symp. quant. Biol.*, **38**, 295–301.

DAVIES, D. R. (1958). Male parthenogenesis in barley. *Heredity*, **12**, 493–8.

DHIR, N. K. and MIKSCHE, J. P. (1974). Intraspecific variation of nuclear DNA content in *Pinus resinosa*. Ait. *Can. J. Genet. Cytol.*, **16**, 77–83.

DOBZHANSKY, T. (1943). Genetics of natural populations IX. Temporal changes in the composition of populations of *Drosophila pseudoobscura*. *Genetics, Princeton*, **28**, 162–86.

DOBZHANSKY T. (1948). Genetics of natural populations. XVIII. Experiments on chromosomes of *Drosophila pseudoobscura* from different geographic regions. *Genetics, Princeton*, **33**, 588–602.

DODDS, K. S. (1965). The history and relationships of cultivated potatoes, 123–41. In *Essays on Crop Plant Evolution*, ed. J. Hutchinson, Cambridge University Press.

DOVER, G. A. and HENDERSON, S. A. (1975). No detectable satellite DNA in supernumerary chromosomes of the grasshopper *Myrmeleotettix*. *Nature*, **259**, 57–9.

DOVER, G. A. and RILEY, R. (1972). Prevention of pairing of homoeologous meiotic chromosomes of wheat by an activity of supernumerary chromosomes of *Aegilops*. *Nature*, **240**, 159–61.

DUPRAW, E. J. (1970). *DNA and chromosomes*. Holt, Rhinehart and Winston, Inc., New York.

EVANS, G. M. and MACEFIELD, A. J. (1972). Suppression of homoeologous pairing by *B*-chromosomes in a *Lolium* species hybrid. *Nature New Biol.*, **236**, 110–11.

EVANS, G. M., DURRANT, A. and REES, H. (1966). Associated nuclear changes in the induction of Flax Genotrophs. *Nature*, **212**, 697–9.

EVANS, G. M., REES, H., SNELL, C. and SUN, S. (1972). The relationship between nuclear DNA amount and the duration of the mitotic cycle. *Chromosomes Today*, **3**, 24–31.

EVANS, H. J. and SUMNER, A. T. (1973). Chromosome architecture; morphological and molecular aspects of longitudinal differentiation. *Chromosomes Today*, **4**, 15–33.

FINCHAM, J. R. S. and DAY, P. R. (1971). *Fungal Genetics*. Blackwell, Oxford.

FLAVELL, R. B. and RIMPAU, J. (1975). Ribosomal RNA genes and supernumerary *B*-chromosomes of rye. *Heredity*, **35**, 127–31.

FLAVELL, R. B. and SMITH, D. B. (1974). Variation in nucleolar organiser rRNA gene multiplicity in wheat and rye. *Chromosoma*, **47**, 327–34.

FOGEL, S. and MORTIMER, R. K. (1969). Informational transfer in meiotic gene conversion. *Proc. Natn. Acad. Sci. U.S.A.*, **62**, 96–103.

FOX, D. P. (1971). The replicative status of heterochromatic and euchromatic DNA in two somatic tissues of *Dermestes maculatus* (*Dermestidae: Coleoptera*). *Chromosoma*, **33**, 183–95.

FRÖST, S. (1958). Studies of the genetical effects of accessory chromosomes in *Centaurea scabiosa*. *Hereditas*, **44**, 112–22.

GALL, J. G. (1956). *Brookhaven Symp. Biol.*, **8**, 17–32.

GALL, J. G. (1968). Differential synthesis of the genes for ribosomal RNA during amphibian oogenesis. *Proc. Natn. Acad. Sci. U.S.A.*, **60**, 553–60.

GALL, J. G. and CALLAN, H. G. (1962). H^3 uridine incorporation in lampbrush chromosomes. *Proc. Natn. Acad. Sci. U.S.A.*, **48**, 562–70.

GALL, J. G., COHEN, E. H. and ATHERTON, D. D. (1973). The satellite DNAs of *Drosophila virilis*. *Cold Spring Harb. Symp. quant. Biol.*, **38**, 417–21.

GERSTEL, D. U. and BURNS, J. A. (1966). Chromosomes of unusual length in hybrids between two species of *Nicotiana*. *Chromosomes Today*, **1**, 41–56.

GRANT, V. (1959). The regulation of recombination in plants. *Cold Spring Harb. Symp. quant. Biol.*, **23**, 337–63.

GRELL, R. F. and CHANDLEY, A. C. (1965). Evidence bearing on the coincidence of exchange and DNA replication in the oocyte of *Drosophila melanogaster*. *Proc. Natl. Acad. Sci. U.S.A.*, **53**, 1340–6.

GRIFFITH, F. (1928). Significance of pneumococcal types. *J. Hyg., Camb.*, **27**, 113–59.

GURDON, J. B. (1973). Nuclear transplantation and regulation of cell processes. *Br. med. Bull.*, **29**, 259–62.

HAMERTON, J. L. (1971). *Human Cytogenics*, Vol. I. Academic Press, Inc., New York and London.

HARINARAYANA, G. and MURTY, B. R. (1971). Cytological regulation of recombination in *Pennisetum* and *Brassica*. *Cytologia*, **36**, 435–48.

HARRIS, H. and WATKINS, J. F. (1965). Hybrid cells derived from mouse and man: artificial heterokaryons of mammalian cells from different species. *Nature*, **205**, 640–6.

HAYES, W. (1968). *The Genetics of Bacteria and their Viruses*, 2nd ed. Blackwell, Oxford.

HAZARIKA, M. H. and REES, H. (1967). Genotypic control of chromosomes behaviour in rye X. Chromosome pairing and fertility in autotetraploids. *Heredity*, **22**, 317–32.

HENDERSON, S. A. (1963). Chiasma distribution at diplotene in a locust. *Heredity*, **18**, 173–90.

HENDERSON, S. A. (1970). The time and place of meiotic crossing-over. *A. Rev. Genet.*, **4**, 295–324.

HERSHEY, A. D. (1955). An upper limit to the protein content of the germinal substance of bacteriophage T2. *Virology*, **1**, 108–27.

HERSHEY, A. D. and CHASE, M. C. (1952). Independent functions of viral protein and nucleic acid in growth of bacteriophage. *J. gen. Physiol.*, **36**, 39–56.

References 141

HEWITT, G. M. (1973). The integration of supernumerary chromosomes into the Orthopteran genome. *Cold Spring Harb. Symp. quant. Biol.*, 38, 183–94.

HEWITT, G. M. and JOHN, B. (1972). Inter-population sex chromosome polymorphism in the grasshopper *Podisma pedestris* II. Population parameters. *Chromosoma*, 37, 23–42.

HOLLIDAY, R. (1964). A mechanism for gene conversion in fungi. *Genet. Res.*, 5, 282–304.

HOWELL, S. H. and STERN, H. (1971). The appearance of DNA breakage and repair activities in the synchronous meiotic cycle of *Lilium*. *J. molec. Biol.*, 55, 357–78.

HUTTO, J. (1974). The response of diploid and tetraploid rye genotypes to phosphate treatments and to cold temperature. *J. agric. Sci., Camb.*, 82, 353–6.

HUTTON, J. R. and THOMAS, C. A. (1975). The origin of folded DNA rings from *Drosophila melanogaster*. *J. molec. Biol.*, 98, 425–38.

JOHN, B. (1973). The cytogenetic systems of grasshoppers and locusts II. The origin and evolution of supernumerary segments. *Chromosoma*, 44, 123–46.

JOHN, B. and HEWITT, G. M. (1965). The *B*-chromosome system of *Myrmeleotettix maculatus* (Thunb.) I. The mechanics. *Chromosoma*, 16, 548–78.

JOHN, B. and HEWITT, G. M. (1968). Patterns and pathways of chromosome evolution within the *Orthoptera*. *Chromosoma*, 25, 40–74.

JOHN, B. and LEWIS, K. R. (1957). Studies on *Periplaneta americana*. II. Interchange heterozygosity in isolated populations. *Heredity*, 11, 11–22.

JOHN, B. and LEWIS, K. R. (1965). *The Meiotic System, Vol. VI F1, Protoplasmatologia*. Springer-Verlag, Vienna.

JONES, G. H. (1967). The control of chiasma distribution in rye. *Chromosoma*, 22, 69–90.

JONES, G. H. (1971). The analysis of exchanges in tritium-labelled meiotic chromosomes II. *Stethophyma grossum*. *Chromosoma*, 34, 367–82.

JONES, K. (1974). Chromosome evolution by Robertsonian translocation in *Gibasis* (*Commelinaceae*). *Chromosoma*, 45, 353–68.

JONES, K. W. (1970). Chromosomal and nuclear location of mouse satellite DNA in individual cells. *Nature*, 225, 912–15.

JONES, K. W. and ROBERTSON, F. W. (1970). Localisation of reiterated nucleotide sequences in *Drosophila* and mouse by *in situ* hybridisation of complementary RNA. *Chromosoma*, 31, 331–45.

JONES, R. N. (1975). *B*-chromosome systems in flowering plants and animal species. *Int. Rev. Cytol.*, 40, 1–100.

JONES, R. N. and BROWN, L. M. (1976). Chromosome evolution and DNA variation in *Crepis*. *Heredity*, 36, 91–104.

JONES, R. N. and REES, H. (1968). Nuclear DNA variation in *Allium*. *Heredity*, 23, 591–605.

JONES, R. N. and REES, H. (1969). An anomalous variation due to *B*-chromosomes in rye. *Heredity*, 24, 265–71.

JUDD, B. H. and YOUNG, M. W. (1973). An examination of the one cistron:one chromomere concept. *Cold Spring Harb. Symp. quant. Biol.*, 38, 573–9.

KARPECHENKO, G. D. (1928). Polyploid hybrids of *Raphanus sativa* L. × *Brassica oleracea* L. *Zeitschr. ind Abstamm. Vererbungsl.*, 39, 1–7.

KASHA, K. J. and KAO, K. N. (1970). High frequency haploid production in barley (*Hordeum valgare* L.). *Nature*, 225, 874–5.

KAVENOFF, R., KLOTZ, L. C. and ZIMM, B. H. (1973). On the nature of chromosome-sized DNA molecules. *Cold Spring Harb. Symp. quant. Biol.*, 38, 1–8.

KAYANO, H. (1971). Accumulation of *B*-chromosomes in the germ line of *Locusta migratoria*. *Heredity*, 27, 119–23.

KEYL, H. G. (1965). Duplikationen von unterreinheiten der chromosomalen DNA während der evolution von *Chironomous thummi*. *Chromosoma*, 17, 139–80.

KIRK, D. and JONES, R. N. (1970). Nuclear genetic activity in B-chromosome rye in terms of the quantitative interrelations between nuclear protein, nuclear RNA and histone. *Chromosoma*, **31**, 241–54.

KIRK, J. T. O., REES, H. and EVANS, G. M. (1970). Base composition of nuclear DNA within the genus *Allium*. *Heredity*, **25**, 507–12.

KITANI, Y., OLIVE, L. S. and EL-ANI, A. S. (1962). Genetics of *Sordaria fimicola* V. Aberrant segregation at the g locus. *Am. J. Bot.*, **49**, 697–706.

KORNBERG, R. D. and THOMAS, J. O. (1974). Chromatin structure: oligomeres of the histones. *Science, N.Y.*, **184**, 865–8.

LAWRENCE, C. W. (1961). The effect of radiation on chiasma formation in *Tradescantia*. *Radiation Bot.*, **1**, 92–6.

LEWIN, B. (1975). The nucleosome: subunit of mammalian chromatin. *Nature*, **254**, 651–3.

LEWIS, K. R. (1967). Polyploidy and plant improvement. *The Nucleus*, **10**, 99–110.

LEWIS, K. R. and JOHN, B. (1963). *Chromosome Markers*. J. and A. Churchill, Ltd., London.

LEWIS, K. R. and JOHN, B. (1970). *The Organization of Heredity*. Edward Arnold, London.

LIMA-DE-FARIA, A. (1954). Chromosome gradient and chromosome field in *Agapanthus*. *Chromosoma*, **6**, 330–70.

LIMA-DE-FARIA, A. (1959). Differential uptake of tritiated thymidine into hetero- and euchromatin in *Melanoplus* and *Secale*. *J. Biophys. Biochem. Cytol.*, **6**, 457–66.

LISSOUBA, P., MOUSSEAU, J., RIZET, G. and ROSSIGNOL, J. L. (1962). Fine structure of genes in the ascomycete *Ascobolus immersus*. *Adv. Genet.*, **11**, 343–80.

LOENING, U. E., JONES, K. W. and BIRNSTIEL, M. L. (1969). Properties of the ribosomal RNA precursor in *Xenopus laevis*; comparison to the precursor in mammals and in plants. *J. molec. Biol.*, **45**, 353–66.

LONGLEY, A. E. (1927). Supernumerary chromosomes in *Zea mays*. *J. Agric. Res.*, **35**, 769–84.

LYON, M. F. (1963). Attempts to test the inactive-X theory of dosage compensation in mammals. *Genet. Res.*, **4**, 93–103.

MCPHEE, C. P. and ROBERTSON, A. (1970). The effect of suppressing crossing-over on the response to selection in *Drosophila melanogaster*. *Genet. Res.*, **16**, 1–16.

MARTIN, P. G. (1966). Variation in the amounts of nucleic acids in the cells of different species of higher plants. *Expl. Cell. Res.*, **44**, 84–94.

MATHER, K. (1936). Segregation and linkage in autotetraploids. *J. Genet.*, **31**, 287–314.

MATHER, K. (1937). The determination of position in crossing over. II. The chromosome length chiasma frequency relation. *Cytologia* (Tokyo) Fujii Jub. vol., 514–26.

MATHER, K. (1938). Crossing over. *Biol. Rev.*, **13**, 252–92.

MATHER, K. (1973). *Genetical Structure of Populations*. Chapman and Hall, London.

MATSUDA, T. (1970). On the accessory chromosomes of *Aster*. I. The accessory chromosomes of *Aster ageratoides* group. *J. Sci. Hiroshima Univ.*, **13**, 1–63.

MIRSKY, A. E. and OSAWA, S. (1961). *The Cell*, Vol. II, Academic Press.

MILLER, O. L. (1965). Fine structure of lampbrush chromosomes. *Natn. Cancer Inst. Monograph*, **18**, 79–89.

MITTWOCH, U. (1967). *Sex chromosomes*. Academic Press, Inc., New York.

MOENS, P. B. (1969a). The fine structure of meiotic chromosome polarization and pairing in *Locusta migratoria* spermatocytes. *Chromosoma*, **28**, 1–25.

MOENS, P. B. (1969b). The fine structure of meiotic chromosome pairing in the triploid, *Lilium tigrinum*. *J. Cell Biol.*, **40**, 273–9.

MORESCALCHI, A. (1973). Amphibia. In *Cytotaxonomy and Vertebrate Evolution*, ed. A. B. Chiarelli and E. Capanna, Academic Press, Inc., New York and London, pp. 233–348.

MORRIS, T. (1968). The *XO* and *OY* chromosome constitutions in the mouse. *Genet. Res.*, **12**, 125-37.

MOSES, M. J. (1956). Chromosomal structures in crayfish spermatocytes. *J. Biophys. Biochem. Cytol.*, **2**, 215-17.

MOSS, J. P. (1966). The adaptive significance of *B*-chromosomes in rye. *Chromosomes Today*, **1**, 15-23.

MULLER, H. J. and KAPLAN, W. D. (1966). The dosage compensation of *Drosophila* and mammals as showing the accuracy of the normal type. *Genet. Res.*, **8**, 41-59.

MÜNTZING, A. (1963). Effects of accessory chromosomes in diploid and tetraploid rye. *Hereditas*, **49**, 371-426.

NARAYAN, R. K. J. and REES, H. (1976). Nuclear DNA variation in *Lathyrus*. *Chromosoma*, **54**, 141-54.

OHNO, S. (1969). Evolution of sex chromosomes in mammals. *Ann. Rev. Genetics*, **3**, 495-524.

OHNO, S. (1970). *Evolution by Gene Duplication*. Allen and Unwin, London.

OLINS, A. L. and OLINS, D. E. (1974). Spheroid chromatin units (v bodies). *Science, N.Y.*, **183**, 330-2.

OUDET, P., GROSS-BELLARD, M. and CHAMBON, P. (1975). Electron microscopic and biochemical evidence that chromatin structure is a repeating unit. *Cell*, **4**, 281-300.

PARDUE, M. L. and GALL, J. G. (1970). Chromosomal localisation of mouse satellite DNA. *Science, N.Y.*, **168**, 1356-8.

PARKER, J. S. (1975). Chromosome-specific control of chiasma formation. *Chromosoma*, **49**, 391-406.

PEACOCK, W. J. (1970). Replication, recombination, and chiasmata in *Goniaea australasiae* (Orthoptera: Acrididae). *Genetics*, **65**, 593-617.

PEACOCK, W. J., BRUTLAG, D., GOLDRING, E., APPELS, R., HINTON, C. W. and LINDSLEY, D. L. (1973). The organisation of highly repeated DNA sequences in *Drosophila melanogaster* chromosomes. *Cold Spring Harb. Symp. quant. Biol.*, **38**, 405-16.

PEARSON, G. G., TIMMIS, J. N. and INGLE, J. (1974). The differential replication of DNA during plant development. *Chromosoma*, **45**, 281-94.

PERRY, P. E. and JONES, G. H. (1974). Male and female meiosis in grasshoppers. I. *Stethophyma grossum*. *Chromosoma*, **47**, 227-36.

PONTECORVO, G. (1953). The genetics of *Aspergillus nidulans*. *Adv. Genet.*, **5**, 141-238.

PONTECORVO, G. and KAFER, E. (1958). Genetic analysis based on mitotic recombination. *Adv. Genet.*, **9**, 71-103.

RANDOLPH, L. F. (1928). Types of supernumerary chromosomes in maize. *Anat. Rec.*, **41**, 102.

REDDI, K. K. (1959). The arrangement of purine and pyrimidine nucleotides in tobacco mosaic virus nucleotides. *Proc. Natn. Acad. Sci. U.S.A.*, **45**, 293-300.

REES, H. (1964). The question of polyploidy in the *Salmonidae*. *Chromosoma*, **15**, 275-9.

REES, H. (1972). DNA in higher plants. *Brookhaven Symp. Biol.*, **23**, 394-418.

REES, H., CAMERON, F. M., HAZARIKA, M. H. and JONES, G. H. (1966). Nuclear variation between diploid angiosperms. *Nature*, **211**, 828-30.

REES, H. and DALE, P. J. (1974). Chiasmata and variability in *Lolium* and *Festuca* populations. *Chromosoma*, **47**, 335-51.

REES, H. and HAZARIKA, M. H. (1969). Chromosome evolution in *Lathyrus*. *Chromosomes Today*, **2**, 158-65.

REES, H. and HUTCHINSON, J. (1973). Nuclear DNA variation due to *B*-chromosomes. *Cold Spring Harb. Symp. quant. Biol.*, **38**, 175-82.

REES, H. and JONES, G. H. (1967). Chromosome evolution in *Lolium*. *Heredity*, **22**, 1-18.

REES, H. and JONES, R. N. (1971). Chromosome gain in higher plants, in: *Cellular Organelles and Membranes in Mental Retardation*, ed. P. F. Benson. Churchill Livingstone, Edinburgh and London, pp. 185–208.

REES, H. and JONES, R. N. (1972). The origin of the wide species variation in nuclear DNA content. *Int. Rev. Cytol.*, **32**, 53–92.

REES, H. and SUN, S. (1965). Chiasma frequency and the disjunction of interchange associations in rye. *Chromosoma*, **16**, 500–10.

REES, H. and THOMPSON, J. B. (1956). Genotypic control of chromosome behaviour in rye. III. Chiasma frequency in homozygotes and heterozygotes. *Heredity*, **3**, 409–24.

RHOADES, M. M. (1968). Studies on the cytological basis of crossing over, in: *Replication and Recombination of Genetic Material*, ed. W. J. Peacock and R. D. Brock. Australian Academy of Sciences, Canberra, pp. 229–41.

RHOADES, M. M. and DEMPSEY, E. (1973). Chromatin elimination induced by the *B*-chromosome of maize. *J. Heredity*, **64**, 13–18.

RICK, C. M. and BARTON, D. W. (1954). Cytological and genetic identification of the primary trisomics of the tomato. *Genetics*, **39**, 640–66.

RILEY, R. and CHAPMAN, V. (1958). Genetic control of the cytologically diploid behaviour of hexaploid wheat. *Nature*, **182**, 713–15.

RITOSSA, F. M. and SCALA, G. (1969). Equilibrium variations in the redundancy of rDNA in *Drosophila melanogaster*. *Genetics Suppl., Princeton*, **61**, 305–17.

RITOSSA, F. M. and SPIEGELMAN, S. (1965). Localisation of DNA complementary to ribosomal RNA in the nucleolus organizer region of *Drosophila melanogaster*. *Proc. Natn. Acad. Sci. U.S.A.*, **53**, 737–45.

ROBERTSON, W. R. B. (1916). Chromosome studies. I. Taxonomic relationships shown in the chromosomes of *Tettigidae* and *Acrididae*: V-shaped chromosomes and their significance in *Acrididae*, *Locustidae*, and the *Gryllidae*: chromosomes and variation. *J. Morph.*, **27**, 179–331.

ROMAN, H. (1948). Directed fertilisation in maize. *Proc. natl. Acad. Sci. U.S.A.*, **34**, 36–42.

ROSSEN, J. M. and WESTERGAARD, M. (1966). Studies on the mechanism of crossing-over. II. Meiosis and the time of chromosome replication in the Ascomycete *Neottiella rutilans* (Fr.) Dennis. *Comp. Rend.*, **35**, 233–60.

RUDDLE, F. H. (1972). Linkage analysis using somatic cell hybrids. *Adv. Hum. Genet.*, **3**, 173–235.

RUDDLE, F. H. and KUCHERLAPATI, R. S. (1974). Hybrid cells and human genes. *Scientific American*, **231**, 36–44.

SAGAR, J. and RYAN, F. J. (1963). *Cell Heredity*. John Wiley, Chichester.

SANNOMIYA, M. (1962). Intra-individual variation in number of *A*- and *B*-chromosomes in *Pantanga japonica*. *Chromosome Information Service*, **3**, 30–2.

SCHACHAT, F. H. and HOGNESS, D. S. (1973). Repetitive sequences in isolated Thomas circles from *Drosophila melanogaster*. *Cold Spring Harb. Symp. quant. Biol.*, **38**, 371–81.

SCHINDLER, A. M. and MIKAMO, K. (1970). Triploidy in man. Report of a case and a discussion on etiology. *Cytogenetics*, **9**, 116–30.

SELIGY, V. and MIYAGI, M. (1969). Studies of template activity of chromatin isolated from metabolically active and inactive cells. *Exp. Cell Res.*, **58**, 27–34.

SHARMAN, G. B. and BARBER, H. N. (1952). Multiple sex chromosomes in the marsupial *Potorus*. *Heredity*, **6**, 345–55.

SIMONSEN, Ø. (1973). Cytogenetic investigations in diploid and autotetraploid populations of *Lolium perenne* L. *Hereditas*, **75**, 157–88.

SIMONSEN, Ø. (1975). Cytogenetic investigations in diploid and autotetraploid populations of *Festuca pratensis*. *Hereditas*, **79**, 73–108.

SMITH, D. B. and FLAVELL, R. B. (1975). Characterisation of the wheat genome by renaturation kinetics. *Chromosoma*, **50**, 223–42.

SOUTHERN, D. I. (1970). Polymorphism involving heterochromatic segments in *Metrioptera brachyptera*. *Chromosoma*, **30**, 154–68.

SOUTHERN, E. M. (1970). Base sequence and evolution of guinea-pig satellite DNA. *Nature*, **227**, 794–8.

SOUTHERN, E. M. (1974). Eukaryotic DNA, in: *Biochemistry of Nucleic Acids*, ed. K. Burton. Butterworths, London, pp. 101–39.

SPARROW, A. H., PRICE, H. J. and UNDERBRINK, A. G. (1972). A survey of DNA content per cell and per chromosome in prokaryotic and eukaryotic organisms; some evolutionary considerations. *Brookhaven Symp. Biol.*, **23**, 451–94.

STEBBINS, G. L. (1950). *Variation and Evolution in Plants*. Columbia University Press.

STEBBINS, G. L. (1966). Chromosomal variation and evolution. *Science, N.Y.*, **152**, 1463–9.

STERN, C. (1936). Somatic crossing over and segregation in *Drosophila melanogaster*. *Genetics*, **21**, 625–730.

STERN, C. (1970). Somatic recombination within the white locus of *Drosophila melanogaster*. *Genetics*, **62**, 573–81.

STERN, H. and HOTTA, Y. (1973). Biochemical controls of meiosis. *Ann. Rev. Genet.*, **7**, 37–66.

STONE, W. S. (1955). Genetic and chromosomal variability in *Drosophila*. *Cold Spring Harb. Symp. quant. Biol.*, **20**, 256–70.

SUEOKA, N. (1961). Variation and heterogeneity of base composition of deoxyribonucleic acids: a compilation of old and new data. *J. molec. Biol.*, **3**, 31–40.

SUN, S. and REES, H. (1964). Genotypic control of chromosome behaviour in rye VII. Unadaptive heterozygotes. *Heredity*, **19**, 357–67.

SWANSON, C. P. (1960). *Cytology and Cytogenetics*. Macmillan, London.

TAYLOR, J. H., WOODS, P. S. and HUGHES, W. L. (1957). The organisation and duplication of chromosomes as revealed by autoradiographic studies using tritium-labelled thymidine. *Proc. Natn. Acad. Sci. U.S.A.*, **43**, 122–8.

THOMAS, C. A. (1970). The theory of the master gene, in: *The Neurosciences: Second Study Program*, ed. F. O. Schmitt. Rockefeller University Press, New York, pp. 973–98.

THOMAS, C. A., PYERITZ, R. E., WILSON, D. A., DANCIS, B. M., LEE, C. S., BICK, M. D., HUANG, H. L. and ZIMM, B. H. (1973). Cyclodromes and palindromes in chromosomes. *Cold Spring Harb. Symp. quant. Biol.*, **38**, 353–70.

THOMPSON, J. B. (1956). Genotypic control of chromosome behaviour in rye. II. Disjunction at meiosis in interchange heterozygotes. *Heredity*, **10**, 99–108.

TIMMIS, J. N., SINCLAIR, J. and INGLE, J. (1972). Ribosomal-RNA genes in euploids and aneuploids of Hyacinth. *Cell Differentiation*, **1**, 335–9.

TIMMIS, J. N. and REES, H. (1971). A pairing restriction at pachytene upon multivalent formation in autotetraploids. *Heredity*, **26**, 269–75.

TIMMIS, J. N. and INGLE, J. (1973). Environmentally induced changes in rDNA gene redundancy. *Nature New Biol.*, **244**, 235–6.

TOBGY, H. A. (1943). A cytological study of *Crepis fuliginosa*, *C. neglecta* and their F_1 hybrid and its bearing on the mechanism of phylogenic reduction in chromosome number. *J. Genet.*, **45**, 67–111.

VOSA, C. G. and BARLOW, P. W. (1972). Meiosis and B-chromosomes in *Listera ovata* (Orchidaceae), *Caryologia*, **25**, 1–8.

WAHRMAN, J. and GOUREVITZ, P. (1971). Extreme chromosomal variability in a colonising rodent. *Chromosomes Today*, **4**, 399–424.

WALTERS, J. L. (1942). Distribution of structural hybrids in *Paeonia californica*. *Am. J. Bot.*, **29**, 270–5.

WARD, E. J. (1973). Nondisjunction: localisation of the controlling site in the maize B-chromosome. *Genetics, Princeton*, **73**, 387–91.

WASSERMAN, M. (1963). Cytology and phylogeny of *Drosophila*. *Am. Nat.*, **97**, 333–352.

WESTERGARRD, M. and VON WETTSTEIN, D. (1972). The synaptonemal complex. *Ann. Rev. Genet.*, **6**, 71–110.

WHITE, M. J. D. (1956). Adaptive chromosomal polymorphism in an Australian grasshopper. *Evolution*, **10**, 298–313.

WHITE, M. J. D. (1957a). Cytogenetics of the grasshopper *Moraba scurra*, I. Meiosis of interracial and interpopulation hybrids. *Aust. J. Zool.*, **5**, 285–304.

WHITE, M. J. D. (1957b). Cytogenetics of the grasshopper *Moraba scurra*. II. Heterotic systems and their interaction. *Aust. J. Zool.*, **5**, 305–37.

WHITE, M. J. D. (1957c). Cytogenetics of the grasshopper *Moraba scurra*. IV. Heterozygosity for 'elastic constrictions'. *Aust. J. Zool.*, **5**, 348–54.

WHITE, M. J. D. (1973). *Animal Cytology and Evolution*. Cambridge University Press.

WHITEHOUSE, H. L. K. (1963). A theory of crossing-over by means of hybrid deoxyribonucleic acid. *Nature*, **199**, 1034–40.

WHITEHOUSE, H. L. K. (1969). *Towards an Understanding of the Mechanism of Heredity*, 2nd ed. Edward Arnold, London.

WIMBER, D. E. and STEFFENSEN, D. M. (1973). Localisation of gene function. *Ann. Rev. Gent.*, **7**, 205–23.

WURSTER, D. H. and BENIRSCHKE, K. (1968). Chromosome studies in the super-family *Bovoidea*. *Chromosoma*, **25**, 152–71.

ZARCHI, Y., SIMCHEN, G., HILLEL, J. and SCHAPP, T. (1972). Chiasmata and the breeding system in wild populations of diploid wheats. *Chromosoma*, **38**, 77–94.

ZHDANOV, V. M. (1975). Integration of viral genomes. *Nature*, **256**, 471–3.

Index

The Bulldog

Gordon Galloway

Also by Gordon Galloway

SCARS OF A SOLDIER
HILLBILLY POET

Library of Congress Catalog Card Number: 96-85730
Copyright © 1996 by Gordon Galloway
All rights reserved
Printed in
The United States of America
Published by Deerfield Publishing Company
1613 130th Avenue
Morley, MI 49336
ISBN 0-9644077-2-8

ACKNOWLEDGMENTS

The art work for our cover was done by Julie Martin. We thank her for another outstanding job.

My wife, Mary Ann, deserves great credit for her solid support on this project. She spent long hours at the word processor, was patient with our changes, and, most of all, provided an editing opinion I could trust.

We also thank Lonnie Deur, Mike Seelhoff, Count Fisher and many others for recalling the Lake County years from their perspective.

With deep admiration I thank Bob and Dee Blevins for having the courage to share their story with us. They are two Americans we can all be very proud of.

From the Author

They needed some National Guard officers to escort the dignitaries at Governor Blanchard's inauguration. I thought it might be interesting to rub shoulders with some of the bigwigs of Michigan politics, so I volunteered to participate. The inauguration would take place on the State Capitol lawn in Lansing, on the 1st of January. One thing for sure, it probably would be damn cold.

This was Governor Blanchard's first inauguration. He wanted the full ceremony, gun salute and all. My brother was a State employee at the time, and I remember hearing him grumble about how much it was going to cost the State to put this shindig on. Although the inauguration was open to the public, over 30,000 invitations were mailed out.

Well, the big day rolled around and fortunately the temperature was up in the thirties. It was a cool crisp day, with a little breeze, but the sunshine made it a pleasant one. We all milled around on the Capitol main floor while they sorted out the pecking order for the grand entrance.

It would be the Democrats' turn to run Michigan's government again. It had been in the hands of the Republicans for 20 years straight. You could sense the excitement of change as immaculately dressed politicians scurried about trying to make sure everything ran perfectly.

I was slated to escort a man who was on the University of Michigan Board of Regents. Both he and his wife seemed like pleasant folks, but I refrained from mentioning that I had graduated from Michigan's other big University at East Lansing.

Before the ceremony started, Governor Milliken and his wife were ushered in. There was a warm and sincere round of applause for them. He had been a popular governor and held office for 14 years, longer than any other Michigan governor.

Five of the last six governors were present. Supreme Court Justice G. Mennon Williams would administer the oath of office. Of all the former governors, Williams was the most familiar to me. When I was a kid, then Governor Williams would walk through the crowds at the Ionia Free Fair and shake hands with folks. He was a big man with a bow tie and a ready smile. I had shaken his hand on two or three occasions.

I didn't recognize many other politicians. Of course they were standing facing the crowd as I was, so it was difficult to see their faces.

The new Lieutenant Governor, Martha Griffiths, stood beside Blanchard. She had a bright red coat on and was acting all excited about the whole thing.

The newspaper stated that Blanchard had already made most of his key appointments in the new administration. One official that should be easy to recognize would be the State Attorney General, Frank Kelley. Mr. Kelley was elected separate from the Governor, and had been Attorney General since 1961. Probably he was looking forward to working with a Democratic Governor for a change. I'd never seen him in person, but from his pictures he looked quite distinguished.

Because of the cold, the ceremony was reasonably short and moved right along. Blanchard's 12-year-old son Jay led the pledge of allegiance. Blanchard's wife Paula held the family Bible, while Justice Williams swore Jim Blanchard in as Michigan's 45th Governor.

Governor Blanchard's remarks lasted about 15 minutes. It was to be a New Alliance between State government and Michigan's citizens. The common enemy was joblessness. The new administration would also restore order out of financial chaos. The crowd was reported to number around 5000, and they gave enthusiastic approval to Governor Blanchard's remarks.

Following the new Governor's speech, four Michigan National Guard 105 millimeter howitzers started blazing away with the 19-gun salute. The noise was startling.

I thought that the whole ceremony was pretty impressive. With the elegance of the State Capitol building in the background, the enthusiastic crowd, the 19-gun salute, this was certainly a fine display of our democratic government.

Thirteen years later I would have a significantly different view of that inauguration. Yes, it had been entertaining, but the ceremony had little to do with quality government or responsible leadership. How well a government serves the people is dependent on the integrity and overall character of the people who are governing. Image and ceremony have nothing to do with it.

This is the story of a man whose diligent pursuit of organized crime became a threat to this new administration. He had to be silenced one way or the other. This is his true story. It must be told.

... willing to stand in the shadows. Always there to defend and support. The patience and loyalty to wait while I served my Country. Never asking for, or expecting more, than I could give. God has truly blessed me with a partner in life that has always given and never asked.

This book is dedicated to my wife and best friend, Dee.

Bob Blevins

FOREWORD

Although human beings are very complex in nature, there is a basic core in each of us that defines our character. In some individuals there is a special spirit and a certain morality that we both respect and admire, for it will have a positive effect on how they react to others and the adversities of life.

The final value judgement of an individual's life is based not on an image, popularity or net worth, but rather on the core quality that makes that individual what he is. In other words, is the human race better off for that individual having been here? In this respect I think Bob Blevins will fare very well. In this age of conflicting role models, Bob seems to be a giant among men. This is his story, you be the judge.

"Far better is it to dare mighty things, to win glorious triumphs, even though checkered by failure, than to take rank with those poor spirits who neither enjoy much nor suffer much, because they live in the gray twilight that knows not victory or defeat."

… Theodore Roosevelt

Table of Contents

Rough Times

My mother, Ruth, was a full blooded Chippewa Indian born on a reservation near Mt. Pleasant, Michigan. When she was about ten years old an epidemic of some sort killed both her parents. She lived with an aunt for awhile, then was sent to live with a wealthy white family in Mt. Pleasant. Actually Mother felt she had been sold into slavery. The Indian relative she had been living with received a sum of about $500, and Mother was cast into the role of a domestic servant.

Mr. and Mrs. Henry Welch owned the Ford dealership in Mt. Pleasant. The Welch's gave my mother a warm place to sleep, good food and treated her well, but they didn't send her to school. Neither did they pay her a wage, so she was their servant for room and board only.

Mother had an independent nature and at seventeen years of age she grabbed at the first opportunity to leave the Welch's control. That opportunity was in the form of my father, Robert Glenn Vaughan. She never did tell me anything about their courtship; I'm not sure how much love was involved, but they got married and we kids started coming along.

My sister Judy was born in February of 1944 and I came along the next February. My earliest memories were of the family always moving. In fact, my mother told me that when my father was driving trucks for a living, she and Judy would ride along, and that she started labor for me passing through Toledo, Ohio. I don't know whether I was born in the truck, a motel room or the hospital, but that's where it all started.

The next thing that I remember most is that Mom and Dad were constantly fighting. I don't remember what they fought about, but there always seemed to be bickering followed by yelling, screaming and then dishes breaking.

I'm not sure how long my father drove truck after I was born, but from what I could find out he was regularly changing jobs, and the family was moving. Maybe the relocation required a different job, or perhaps a new job required the move, but we never stayed in one place long.

I do remember one job my dad had, and I was in the audience to watch him perform. The man that had the band was called Little Jimmy Dickens and Dad played the bass fiddle. Dad also was the clown of the bunch and wore comical looking clothes. His stage name was Hyram P. Barnsmell. He wore bib-overalls and a floppy hat like some hillbilly.

Every so often during the band's performance, Dad would take a tin bucket over to a hose and put some water in it. By the end of the performance and several trips to the hose, the bucket was about full of water. Sometime, when the audience's attention was distracted, the bucket was switched with another full of confetti. At the end of the show Dad would grab the bucket, then rush towards the audience and empty the contents on them. They'd all duck to avoid the water, and then laugh a lot when it turned out to be confetti.

There was one period of a year or so, when we lived in a tent along a stream in a state game area. This was in Lake County. I started kindergarten there. My sister Judy and I walked from the tent to a youth camp on the west side of Wolf Lake, then caught the school bus into Baldwin. We never did have a chance to make many friends there, because soon we were on our next move.

With our parents always fighting, the only stability and happiness Judy and I had seemed to be when we visited our grandparents. They owned a little two room home, with seven acres located on Ten Mile Road outside of Sparta. There was no running water or toilet in their house, but compared to what we had they were rich.

From my earliest memories, there always was love and fun with my grandparents. Between moves my sister and I would spend time with them and they treated us as being very special. This was not the case with my father.

My grandmother was my dad's real mother, but Granddad was his stepfather. Grandpa hated my father and made no bones about letting that be known. More than once, as I grew up, Grandpa told me how worthless my father was. I'm not sure why he felt that way. Maybe it was because Dad was always changing jobs and moving the family around. Perhaps it was something else, but whatever it was he never changed his mind about him.

Once, after Mom and Dad had a big fight, we left Dad and moved into an old chicken coop at Grandpa's. Grandpa butchered all the chickens and cleaned the manure out of the coop. He scrubbed the floor and put some old carpet pieces down, then brought in a bed and an old dresser. We had a two-burner fuel oil stove to cook with plus a small space heater to keep the place a little warm. My most vivid memory of the chicken coop is spending one Christmas Eve there. We had no tree or gifts, but we laid there in bed and I can still hear my mother singing Silent Night.

Grandpa and Mother didn't get along that well either. Mother would willingly stay on grandpa's property, but not under the same roof. She was a hot-tempered Indian, and he was a strong willed German that didn't approve of the way she was raising us. There were lots of clashes between them.

While living in the chicken coop we went to the Englishville one room school. There was just one teacher for all eight grades, plus kindergarten. The teacher I remember most during that time was an older lady named Mrs. Johnson. She was very stern. The boys that caused her trouble, or didn't do well, had to carry wood in for the pot-belly stove setting in the center of the room. It seems like she had me carrying in wood most of the time.

There was a big hill by the school. During recess or noon hour in the winter, we had a great time sliding down it. There were a couple of old junk car hoods for us to pile on, and we'd get some pretty good speed built up on the way down.

The Englishville School had the old fashioned desks with a hole in the top for an ink well. Each grade had their desks in a row and I believe there were probably 20 or 30 kids in school with us.

We moved away and back to Grandpa's twice. Each time we were away we started in a different school, then returned to the old school at Englishville. With all those disruptions in our schooling, we sure weren't learning much. I was getting farther and farther behind in my education.

During those years my father just didn't seem to be with us very much. Finally, we got together and moved to an apartment house in Grand Rapids. For some reason the address is still embedded in my memory; 604 W. Fulton Street.

Another family, named Blevins, lived in an apartment close to us. They had two girls and a boy that Judy and I used to play with. Their oldest kid was a girl named Daisy. She and I didn't get along very well. One day we had a confrontation over something or other and she threw an old glass door knob at me. It hit right on the side of my face, causing a nice cut inside my mouth. That was the first time I ever tasted my own blood. I decided to keep away from that nasty little girl.

Down on Fulton Street things didn't get any better between Mom and Dad. Judy and I had heard the fighting for years, but it was getting worse. We'd crawl into bed and put the covers over our heads to get away from it.

Finally they agreed to a permanent separation. After one final argument, Dad packed up everything he owned in a pillowcase, and went out the door for the last time. I remember leaning out the window and begging him to come back. He didn't stop, but as he continued down the sidewalk he did turn and wave. With that he was out of my life forever. I don't have unpleasant memories of him. Even though he wasn't around much, he treated me well. There were no beatings or abuse, and I would have liked to have known him better.

The father figure in my life was now my grandfather. He had provided love and stability before, and with Dad gone his influence would grow. I know Grandma also thought a lot of Judy and me, though she wasn't real vocal about anything. Grandpa was a very dominant person and Grandma always went along with everything he said and did. I guess she was just a little old fashioned in her ways. She let him be boss, and would not ever even interrupt him if he was talking. There never were any arguments or fights between them. Grandma died when I was seven and probably in his loneliness we became even more important to Grandpa.

After Dad left, Mother decided to retrace her roots and reclaim her Indian heritage. She had been telling Judy and me as much as she could remember about living on the reservation near Mt. Pleasant. Her family didn't have much; the floors were dirt, there was no inside plumbing and not much food to eat, but the Indians were a proud people and we should be thankful for that heritage.

Well, Mother took us all over to Mt. Pleasant with the idea of returning to the reservation. The Chippewa's didn't throw out the welcome mat for us. Maybe they were already too short of resources or maybe they just didn't want the woman with the two half white kids.

Mother was devastated by this rejection and angry at her people. She had reached the end of her rope. With no means to support us, she made a drastic decision. Judy and I were left at St. Johns orphanage in Grand Rapids. Things in the past had been rough, but nothing like what we were about to encounter.

St. Johns orphanage was like a jail. My memories of the place are terrible. At nine years of age I was now separated from all the people I loved. My sister was taken to a different floor and I didn't see her for a month. I lost my mother, father and grandfather and gained a supervising nun that never smiled. She treated us like prisoners.

There were a lot of rejected kids around in the days after WWII, and this orphanage was jam packed. Children born out of wedlock were not well accepted by families or society then. And, there wasn't any generous welfare system to look after them. It's still hard though, to understand why we were treated so badly.

I stayed on a long floor with boys five to twelve years old. There was another floor for the teenage boys and one for the girls. On the ground level they kept the cribs with babies and the ones that were just learning to walk.

We didn't have bunk beds, just long rows of cots. Sister Phyllis was the nun in charge and she looked like she was a hundred years old. The sister never smiled or said a kind word. She ran the place with an iron hand. If you got out of line, punishment was swift and brutal. A little pushing or shoving incident between boys would land you in the punishment room with your pants down. A ping pong paddle or chunk of garden hose was the instrument used for correction and the beating was severe. Crying did no good, in fact the beating continued until you stopped crying. When the sister was finished you pulled up your pants and said, "Thank you Sister," and left.

One boy, that was about eleven years old, had the problem of occasionally wetting his bed. Now each floor had a large grate that allowed heat to come up from the furnace in the basement. To solve this boy's problem, or rather as punishment because he had the problem, he was tied down naked to this grate at bedtime. The metal was not hot enough to burn him; however, when he was let up, his back was all imprinted with criss-cross marks from the grate. After this happened a couple times, the rest of us would take turns waking him up to go to the bathroom whenever we woke up and had to go.

Our dining area was set up like a cafeteria. If you put your tray out for some food, God help you if you didn't eat every scrap.

There wasn't any time allotted for playing at the orphanage. No toys or play yard with swings, just school, sleep and mass. Religion was paramount and we sure did a lot of praying. There was mass when we first got up in the morning, before lunch and again after school. There is nothing wrong with praying except they overdid it by a lot.

We walked to our school, which was along Plainfield Ave. by Three Mile Road. I was not a good student and spent a lot of time in the corner with the dunce cap on my head. Moving around so much, my education had really suffered. I could barely read. The teacher would stand me in front of the class and force me to try and read. I hated that. It was easier to set in the corner.

They failed me in the third grade and that sure was embarrassing. When I showed up in the same grade the next year, I was much bigger than the other kids. It didn't take them long to figure out why I was there.

After several months at the orphanage, Grandpa occasionally started getting us out on a one day pass. That eventually turned into a weekend at his place. It was absolutely great. Grandfather again became our refuge from the bad things in life. Before it had been the constant fighting between our parents, now it was momentary salvation from the orphanage.

Grandpa didn't have a big house. It just had two rooms; a bedroom closed off with a blanket in the doorway, and another room which was both kitchen and living room. I remember that he didn't even have electricity or plumbing in the house. Eventually he did get electricity, a refrigerator and an electric stove he was really proud of. In fact he was one of the first in the neighborhood to get a television.

When we visited, my favorite program on TV then was a western called The Cisco Kid and Poncho. It came on at 6 p.m. every Sunday. As much as I looked forward to watching it, I knew that when it was over we would have to go back to the orphanage, so it was also a sad time.

Sunday also meant setting on Grandpa's knee while he read me the Sunday newspaper funnies. Li'l Abner was my favorite and I loved the way those mountain folks talked. He'd help me pronounce their strange words like "whacha doin?" and "Heeay Paw". It was fun trying to sound out words and pretty soon I progressed to other comics including Dagwood and Blondie. After a few Sundays I was reading to Grandpa instead of him reading to me.

It was as rough for Judy as it was for me and we sure did dread going back to the orphanage on Sunday night. Sometimes we made plans to get together at night when the other kids and the nuns were asleep. We'd sneak around, meet in some corner and talk for a long time about when we were together as family and all the fun things we used to do. Before we separated to go back to our cots we'd hug each other, then get down on our knees and do some serious praying. We'd ask God to bring back our mother and father, to put our family back together again.

Judy and I had been at the orphanage for about three years when one Sunday Grandpa came to pick us up for a visit. He had some terrific news. "I've got something to say that I think will make you happy. Your mother has remarried and wants to see you kids. She and her new husband plan to get you out of here and have you live with them." In all the time we had been at the orphanage, we had only seen our mother twice. Judy and I were ecstatic. After three years we were getting out of our jail.

Free at Last

My stepfather was Finis Blevins. How he and my mother ever happened to get together, I don't know. Finis and his family lived in the same apartment house, at 604 Fulton, as we did. We played with his kids and his girl Daisy is the one that hit me in the mouth with the glass door knob. I guess Finis went through a divorce about the same time as my parents. Maybe my mother and Finis even had something going on before the divorces. But, if they did, we never knew about it. It was 2 years later when they married.

Grandpa seemed to accept Finis. He gave him and my mother a small plot of land on his seven acres. Finis bought a military surplus house and had it moved from Lansing to the lot on 10 Mile Road, south of Sparta. The house was one of those prefab kind produced to satisfy the huge housing demand after WWII. It only cost Finis $600, which included the move.

Our house arrived in sections, then they bolted it together on a concrete slab. There was no electricity, water or toilet, but there was a five gallon fuel oil space heater. With little insulation in the house, once you got very far away from the heater you were cold. Even though it wasn't a very fancy house, it was home and I could walk to Grandpa's anytime I wanted to.

I became Grandpa's favorite. We had been close before, but now we lived next door and I could be with him a lot. Even though we were not blood relatives, he never failed to demonstrate his love for me, to be there when I needed him. Of all the men I had ever been associated with, Grandpa stood out in my mind as a remarkable individual.

My grandfather, Lawrence Friedrich, was the only child of two German immigrants that settled in Wausau, Wisconsin. From his great grandfather on down the men had been cabinetmakers, so his father taught him that trade. All the rest of his education came from his parents at home. He attended no public schools. His parents did a good job teaching him, for he was an excellent reader and a real whiz at math. After he gained experience in his trade he moved to Michigan to build houses. Unfortunately he lost several fingers in a table saw accident, and his building career came to an abrupt end. Grandpa then had to work in a factory to support his family, but he really didn't enjoy it. He was grateful for the job and liked the men he worked with, he just didn't like making parts without knowing what they were to be used for.

The factory work at Applied Arts didn't pay much when he started, because the pay was based on piece work, not salary. His hand didn't heal properly and the cords had almost pulled it closed, so it was hard for him to work very fast. To drive a car he had to force his hand open to grasp the steering wheel. Now at work, he operated a punch press and did spot welding. They nicknamed him Baldy, and he loved that.

My grandfather had lots of stories, advice, and a philosophy that became the foundation of my character. The meaning of some of his stories didn't become apparent until I was much older, and then it began to sink in.

One story he told was about the wise old man that was the leader of his village. Another younger man was jealous of the old man's following and so he told his own small group that he would embarrass the old man in front of the whole village. Out of a nest the young man took a baby bird that was feathered out and almost ready to fly. He said, "I will hold this bird in my hands and ask the old man if what I have in my hands is alive or dead. If he says it is alive, I will crush the bird with my hands and produce a dead bird. If he says it is dead, I will open my hands and the bird will fly away. Either way, he will be wrong."

So all the villagers gathered around as the young man held his hands out in front of the wise old man. "Old man, is what I have in my hands alive or dead?"

The old man slowly replied, "The answer is in your hands, my son."

More than once my grandfather reminded me of this story and the fact that what happened in my life was in my hands. In our day to day chats he would give me advice and tell me what he would do in different situations, but when I chose my own course of action he would always support me even when I fell on my face.

As our family didn't have a lot of possessions, I was always wishing that I could have something that some other kid had, whether it be a toy, a shiny car or even an inside bathroom. Grandpa would say, "Son, if you have your wish in one hand and a pile of shit in the other, you will always have the shit before you ever see the wish." He would reassure me though, that everything I wanted was available if I just made up my mind to do it, there would be a way.

Another of Grandpa's stories became very meaningful in my later years. It was the story about the turtle and the snake. It seems this snake came upon a turtle who was resting on the bank of a river and said, "Mr. Turtle, I need to get to the other side of this river. Could I ride across the river on your back?"

From inside his shell the turtle replied, "You are a snake. You will bite me."

"No , Mr. Turtle, I only want to get to the other side of the river. I will not hurt you."

After some coaxing from the snake, the turtle allowed the snake to ride on top of his shell as he paddled across the river. Once on the other side of the river, the snake crawled off the turtle's back then promptly turned around and bit the turtle in the neck. As the turtle lay there dying he said, "But Mr. Snake, you promised not to bite me!"

The snake replied, "Yes, but you must remember, you knew I was a snake."

Our house was on Ten Mile Road, close to the Bow Tie Tavern, the Englishville School and a set of railroad tracks. On the other side of the railroad tracks was Stout's Dairy Farm, operated by a man named Bud Kellogg. Well, Bud Kellogg needed some help on his farm. He could see this twelve year old fiddling around next door, so one day he came over and asked me if I would like a job cultivating weeds out of his corn.

"Can you drive a tractor?" he asked.

"I can if you show me," I replied.

That summer I drove Bud's John Deere H, cultivating corn for 25 cents an hour. When there wasn't the need for any more cultivating, Bud asked if I might like to do some different work, so I helped with the milking and other farm jobs.

When school time came in the fall, Judy and I were bused into Sparta to attend junior high. Thanks to Grandpa's help with my reading, I was able to skip ahead and now be in a class with kids my own age.

Things certainly were changing rapidly in my life. I was out of the orphanage, back with my family in a home next to my grandfather's house and I had a job. You would have thought that I would have been a very happy boy, but that was not the case.

Judy and I felt very close to each other during the orphanage years, but now we were fighting regularly. Instead of feeling grateful towards my mother for getting us out of the orphanage, I felt more and more that it was her fault that we had been in there to begin with. It seemed that we were just excess luggage to drop off while she got along with her life. I was full of anger and didn't understand why this had happened.

My stepfather legally adopted us and our last name was now Blevins. I did not allow him to be a father though. Towards him and my mother I became more and more antagonistic. With my grand-father and Bud Kellogg I was a model teenager, but at home I was becoming an unmanageable rebel.

By the following spring, I was milking Bud Kellogg's 60 holstein cows morning and night by myself. Bud's main job was being a super-visor for L.W. Edison, a big contractor that was building much of US 131, the main highway that ran north and south through Grand Rapids. To get all these cows milked and make it out for the school bus, I had to get up at 3:30 a.m. Of course, right after school it was time to feed and milk the cows again, so I was spending less and less time at home. When I did have free time on a weekend, I was with Grandpa.

Grandpa had one of the first televisions in the neighborhood. On Saturday several of his friends would come over and watch the baseball games with him. There weren't a lot of chairs in his house, so I just sat under his big kitchen table and listened to the old guys laugh and talk.

In the fall when the weather was nice, Grandpa would always take me hunting. When I was still in the orphanage, maybe 10 years old or so, he taught me how to handle and shoot a shotgun. He bought a 12 gauge for me, and that sucker would really knock me back when I fired it. I was careful not to show any lack of confidence with the gun, though there was a tiny bit of fear inside. We just hunted pheasants and rabbits, but it was a long time before I could hit anything. Grandpa on the other hand was an excellent shot with his old 20 gauge. I would throw up apples for him to shoot at and he never missed blowing them apart.

Grandpa was always taking me fishing in the summer too, but sometimes, we'd just walk in the woods. He had a lot of respect for the woods. Being a cabinetmaker by trade, he knew all the different kinds of trees and would show me how to tell how old they were. Shucks, he could even reach down, pick up a weed or some grass, tell me what kind it was and what it might be good for. I tried to absorb all this, but I'm afraid Grandpa forgot more than I'll ever remember.

By the time I was 14 years old I had just about been banned from riding the school bus. I had this terrible problem with motion sickness. Both going and coming from school I would quite often throw up. If I couldn't get the window down in time or if the bus driver couldn't stop to let me out, my mess ended up on the bus floor. This usually started a chain reaction and there would be some other kids throwing up too.

I didn't get sick when I drove Bud's pickup truck. He regularly let me take a load of corn to town for grinding on Saturday. I didn't have a license or anything, but he didn't seem to mind and I thought it was great.

Bud knew about my getting sick on the bus, so he spoke to some school official. This resulted in the school superintendent and police chief cutting a deal that allowed me, at 14 years of age, the restricted privilege of driving back and forth to school. I was delighted. I was the only kid allowed that privilege and it sure gave me some status in the freshman class.

What brought me even more recognition, than merely driving to school, was the surprise that Bud had for me one night after chores. He led me out to the garage and showed me the prettiest little 1946 Chevy pickup that you could imagine. It had a straight stick shift, a roll out windshield and a nice shiny rebuilt engine under the hood. Boy did the girls like to ride in that.

I was still having some trouble dealing with all the changes in my life. My mother's temper wasn't any better than it had ever been. She was always yelling at me. It seemed like I was depressed a lot and during my freshman year there was a fair amount of hazing going on by the upper classman. Finally, one day in the school cafeteria I decided to take some action. This older kid had been pushing me around and as I sat down with my lunch, he came over and poured my milk all over the food. He laughed, went back over and sat down with his buddies. I waited until the cafeteria pretty well filled up with kids, then I went over to where he sat, hauled him out of his chair and beat the shit out of him.

Of course I got kicked out of school for a week, but I never got into another fight in high school. My reputation had been established. I was treated with new respect. In fact, I decided to go out for football that freshman year, which gained me an even higher status.

Football became my salvation. I was good sized for my age and I discovered that I had a lot of natural ability. Starting out on the freshman team as center, I could hike a perfect spiral to the punter, plus no one got over my position to the quarterback. The opposing ball runners were also stopped. I was aggressive and before long I found myself on the junior varsity team. I played hard on the junior varsity team and again my talent did not go unnoticed. Before the end of the season I was moved up to the varsity squad. I was sure in the limelight now. Not only did I have my own shiny truck out in the school parking lot, I also was on the varsity football team, had a job and spending money. For the first time in my life other kids envied me. In particular, a senior girl took a liking to me and we started dating. I never was sure what this girl saw in me, because I was so much younger than she was. I thought she was about the prettiest thing I had ever seen.

We went together the last half of my freshman year, and one date almost got me in trouble with Bud. The date ended up with us out parked, steamin' up the windows in my pickup. She had lots of patience and before the evening was over I had lost my virginity.

After taking the girl home, it was almost four a.m. when I finally got to the barn. The cows had given up on me feeding them their grain, and had walked back out to pasture to fill up on grass. I thought to myself, "This is all right. I'll just let them go and milk them good tonight."

I filled all the milk cans with water and tried to make it look like chores had been done. For some reason Bud came down to the barn that morning. He looked around for a bit, then took a rubber hammer and knocked one of the milk can lids off, exposing a can full of water.

Without raising his voice, he gave me one of his looks and said, "Guess we better get the cows in and milk them, don't you think?"

"Ya, I think we better," I said.

He didn't ask why it had happened or anything. I knew I was wrong and didn't need any lecture. It never happened again.

The Kelloggs treated me like a son. Not only did Bud buy me the truck, if I needed a new pair of shoes or some clothes, that was all part of the job perks. I didn't get paid a lot, just $15 a week, but I had a job and liked having spending money.

I was getting the milking process pretty well streamlined. I'd put a pile of grain in front of each cow's stanchion, so when the barn door was open they would be eager to walk to their own personal places and put their head through the stanchion to eat. When the stanchions were all closed and locked the cows weren't going any place, so then I could start milking. Bud had two Surge milkers and I would have these two going while I washed the next two cow's teats, stripped out the first milk and hung the strap on their back that would hold the milker. When the first cow was milked I'd pour the milk in a pail and hang the milking machine on the next cow, hook up the vacuum hose, turn it on and let the milking cups suck up on each teat. Then I'd empty the pail of milk in the milk can setting in the milk house cooling tank and move on with the other machine. This job was mine from the time I was 12 years old until I graduated from high school. I did get some relief during football season. On game night Bud either did the chores himself or hired somebody else, so that I was ready for the game. Milking cows morning and night for five years certainly established a strong work ethic within me. During the summer it was even more hectic, because I was involved in the rest of the farm work, from planting to baling hay, filling the silos and harvesting the grains.

Grandpa loved to fish, and whenever I could I went with him. There was less time for it now, and also I had developed another close friendship. Bud had hired a boy to help out with chores during football season. His name was George Stickle. He wasn't particularly interested in sports, but we had a lot of other good times together. Sometimes we'd hop in the truck and drive up to set on the shoreline and watch them build the Mackinac Bridge. It would be late when we returned, with time for only a couple hours of sleep before we started milking. Other times, we'd just get the cows in, milk right away and either get a little rest during the day or skip sleep, going to bed early that night.

At least one night of the week we went out on a double date or hung around with some other guys. In those days our idea of mischief was tipping over an outhouse, cutting someone's clothesline, or rolling some dry corn silk up in tissue and smoking it. We might even have a cherry coke in Wolf's Drug Store and slip in a couple of aspirins. A real buzz could be imagined from that. Our crowd wasn't aware marijuana even existed. I was careful what I did, for I knew they would kick you off the football team for about anything. Football was too important for me to risk losing that.

My stepfather, Finis, tried to be my friend, but I wasn't too receptive. My mother wouldn't let Finis be the disciplinarian. She would take over that role when I got out of line. Normally though, she'd lose control, start yelling and I'd walk away to return in a few days. When I came back she was always glad to see me. I was manipulative, took advantage of her lack of control and just did what I felt like doing.

For all practical purposes I was completely independent. I really didn't know how to deal with my mother's temper anyway. She had a real low flash point. If you were in her presence when she flew off the handle, there was a good chance that you would be hit with something. She even chased me around with a hammer once. When she hit you or broke something, her fury seemed to subside. If she had this tantrum when no one was around, sometimes she would pick up a knife and start slicing her arms. Today they probably would have medication or therapy for her.

Grandpa would never say a bad word to me about my mother. After she screamed at him, he would just give her the cold shoulder for a month. When staying over at his place caused too many confrontations between him and mother, I started spending more nights at Bud's. I was welcome there, and treated more like a son by them, than I was at home.

One thing that did bring Finis and I much closer together was my first deer hunting trip. Grandpa wasn't a deer hunter, but Finis did like to go, so he invited me up north hunting with him. Finis liked to hunt and tell deer stories, though I don't think he ever killed a deer. He told me up front, "This is a 50/50 operation. You make the mess and I'll clean it up." This sounded strange to me. It was my first chance to go deer hunting and I was looking forward to the opportunity no matter how strange it sounded.

Our deer camp was not too far outside of Luther. It was the first time that Finis and I had spent much time alone together, but it was working out fine. He liked to be out in the woods and look after all the camp chores, while I was out trying to shoot a buck.

In those days Luther was known as the Dodge City of Michigan. It was a pretty wild place normally, and then with the influx of deer hunters it was extra active. Finis decided it would be nice to go into town and get something to eat, so we drove to the Totem Pole Bar and Restaurant. It was crowded but we found a place to set down and ordered a sandwich and something to drink.

I was 14 years old and had on a fine pair of Soo Wool hunting clothes. It was nice to be in there with all the other hunters and even though it was my first hunt, I felt I looked like a veteran.

The crowd was pretty noisy. We sat there eating our sandwiches when along comes this guy who trips on a back leg of Dad's chair. The man said something to Dad, which I couldn't hear. Dad said he was sorry and pulled in his chair. It wasn't long before the guy that tripped on the chair was back and again said something to Dad. Dad kinda looked up with a half smile and the guy proceeded to pour his beer on Dad's head. When the guy left this time I said, "Dad it's time for us to leave." He just took a handkerchief out of his pocket and started drying off his head. "Son, it's all right, just eat your hamburger and don't worry about it." I wasn't at all reassured.

My uneasiness was definitely justified, because now the guy was coming back with two buddies. Before Dad could turn around, this guy's friends had a hold of him and hauled him out of his chair so that the guy could start beating on him. I jumped into the middle of the brawl a couple of times, but was no match for these guys. Dad was on the floor now trying to defend himself from this guy's punches.

I had an idea, something that might bring this to a halt. Out the front entrance to the truck I went, and from the back door of the homemade camper top I grabbed our old McCullough chain saw. I knew that when it was cold it would always start. With the choke out I gave the start rope a pull and with a puff of black smoke the old girl roared to life. With the throttle trigger all the way back she was really screaming. With the chain bar up in front of me I ran back into the bar where I'd left Dad.

Boy, did those guys pour out of that bar. The chain saw noise and smoke sobered that place up fast. It was pretty evident that nobody was planning to do battle with my chain saw. Dad got up a little slowly, but he could walk. We headed to the truck with the saw still going full blast. While he got in to drive, I opened the passenger side window and held the saw outside.

Down the road a ways, Dad said, "Shut the saw off now." It was close in the truck and I suppose the noise was getting to him. "I think I'll just keep it running till we get to camp, and I'm sure no one is following us." I knew that once the saw was warmed up and shut off, it would be nearly impossible to restart. Later on we sure had lots of laughs over my first deer hunting trip. It just didn't seem that funny while it was all going on though.

The varsity football coach was named Jim Dickey. He seemed quite fond of me and I enjoyed playing for him. I was always aggressive in both practice and the games. Some of the other players would question why I hit and tackled so hard in practice. I told them that as far as I was concerned the coach was watching all the time. To be on the varsity first string at my age was really something. I didn't plan to lose that status. During the games the opposing players didn't go over or through my position and the coach liked that. I also was very vocal during the games trying to keep everybody fired up. My nickname became Spark Plug.

During the football season my popularity was at an all time high. It was great. I played it to the hilt socially too. When the senior girl I dated graduated, I kinda lost track of her. Soon there were others though, and several were willing to help with my homework. As good as I was in football, I was that bad in my studies. The grades were really poor, the result of low interest and lack of motivation. I didn't pay attention and had a real poor attitude towards school work. Football was the only thing that really seemed important.

Many of my teachers were either coaches, married to coaches or associated with coaches and had an interest in football. This certainly did give me an advantage in achieving passing grades. One teacher didn't go along with that program, the typing teacher Mrs. Ree.

The only reason I took typing was that I figured it would be easy, plus there were lots of girls in the class. It turned out to be one of my hardest classes. Up front, Mrs. Ree told me if I was there to coast, I might just as well leave. I stayed and she made me really work, which was good. Today I can type 40 plus words a minute.

I never did buy any school books. We had to pay for our own then, and I wasn't about to spend my money on books. The folks didn't have much money and I wasn't going to ask them for any. My reading was fast though, so I'd borrow some books to read in class or study hall. That didn't help with take-home assignments. I just didn't get many of those done.

Occasionally Dad would bring his children up from Grand Rapids to visit. Being busy with all the farm work, I didn't spend too much time with them, but I did notice that the girl who hit me in the mouth with the door knob was rather cute now. She didn't seem at all like the nasty girl I remembered. Over time we became quite well acquainted, and even started dating some.

During the last part of my junior year in high school, a Marine recruiter came to Sparta and talked to boys that might be interested in enlisting when they graduated. There was also a reserve unit in Grand Rapids that you could join, to attend training drills once a month during your senior year in high school. You would even get paid for attending the drills. Wow, this recruiter really had a sharp dress blue uniform, and that training sounded neat. I felt that this program was definitely for me and I could just see myself in one of those uniforms. I joined up.

My senior year in football turned out great. I was an all city, all conference center and the college recruiters were sniffing around. When the football season was over, all I could think about was the Marines, and there didn't seem that much reason to hang around high school any longer. I thought I could probably even play football for the Marines.

Football had kept up my interest in high school and because I was good the teachers allowed me to obtain passing grades. There were those who were hoping that football might turn my life around and help me develop a more serious attitude towards school. That didn't happen. The season was over and I seriously considered dropping out. I really didn't need to graduate to go into the Marines anyway. Someone finally brought it to my attention that I had earned enough credits to graduate, so reluctantly I decided to stay in and graduated in June. They really didn't see a lot of me in class though, as I managed to miss over 50 days of school.

One girl had caught my eye during the last year and it was turning into a real serious relationship. Knowing that I would be going for Marine basic training soon after graduation, I started making plans for the future. We were talking about marriage.

Bud Kellogg had given me a calf each year that I could either sell or keep in the herd. I kept them, and had been getting a share of the milk check. I now owned six cows. I decided it was time to turn these assets into cash. Bud bought the cows from me and with the money I bought a real sharp 1960 Mercury. I also put $1000 in the bank. The Mercury lasted only a week.

The girl friend and I were visiting another couple on Lyon Street in

Grand Rapids. When we got to the apartment the other girl asked me on which side of the street I had parked. When I told her, she said I had parked on the wrong side and because of the snow removal schedule I would have to park it on the opposite side, or get a ticket. Well, I started my nice car up and backed right out in front of another guy. He piled into my backend and the car was totaled. I still had the $1000 in the bank though, and that was a lot of money in 1963.

By graduation time, Beverly Hammond and I had definitely decided on getting married when I finished basic training. I put her name on my savings account too, and that's where my extra money went.

Out of about a hundred graduates, I had the distinction of graduating dead last. Even with that standing, I had been offered a football scholarship at Michigan State University. College was not my goal, being a Marine was.

Sparta's Star Center

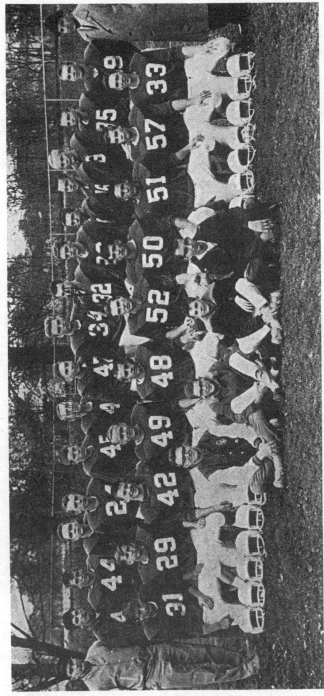

Assistant Coach, Hess Wever, D. Willis, N. Berg, T. Carpenter, T. Cummings, S. Tindel, J. Hass, L. Cumings, G. Brown, P. Bayes, J. Hickman, N. Fuller, K. Thayer, B. Ekonen, D. Fairchild, T. Fryear, Head Coach, James Dickey, T. Murry, M. Patterson, D. Pitchure, K. Stoppenhagen, B. Blevins, J. Hamacher, D. Whitney, J. Koning, K. Erickson, C. Ryan, Student Trainer, John Lauffer, Mgrs., D. Davis, J. Davis, D. Wolters.

Bob Blevins #48

A Marine

Two weeks after graduation from high school, I was at Paris Island, South Carolina, training to become a full fledged Marine. This was the 1960's and they were making Marines the same way they had for a long time, with severe discipline. Those that were in training were volunteers, and it was no secret that the training would be tough and that becoming a Marine would not be easy.

Some of the recruits had an attitude similar to what mine had been in high school. They were independent, rebellious and not used to being told what to do. It was just a matter of time before they either changed their attitude or faced an even more difficult time. For those that were openly defiant, there was always the possibility of ending up in the shower room with your blood and maybe a tooth or two washing down the drain.

Compared to the orphanage, Paris Island was like a cakewalk for me. There were several things in my favor that helped make it easier for me than it was for my fellow recruits. First, I had been attending training weekends with a Grand Rapids Marine reserve unit for almost a year. I knew how to march, handle an M-1 rifle and I learned about the Code of Conduct. My service number was committed to memory. All these things seemed to be stumbling blocks for the others.

From the first day we arrived at Paris Island, we were screamed at and verbally abused in the most intimidating way possible. Fortunately, I understood that all this yelling in my face was only aimed at creating a Marine that could accept discipline and willingly follow orders. The DI's scared the piss out of about everybody else, but they didn't bother me that much. I didn't think that underneath that front they were that bad.

Even though I had been independent and rebellious in high school, I soon realized that for things to go smooth here, one needed to do what you were told and certainly without showing disgust or hostility. My memories of the orphanage were still vivid and I remembered that when you didn't do as you were supposed to do there, some tough old Catholic Nuns would swiftly deliver some severe punishment. As bad as it was, I guess at least the orphanage was good preparation for Marine boot camp.

Others had a more difficult time adjusting to the Marine way of doing things. Consequently their lives became rather miserable. If the DIs couldn't correct bad recruit behavior with their yelling or physical blows they always had the Corrective Custody (CC) Camp or the Motivational Platoon. Both were bad. In the camp they had this big pile of gravel and each man had two buckets to carry the gravel to the top of a small hill. When they got it all up there, then they were required to carry it back down. When they came back to us, if they came back to us, a good share of the skin was missing from their hands. I heard some even had heart attacks and went home.

A few of the boys wanted to wash out so bad they tried to create their own injuries. One guy took his boots off in the chow hall and was beating his heels against the cement floor to damage his feet. Another took a knife and tried to cut the tendons in his leg.

At 5:00 a.m. every morning, our DI would come into the barracks and beat on a garbage can with his swagger stick to wake everybody up. He would always find me up and dressed, with my bunk made. I had been getting up at 3:30 for the last 5 years to milk cows, so 5:00 a.m. to me was like sleeping in. By that time I also had shined both my boots and shoes, and was wishing I had something else to do to keep busy. The DI would always have a strange look on his face when he saw me. It was like he was thinking to himself, "No matter how hard I work them the night before, it just doesn't seem to phase this guy."

Things were going very well for me. I shot a high expert on the rifle range, plus I had the honor of being the right guide when we marched. The right guide carried the flag with the company designation and the platoon number on it. He was considered the senior man.

One day after a lengthy marching drill this scroffty old sergeant was giving us hell. I stood there at attention with this chessey cat grin. In short order the old sergeant was right in my face, "Private, I just don't scare you, do I?"

"Sir, I spent three years in a Catholic orphanage. There ain't nothin' you guys can do to me."

Well, we were halfway through training and for the first time I thought I saw a smile start to form in the corner of his mouth. Seemed like he might really be human after all. Maybe he knew something about Catholic orphanages, or just imagined they were bad as I had said. I thought my smart remark would probably have me doin' a few hundred pushups, or at least gain me a hard whipping from him to prove how bad he could be. Neither happened, he just walked away.

One of the requirements to join the Marines was to be able to swim well. I was a good swimmer and felt comfortable and confident in the water. They arranged this little exercise to test your swimming ability. Several didn't pass.

Dressed in our utility uniform with boots, a pack and helmet, we entered one end of an olympic size pool carrying a rifle without a sling. The object was to swim or float to the other end without touching the pool sides or losing the rifle. All the DIs would be lined up along the sides of the pool and stomp on anybody's fingers that touched there. Of course they wouldn't let anyone drown, but if you panicked and went under they would let a lot of bubbles come up before they snagged you with one of their big wire hoops. When they got one of these boys out, they would strip off their clothes and resuscitate them, then tear the Marine emblem off their swimming shorts, give them a light blue suit and send them home.

I really had a knack for the military. This training was almost enjoyable, and I did well at everything. It looked like I would be among the four company honor graduates and receive my first stripe at the end of boot camp. With only two weeks left, things were really going great, until the letter arrived from home.

It was quite a thick letter, with the address in Beverly's handwriting. I saved it until the end of the day when I could set down and enjoy it.

Along with the letter was a news clipping. I briefly glanced at the article, something about police breaking up a teen drinking party, then started reading Beverly's letter. She explained that she had gone out with this other fellow to the party that the police had raided. She wanted to tell me about it up front, so I wouldn't hear it from someone else. Furthermore, she had been dating this other fellow some and thought it would be best if we broke up. This was the girl I loved; the one I had put all my trust in; the one whose name I had put on my savings account. I snapped. I went bananas and started tipping over lockers, beds and throwing things through the windows. They literally had to tie me down to stop this craziness.

The DIs were undecided what to do with me. If they sent me to the correctional platoon for attitude adjustment, I wouldn't graduate with my company. I had lost everything I had worked so hard for. The right guide position was gone, I certainly wouldn't get my first stripe at graduation.

When the platoon marched, I was made to bring up the rear with my head down, like some puppy dog. How demoralizing that was. They could have beat me half to death and it wouldn't have been any worse than that.

Finally after a week of this, the DI, old P.I. Porter called me aside and told me to report to his hut. We had decided his initials stood for Paris Island. Anyway, old Paris Island Porter wanted me in his hut. When I stepped inside he said, "Set down there and start reading." He shoved a box over that looked about the size of a recipe box. The box was full of "Dear John" letters, that P.I. had collected over the years. I read them all, including mine. It was comical. I swear they all sounded exactly the same, just like they had all copied the same letter. "Dear Johnny, he's here, you're there, I'm going to get married, Good-by." It was therapeutic and I almost laughed out loud. Before, I'd felt that I was the only one this had ever happened to.

They gave me a second chance. I'm not sure I deserved it, but I was right guide marching in front of my platoon again. I also was one of the four out of ninety that was honored with their first stripe at graduation.

After basic I had the choice of coming home to rejoin my reserve unit or enlisting for an active duty tour of two or four years. I chose the four year enlistment and was sent to Infantry training at Camp Geiger, North Carolina.

Camp Geiger will stand out in my mind because of only one vivid memory. It was cold and rainy that November day in 1963. The Marine in the foxhole next to mine had been listening to the news on his battery operated radio. This was against regulations, but the infraction was quickly forgotten when he jumped out of his foxhole and started screaming, "They killed the President!!"

Following Geiger, I was assigned to the 1st Marine Division at Camp Pendleton, in California.

Back in September, I had received my "Dear John" letter from Beverly. Since then I had received some correspondence from her and I had the impression that she cared for me and was not dating that fellow anymore. Time and rationalization had eroded my original anger with her. There seemed to be the possibility of patching things up, so I took a couple weeks leave over Christmas. No one knew that I was coming home, as I meant to surprise them all.

When I arrived at Beverly's home her mother said that she was out for a pajama party that night. I had the distinct impression that she was lying, so I hung around there for awhile. Sure enough it wasn't long before Beverly came home with the guy she wasn't supposed to be dating anymore. This time I wasn't devastated and there was no confrontation. I just knew it was over.

It was nice being home even if the Beverly thing was over. Seeing Grandpa was great. He, Bud, Finis and Mother all seemed quite proud of me and the fact I had survived Marine boot camp. They thought my Marine uniform looked pretty spiffy too. Once again, I was in the limelite.

I had the opportunity to see Finis's daughter Daisy again. She certainly did look attractive. We had dated some while I was in high school, but she lived in Grand Rapids and the relationship with Beverly had become quite intense. After Beverly's " Dear John" letter, Daisy and I corresponded some and I planned to see her when I came home.

On my Christmas leave I started dating Daisy again. Everyone called her Dee for short and it wasn't long before I felt Dee was the girl for me. When my leave was about over I asked Dee if she would marry me. She said that she would. That nasty girl that hit me in the mouth with a glass door knob would now share my life.

In the Marines you didn't get married without receiving permission from your commanding officer. If you didn't get permission, you wouldn't get base housing for you and your wife. If you had your wife living off base, they wouldn't allow you to leave the base during the week to visit her. In short, they could make life difficult for you if you didn't follow the rules. I learned that very early in the Marine Corp program.

Back at Camp Pendleton, I told the base First Sergeant that I wanted to get married. He relayed this request to the base commander. The Captain gave his permission, and they granted me 15 days advance leave to go home again to marry Dee. We were married on February 1st, 1964. With a newly purchased 1953 Chevrolet, Dee and I headed back to California.

Back while I was finishing Boot Camp, the Soviet Prime Minister, Nikita Kruschev, was attempting to send missiles to Cuba's dictator, Fidel Castro. President Kennedy wasn't going to let that happen. He formed a naval blockade, risking war with Russia, and turned back the missiles. Since all the 1st Marine Division was deployed in support of that mission, I reported to the Marine base with little to do until my assigned unit came back. After receiving Christmas leave, and another 15 days of advance leave in February to get married, I was certain I would be contacted with arrangements to join my unit. I reported again to the base First Sergeant and was told my unit was still deployed with no certain date of return. I also received the news that there would be no provision for me to rejoin the Division.

I would be kept on duty in California. I was uneasy about what was going to happen now, because no one seemed to know how long the Division would be gone.

The married housing for a low ranking enlisted man was a little house trailer in the South Mesa trailer park. That's where Dee and I began our married life together and started to get to know each other.

There was a lot I didn't know about Dee. She seemed real receptive to marriage and didn't mind leaving home at all. For some reason, she had not gone to school past the seventh grade and had stayed home to help her mother. I wondered how happy that home life had been. Dee was quite young when we got married. She had just turned seventeen, but she seemed quite mature and was intent on making our marriage a good one. Dee was easy to get along with, because she agreed with anything I wanted to do.

The base First Sergeant said that I would have to find something to do or they would be detailing me to do trivial work, picking up cigarette butts and things like that. I kept a close eye on the base bulletin board and occasionally the First Sergeant would suggest something I might be interested in.

They were having tryouts for the Navy football team that spring and I thought that would be a great way to spend my Marine Corp career, playing football. I had been a hot shot high school player, more than once bragging a little about that to my fellow Marines.

Well, the First Sergeant told me to report down to special services and get equipment. The guy at special services looked at me with hesitation, "I don't know whether we've got equipment small enough to fit you Marine." I thought he was just joking, but the shoulder pads he issued me did seem rather large.

When I reported down at the playing field with my equipment on, they let me form up on the line as center. This was my position in high school and it seemed good to be back at it. The quarterback called for the ball and after a perfect hike I looked up to see this huge forearm coming at me. The next thing I remember is waking up on the sidelines. The extra 100 pounds and forearm of the opposing lineman had caused my head to rattle around in my helmet like a marble in a barrel. Right then and there, I knew that I wasn't going to be making a career out of playing football.

They were looking for volunteers to be on the Marine rodeo team too. Gee, I had been a farm kid and had been around horses and cattle. I never did ride horses much, but occasionally I'd jump on a cow's back and ride her out of the barn after milking. When they left the barn some would start running to pasture; others might kick a little. I'd either slide off or be thrown off in the mud. Seemed like I might be a pretty good bronc rider. Still, after my distasteful tryout for the football team, I wasn't beaming with confidence.

The man in charge of the Marine rodeo operation was all dressed up in cowboy clothes. His name was Chuck Wilson. I told him I'd like to ride bucking broncos for the Marines. He gave me a once over look and said, "I think your legs are too short for bronc riding, but if you got heart and guts you might be a good bull rider." I was willing to try anything.

Camp Pendleton had their own rodeo grounds and arena. They also had all the horses, calves for roping and some pretty mean bulls to practice on. The bulls got even meaner when they worked them over with an electric cattle prod.

The bull riding went much better than the football tryout. Chuck thought after watching me that I had the balance and rhythm to make a reasonable rider, it would just take some practice. I hoped that I could survive all the practices to actually make the team and be able to compete in some rodeos.

The rodeo team was only a part-time activity. I also wanted to be something in addition to being a Marine rifleman. In those days the Marines had what they called pathfinders. They were jump qualified and the First Sergeant said that they had a quota to fill for jump school at Ft. Benning, Georgia. To qualify for jump school you were required to pass a physical readiness test with a perfect score of 300. The test involved a three mile run, 20 pull ups, 80 set ups and 60 push ups, all done in full utility uniform with boots.

I scored the required 300 points and stood in formation as they called out the names of the boys that had been picked for jump school. When they got down to the last position my name hadn't been called. The sergeant said, "We have one billet left for jump school and two men are tied for it." He then read my name and another.

"We're going to settle this over at that pull up bar. Give me your best, Blevins, you're first."

Marine pull ups were done with hands over the bar and no body swinging to help with the lift. I did 25 and tried for one more, but couldn't get my chin high enough. Lowering my body slightly to rest, I tried again and again to get another one, but I just couldn't do it. Finally my grip on the bar gave out. I fell in a heap. While I took my place back in formation the other guy started his pull ups. He easily did 25, then one more to beat me and dropped off the bar with a satisfied look on his face. As he returned to the formation the sergeant announced, "The last jump school slot will be filled by Private First Class Blevins." He then looked at the boy that had beaten me and said, "Son, you're good, but I don't send quitters to jump school."

The Marines sent their men to be jump qualified by the Army at Ft. Benning, Georgia. They prided themselves in the fact that no Marines thus far had washed out of the Army program. At Pendleton they set up a junior jump school to weed out those that might have problems at Ft. Benning. We practiced our PLFs (parachute landing falls) from the back of pickup trucks, from a cable into a sand pit and in a harness from a tower. They also made sure we were at the peak of physical fitness.

At Ft. Benning everything went fine with our pre-jump preparation training; lots of PLF practice and numerous safety drills. Our first jump was from the back of a C-130 and I took my place in line with the rest. My safety strap was hooked to the aircraft static line and I followed the rest to the open door ramp at the rear of the aircraft. When it was my turn to go I looked out at the wide open space and froze. The momentary thought that went through my mind was, "Oh God, what have I gotten myself into now?" It was certainly just a split second thought, because the guy behind gave me a shove and out I went. My eyes were closed until I felt the tug of the chute opening. On the way down I just stared up at the open chute until I could hear voices below, then I prepared myself for the landing.

After that first jump, we did four more for our qualification and received our jump wings. I was extremely proud to wear those wings and still have them on my uniform.

Dee wasn't real happy when I left for Georgia and jump training. We hadn't been married that long to be separated right away, but she went along with it anyway. One thing that did brighten things up was that Grandpa had decided to come out and live with us.

Grandpa and Grandma had spent years planning how they would spend their retirement. Like most folks that had a hard time during the Great Depression, they were always saving and went without a lot of things in the present so they wouldn't be caught short in the future. Unfortunately, Grandma had died several years ago and now Grandpa was retired. He said that he would never consider remarrying, for he had been married to one fine woman and considered her still with him. Her last request of him was to take care of her grandkids.

Grandpa asked if we'd like him to join us and that he would be happy to help out with the rent. Of course, we wholeheartedly welcomed his presence and rented a small two bedroom house outside of Oceanside. Grandpa provided company for Dee when I was away training, and now that was SCUBA school down at Del Mar, California.

SCUBA (self contained underwater breathing apparatus) training was very intense and difficult. As a kid, I had learned to swim in Long Lake, by Sparta. I was a good swimmer and felt very comfortable and secure in the water. It was not a place of danger for me, but this training pressed me to my absolute limits. It was taught by Navy Seals and they were tough. They thought nothing of running us for 10 miles before we hit the pool for some strenuous swimming. It was by far the most difficult training course that I attended.

Now it was time for me to get down to some serious bull riding and Chuck Wilson was providing all the pointers he could. I made the team and we started competing at other west coast military installations and civilian rodeos too. Here again Dee went along with my bull riding because I wanted to do it. If I was leaving for a contest at another base, she'd see me off on the bus and tell me she loved me, but I had the feeling she worried a lot. When a rodeo was close, she'd come and set on the hood of the car to watch me either ride or get thrown off the old Brahmas.

Once while Dee was watching me, I was beat up good by an old bull called Jack Daniels. The clowns couldn't get him off me and he just kept grinding me into the ground with his head. My knee got tore up pretty bad on that one too. Dee got shook up over the incident, but it could have been worse for her and me. We found out later that she was two months pregnant and could have lost the baby from that stress.

It was late in the summer, when I noticed the opening for a billet in Kodiak, Alaska on the bulletin board. It was for a multi-service search and rescue team. Both jump and SCUBA qualification was a requirement. Kodiak, Alaska really sounded neat. There was bound to be an opportunity for some hunting and fishing there, plus, I would have an opportunity to use my training. The only bad thing about this opportunity was that it was an unaccompanied one year tour. Dee couldn't go. Dee and I discussed Kodiak and she went along with my decision to apply. I submitted an application, and within a couple weeks I was notified of my acceptance and orders soon arrived. Dee was now about six months pregnant. I drove her and Grandpa back to Michigan. From there I would catch a plane to Kodiak.

The tour at Pendleton had been a strange one. I was assigned to the 1st Marine Division, but when I reported, in the summer of 1963, the Division had already been deployed in support of the Cuban Missile Crisis. Now, a year later, I was leaving the 1st Marine Division, and they still hadn't returned. Fortunately, because I had nothing else to do, I was able to receive some great training and also capture a nice trophy belt buckle for my 10th place on the bull riding circuit. Probably, if the 1st Division had been at the base, I would have had much more competition for the Kodiak assignment too.

The commander at Kodiak introduced himself. He had a full beard and wore civilian clothes. Because of the cold climate, there were different rules here. Facial hair was allowed and uniforms did not have to be worn unless one wanted to wear them. I must be available at a moment's notice to go out on a rescue mission, but unless I wanted to set in the barracks and read magazines and watch TV, I could get a job in town. He said the pay scale in town was good and that most of the team members had additional jobs there.

Kodiak is the biggest island of the Aleutian chain. It's about 300 miles long and 100 miles wide. The Japanese current surrounds the island, so it is ice free year-round and an excellent fishing port. Most of the people on Kodiak are either fisherman or associated with the fishing industry. Probably when I was there the whole island population was not more than 3000. Of course Kodiak was famous for its huge Kodiak bears and King Crab.

On March 27th of 1964, a severe earthquake hit Alaska. The quake produced three tidal waves, the third one being the most damaging.

When it hit Kodiak Island it was reported to be 40 feet high. The wave traveled 10 miles inland, destroying the town of Kodiak and killing hundreds of people.

A certain psychic or fortune teller made news when he correctly predicted President Kennedy's assassination and the March 27th earthquake that hit Alaska. He also predicted that another quake would hit Kodiak a year later. The island would be split in half with one half sliding into the sea. I don't know how many people believe in that stuff, but I was aware of the prediction when I put in my bid for the assignment, and wondered if it had narrowed the field of applicants.

The village of Kodiak was small with dirt streets and a few basic looking buildings spared by the '64 Quake. I wandered around and noticed a sign in this shop window that said WELDER WANTED. In high school I took a welding course. "Maybe I could get a job in here," I thought. Inside I introduced myself to a guy that looked like the owner. He said that there was a job available, but that it was for an underwater welder to repair fishing boats. There just weren't enough dry docks to need an above ground welder, he explained. I told the man I was dive qualified and would be interested in the job. Even though I was dive qualified he said I would still have to start as an apprentice and get some training before he could pay scale wages. That was fine with me. I asked when I should come in.

"If you can come in this weekend, I'll pay you time and a half on Saturday and double time on Sunday. There's a time clock right over there."

The weekend work was welding seams on fuel oil drums. It was tedious, but I had nothing else to do so I worked the whole weekend. I didn't know what my pay would be. Grandpa taught me that it was poor manners to ask right out what a man was gonna pay you. You were just supposed to be appreciative of getting the job. Minimum wages in those days was somewhere around a dollar an hour.

When I went in to work on Monday, there was a pay envelope with my time card. Holy cow! There was a check for several hundred dollars in there. With the check I walked right into the owner's office, and he could see the worried look on my face.

"I told you that you would have to start at apprentice wages. That's $9.20 an hour. Within a month or so I will have you trained and you'll get $25 an hour. I just can't afford to have you welding on somebody's boat and not be sure you're doing it right. Their life may depend on that work."

I just said, "Yes sir," and left to weld some more fuel drums. I didn't tell him I thought he made a mistake and paid me too much.

After two weeks on Kodiak, I was told to show up at the airstrip in uniform to meet the new commander that was coming in, a Major Wilson. Well I was there all right, and who do you suppose got off that plane. It was Major Chuck Wilson, the rodeo team leader. Hell, I didn't even know he was in the Marine Corps let alone an officer. Every time I'd seen him, he had on Levis and a cowboy hat and shirt.

I saluted and we shook hands. Neither of us smiled or displayed any recognition of the other. Officer and enlisted fraternization was strictly forbidden in the Marines. He was a major and I was a lowly corporal.

Later that day, he called me into the commander's office and we had a chance to relax and visit. The first thing he asked was where Dee was. I said that it was supposed to be an unaccompanied tour and that she was 8 months pregnant back in Michigan.

"Why don't you bring her up here? The government won't pay for it, but I'll give you permission. My wife will be here," he said. That sounded darn good to me, so I got on the phone to Michigan and told Dee to get on a plane as soon as she could, and fly to Kodiak. I was sending my weekend welding money. She should sell the '53 Chevrolet and borrow whatever else she needed to get here. This assignment was turning into a gold mine.

The Marines were on Kodiak to guard a Navy weapons storage area. I suppose the real need for the search and rescue team was for a military emergency, but its day-to-day operation was to rescue capsized fisherman and look for missing hunters. Unfortunately we retrieved more bodies than we rescued live people. The team was made up of Navy corpsman, Marine scuba divers, two Air Force navigators and the pilot of our seaplane was the last of the Navy enlisted pilots.

The pilot was an old Navy Chief on his last year of a 30 year career.

He could guide that old seaplane in and out of about any area we needed to go into. He used to grumble when we brought our parachutes along. We were just unnecessarily cluttering up his airplane, because he could take us wherever we needed to go without parachutes.

You would not have thought we were from any branch of the service, because we wore mostly civilian clothes and a mixture of baseball and cowboy hats. Some even on occasion wore beards, except for me. I was young enough that it would have taken forever to grow one, so I didn't bother. Only part of the team was dive qualified, but all were jump trained.

Dee finally arrived in Kodiak. By that time her belly was swollen up with baby. We had a room in Green Apartments and it wasn't too fancy. Three families had to share one bathroom on each floor of the old house. Furniture was sparse, with only a bed, table and three chairs. A wood stove was used for heat and cooking. At least we were together again.

I did work a lot of hours at the ship repair shop. My pay as a corporal was about $400 a month and I easily made that on a weekend welding. We just lived on the Marine pay and stashed away the rest.

We didn't have a lot of call for rescues that winter. There were the three hunters that couldn't be accounted for. They were supposed to be picked up with an airplane. The weather was too bad on the designated day and when the pilot went in the next time, they weren't there. Our plane landed on skis and our team hiked up to the cabin they were supposed to be staying in. They were in the cabin all right, frozen in their sleeping bags. All the fire wood on the porch had been burned up, so they just crawled in their bags to keep warm and froze to death. That was a real shame, for they could have burned the furniture or even the siding on the cabin to stay warm. We left them in their bags and lashed their bodies under the wings for the trip back.

On March 23, 1965, Robert Jr. was born in the Naval hospital on Kodiak. A few weeks later, in April, we were in bed when the whole house started shaking. It was physically impossible to stand up and I was thrown off my feet several times. The seconds seemed like several minutes. To my horror, the prediction of the second earthquake was about to come true. I experienced sheer panic, and thought how could I have brought my family here only to be killed.

The house was still pitching and shaking as I crawled to the window overlooking the fishing port. The boats were still there ... that meant no tidal wave. Or, maybe it meant that a warning hadn't been issued. Taking no chances, I gathered my family and fled the house with the rest of the tenants. We began climbing Old Woman Mountain which was across from the rear of the tenant house. As we climbed, the earth shook with after-shocks. None of us stopped until we were several hundred feet above sea-level, safe from the danger of a tidal wave.

Looking around in the dark, it was easy to see that most of the town's population was perched on the side of that mountain. Now, the thing on most everyone's mind was ... did the island split in half as predicted. And, were we going to slide into the sea?

At daybreak, the fear started to subside and the realization hit that there was once again earthquake damage and there would probably be several casualties. Our search and rescue team was assembled and we began to dig out and search for survivors over the next three days.

This was Dee's first exposure to danger, and for the first time in our marriage she spoke out. She wanted to go home. I assured her that if the military felt the island was going to sink, they would get the families back to the lower forty eight. However, a large area did in fact experience a serious rupture and was under several feet of water during high-tide. Many families and native-born alike crowded the small airport during the next several weeks, filling every flight out of Kodiak. The rumors of the prediction were rampant, and people were leaving everything they owned behind. I convinced Dee to stay, and promised her we would leave at the end of the summer when my tour was finished.

It was a couple of months before things started to return to normal. During this time frame I didn't do any outside work welding, but later I started again and there was plenty of work. I was earning $25 an hour working underwater patching fishing boat hulls and changing damaged propellers. To remove the old props, we'd attach a small charge and blow them off. Then, we'd replace them with big heavy brass ones.

Chuck Wilson and I had a strange friendship going. On the rodeo team we all just called him Chuck, for he never let on he was an officer. Even though beards and civilian clothes were allowed on Kodiak, everyone knew he was the Commander. I just didn't call him anything except Major, or Sir when we were on base. I didn't abuse our friendship nor did he. Off base, away from the military environment, we were Chuck and Bob. My fondest memories of Kodiak are the hunting and fishing trips we took together.

Each of our expeditions started in a rather sneaky manner. Chuck would drive a jeep to Green Apartments and we'd load all my hunting gear in the back next to his. Then, I'd crawl in the back with our stuff and he'd cover me up with a tarp. We had to drive right past the base, so I'd stay hidden until we were past the main gate.

The south part of the island had some high hills on it. It was the wilder region where all the deer, elk and bear lived. I saw several of the big Kodiak bears, but we never hunted them. Mostly, we hunted the Black Tail deer, which were quite small. I did get one that dressed out at 170 lbs. and that held the island record for quite awhile. There was a farm where we usually rented a couple of horses and from there we would ride up in the hills to hunt and fish for two or three days.

Besides the deer, there was a pheasant-like bird that we hunted. It was white and had white feathers on its feet, like leggings. There was also the Dolly Vardin trout, which were plentiful. They were a salt water fish that swam up in the streams. Their bodies were very soft. On one of my first trips with Chuck, I caught a nice bunch of trout and threaded them on a forked stick tied to my saddle. Chuck said that it would be better to put them in the saddle bags, but I ignored his suggestion. After all, that's the way I carried them as a kid. I kept checking them on the way back to camp; however, when we got there, on my stick was nothing left except heads. They were so soft that the body simply fell away from the head. Most of the time Chuck and I had these guilt feelings about leaving our wives back in town, so occasionally we'd bring them along for some fishing or a picnic. We both loved the outdoors and this lifestyle. Chuck assured me that if I wanted to, I could spend the rest of my Marine tour here on Kodiak.

Dee and I were on Kodiak for almost a year after the earthquake. I spent most of my free time either welding or hunting with Chuck. We did have several search missions too. Most were for capsized fishermen, but there was one incident where a hunter crashed his private plane up in the hills. He had survived awhile and then froze to death. If we found the fishermen they were usually floating on the bottom.

The decision to leave Kodiak was one that Dee and I made together. Since the earthquake she just didn't feel comfortable on the island. Dee preferred California or Michigan, just someplace else. She had been left alone a lot with the baby, in a little apartment, while I hunted or welded. Still it had been a good tour. Although it was supposed to be unaccompanied, Dee had been able to be there with me when Robert Jr. was born and we had saved around $15,000 from my part-time job.

Admittedly, I had been like a "kid-in-the-melon-patch". I had worked hard earning extra money for our future, but I had spent most of my free time getting in as much hunting and fishing as possible. I hadn't been very considerate of Dee, alone with my baby son in that one-room apartment. I felt guilty then, as I do now. But, much like Tom Sawyer and Huck Finn, this was a once-in-a-lifetime opportunity. Thankful for her understanding and forgiveness, it was time to put my priorities in perspective and move on.

To be a drill instructor on Paris Island had always held some appeal for me. To wear the Smokey Bear hat and yell at recruits might be a fun thing to do. I applied and was accepted for that assignment. We said a sad good-bye to my good friend Chuck Wilson and his wife, and vowed we'd keep in touch. Chuck stayed on Kodiak and later retired from the Marines there. He became a master hunting and fishing guide.

We purchased a nice house trailer and lived in that on Lady's Island, while I started my DI training. Grandpa put the lock back on his door up in Michigan, and moved in with us.

Grandpa really fit in fine with our family. He enjoyed his new great-grandson and was helpful taking Dee for her shopping trips. We had purchased a 1963 Volkswagon Bug, but Dee had never had the opportunity to get a driver's license, so now Grandpa could take her wherever she wanted to go.

Grandpa seemed to enjoy himself too. He liked his moments of solitude. He might just take off to go fishing by himself or relax next to his radio and listen to his favorite team, the Detroit Tigers, play baseball.

My DI training wasn't finished yet, but already I was having second thoughts about it. I found out from others that the lifestyle of a DI was not conducive to a happy marriage. They had what they called the port and starboard shift. You were on duty from 5 a.m. one day until the evening of the next day, and then you started again at 5 a.m. that following morning. In between classes there were a few extra days off, but basically this was how the whole tour was going to be. You slept at home with your wife every other night. For the first time, I was really concerned about keeping my family together. I had been pretty footloose and fancy-free, doing my own thing through most of our marriage, but now I felt it being threatened with this DI assignment.

Each Marine has a monitor in Headquarters Marine Corps, Wash. D.C. The monitor can counsel on career decisions and problems. I called my monitor up and explained that I felt my current assignment might present a dead end to my marriage. In these times there was a well known saying around, "If the Marines had wanted for you to have a wife, they would have issued one." That attitude might have been popular in some places, but my monitor was sympathetic to my concerns. Within two weeks I had orders for independent duty at the 12th Marine District Headquarters, 100 Harrison in San Francisco. I was to be in charge of the printing office. It was to be a beautiful time together for the family.

The Marines moved my trailer to the West Coast and we climbed into the Volkswagon and headed in the same direction. Grandpa went back to Michigan for a couple months, then joined us again when we were settled in California.

The last place I wanted to be living was in some trailer park. That was like living in a beehive. I drove around looking for an area out in the open to put the trailer. North of the Golden Gate Bridge in Novato, I found the C Bar N Ranch and Training Stable. I located the guy that owned the ranch and told him that I'd like to have a place on his ranch to put my trailer. I explained that I'd pay rent or work it off, whatever he wanted. He asked if I knew much about horses. I said that I'd been around horses a lot and had been on the Marine rodeo team.

Our trailer was parked right next to one of the old bunk houses and we hooked into its sewer line. We were also tied into the ranch's water and electricity. It was all going to be free in exchange for me working with his horses.

Now I kinda bluffed my way into this agreement. True, I had been around horses quite a bit. My Grandfather owned a pair of Belgians that he used for farming a rented forty acres across from his house. Also in the rodeo I had been on and off them a few times, but I had no training experience at all. The C Bar N Ranch had about 75 horses and they catered to renting out to the college kids for weekend trail rides and lessons. A lot of these horses would develop bad habits with these inexperienced riders. They would start doing their own thing. Some just wouldn't move when they were supposed to, others would buck, bite and even start on an uncontrolled run. My job was to ride these horses during the week and reinstate discipline and obedience so they were safe with the inexperienced riders.

I felt that if I could ride bulls I ought to be able to master this horse thing too. After being rolled on, bit, kicked and thrown off a few times, I did learn a little on how to outwit these critters. Also, I gained a lot from watching how the experienced riders went about working with these unruly horses. Eventually I developed some techniques and a real talent for handling horses.

My duties at the printing office didn't amount to much. The most stressful thing in life was playing handball over on Treasure Island or riding the horses with the bad attitudes up at the ranch. Grandpa was with us again and we were having some real happy times together.

In June of 1967 our daughter Karen was born. That fall I made Staff Sergeant and in January of 1968 my enlistment was up. I decided not to reenlist. It was a decision I struggled with, but I thought that it was in the whole family's best interest. Things had changed in the Marines and in the country; it was the Vietnam era.

The Marines had always been a volunteer outfit. Men and women were there because they wanted to be. It meant something to be a

Marine. Now, with Vietnam, there weren't enough volunteers to fill the quotas. Draftees were being taken in and I thought they were lowering the standards. That was a mistake. There were some real undesirable characters out there and at that time they didn't have the drug tests to weed them out. The Marines was a close knit family. You were like brothers and took care of each other. It seemed this unity and comradery was being seriously threatened.

I detested the hippies and war protestors that congregated around San Francisco. It was hard to believe that there were those that would curse and spit on their men in uniform, let alone burn their country's flag.

I was 22 years old when I left the Marines in January of 1968. I had not been tested in combat, but I had no fear of that. I still felt tough and bulletproof, and even though there was the need within me to join the other Marines in Vietnam, there also was a new feeling of responsibility to be with my wife and two young children. I had served four years in the military. Perhaps I could do something worthwhile for this country in civilian life. We packed up the trailer and started our drive back to Michigan.

We paid cash for our three bedroom home in Kentwood. It was the first home we ever owned and it was a dream come true. Still, we had money in the bank and a new car too.

Right away I went over to the Kentree Stables in Ada, and got a job breaking horses. They had an ad in the paper, offering to pay me $75 apiece to break six horses a month.

From the Kentree Stables, wearing my Levis and with shit on my boots, I walked into the Kent County Sheriff's office and tossed my Marine discharge on the desk. You would have thought by my manner I was doing them the biggest favor in the world when I asked for a job. Captain Bill Rector was at the desk. He glanced at my discharge papers, looked up and said, "Can you start in the morning?" I said to myself, "Damn, these Marine discharge papers are powerful. They can really open doors."

Boot Camp Graduation — Bob and Mother

Wedding Day Feb. 1, 1964

Major Chuck Wilson — Bob's rodeo coach, then his Commanding Officer and hunting partner on Kodiak

Top: Search and Rescue Team at Kodiak — Bob center

Below: Main Street Kodiak in 1964

Top: Green Apts — Our home on Kodiak

Bottom: Dee and son "Buck"

Grandfather on the left, Bob's stepfather Finis on the right with Dee and Buck

County Deputy

For the last couple of years, before I left the Marines, I had been con-templating a career in law enforcement. It seemed that belonging to a police force would provide a continuation of what the Marines had already given me, a sense of belonging and unity. That was something terribly lacking when I grew up. I also thought that law enforcement would be exciting, and I could picture myself racing around in a patrol car, catching all the bad guys.

Things certainly were going well. I had a nice house for the family, at 212 Majestic Street in Kentwood, and a job at the Kentree Stables in Ada breaking horses. Now, the very next morning after I had applied, I was reporting for duty at the Kent County Sheriff's Department.

It quickly became apparent that I wasn't going to be racing around in any black and white patrol car right away. Instead I was to be a floor guard with daily exposure to these scruffy, low-life inmates. It was customary for a deputy to be working in the jail for five to seven years before he had patrol car duty. What a bummer.

These inmates represented everything I detested. In my mind, these unshaven, dirty drunks were just an extension of the hippies and war protestors I had seen on the West Coast.

I was young (22) and I looked young. As I walked down the catwalk between the cells, these dudes would try their best to scare and intimi-date me. After the Marines and the orphanage, there wasn't a man alive that could scare me. It wasn't too long before these guys learned that.

When a fight broke out in a cell between inmates, I never missed an opportunity to wade in and break it up. If somebody wanted a piece of me in the process, he was quickly educated to the fact that I could whip his ass. If one got real unruly it was fun just to drag him down the hall to the dark hole for a little solitary.

The inmates didn't like me and I surely had no use for them. If they had been arrested and put in here, surely they were guilty and deserved to be here. My duty as a jailer did not go well. I wanted the keys to a patrol car and instead got the keys to these cells.

Confrontations between me and the prisoners became more numerous, and after a couple of months they were occurring on a daily basis. The shift supervisor could see I just wasn't working out well in this capacity, so he put me in charge of booking and receiving. This didn't last long, because after Martin Luther King was assassinated the riots broke out around the country, and in downtown Grand Rapids.

One night the Lieutenant pulled me off desk duty. He handed me a set of patrol car keys, "You'll find a car in the lot with the same number that's on this key tag. The overhead light switch is on the right side of the steering wheel. Do you know how to use a radio?"

"Yes sir," I replied.

"Well, get going downtown. Report to Frank Pierce."

I was in my glory. Had my own patrol car, no partner, no training and heading towards the action.

Frank Pierce was in charge of trying to control the riots. He was a Congressional Medal of Honor winner; a navy corpsman in Korea. All kinds of police cars were showing up from different directions with their flashing lights all going. The whole thing was a show of force to contain the crowds. There was a lot of looting, destruction of property and unnecessary loss of life, but things finally quieted down.

The decision was made to leave me in a patrol car, so I didn't have to return to desk duty again. Luckily, I had circumvented several years of being a jailer. Now, I could be out on the streets doing what I wanted to do in the first place. I was happy.

Shortly after the riots, I was patrolling in Comstock Park when a call came over the radio asking for assistance at a bad accident on Plainfield Ave. Only a few minutes away, I responded to the call and the dispatcher cleared me to go. There were no other patrol cars even close, so I would be the first to arrive. I was in my glory, the only show in town. North Park Bridge was two lane, however it was a real narrow bridge. Two cars passing had to be vigilant. About a third of the way across there was this old farmer putting along in his pickup truck. My flashing lights were on but obviously he hadn't seen me yet. I was a little reluctant to pass him until he knew that I was there. No telling what he might do. Reaching down on the dash panel I flipped on the siren. Those old federal sirens were really loud and this one started, "Yelp, Yelp." All I could see in his mirror were two saucer-sized, fearful eyes. Since he started driving, he had been programmed to pull over for a cop. But, there was no place to pull over. He tried to anyway, and the first thing the side of the steel bridge brushed off was his big West Coast mirror. The glass came my way in pieces and the frame bounded off my hood. Next he went over the centerline, back to try it again. First came some trim, then the door handle and some other pieces.

I was petrified. In panic, I couldn't locate the siren switch to turn it off. I could just see another accident developing as again and again he hit the right side of his truck against the bridge. Finally we reached the other end, so beyond the bridge he could finally pull over and stop. He looked terrified as I went by, but I could see he wasn't injured.

For days and weeks I was afraid that the old man would show up at the sheriff's office and bitch about the idiot that turned the siren on behind him. He never came in to complain or claim damages for his truck. About a year later, I knew he wasn't ever going to be there. It seemed humorous then.

When I started with Kent County, there were probably only 50 deputies or so. They issued us a shield with our name on the front; then a couple years later we turned these in and got a star with our own badge number on it. My number was 69. I really liked that number. When a bad actor demanded my badge number, I always said he could read it lying down or standing up, whatever he wanted.

It seemed like the deputies were either very young or quite old. For the most part, they were big. The police force recruiting philosophy seemed to be that the tougher, the bigger, the harder they were, the longer they'd last on the street. At 185 pounds I was the little guy in the squad room. The other deputies never failed to remind me of this.

Many of the deputies were ex-military and a lot of them had an active youth. When I say active, I mean as they grew up they were in fights, scrapes with the law and I know some on other departments that had been in jail at one time or another. That was a little different then it is today. Nowadays your record needs to be spotless if you want a job in law enforcement.

When I started, things were different in a lot of ways. There weren't as many deputies around and the laws allowed them more power to make on-the-spot decisions in settling problems and taking people to jail. If an officer was big, and had grown up on the streets, he knew how to talk the talk and walk the walk. He could intimidate the bad guys and knew where to find them too. A deputy did have to be careful of what kind of situation he got himself into, as backup might not be readily available.

For many years the County only had one car on patrol during the night shift, so your backup might be a distant State Police post or a neighboring county patrolman. There was an old saying among the deputies, "Don't write a check with your mouth that your ass can't cover."

The other warning was that backup could be so far away that the bad guy could bump you off, hold services, bury you and grass would be growing on your grave before backup arrived.

I loved the work. Sometimes I had a partner and sometimes I didn't. For a year I patrolled the county, wrote tickets, handled accidents and refereed numerous family disputes, all without any formal training. What I didn't learn on my own I picked up from the deputies that had been around for awhile. I liked them all and they seemed willing to share their experiences with me.

About a year after I started as a deputy, the State passed a law that required Michigan police officers to be formally trained and certified; I had worked long enough as a deputy to be exempt from the requirement, however I really wanted the training. There was so much that I didn't know, that would take too long to learn the way I was now doing it.

At the police academy I learned all aspects of the vehicle and penal code. Our instructors were senior judges, veteran attorneys, Michigan State Police Officers and experienced deputy sheriffs. Their practical experience in law enforcement gave us a realistic approach in handling situational training problems. I listened intently to everything that was said, for I wanted to be as good a police officer as I could possibly be. They also gave us extensive firearms training.

The firearms instructor at the Academy was an FBI resident agent, Dan Bidwell. I had received plenty of firearms training in the Marines and had hunted most of my life, but the only pistol that I had any experience with was a 45. Revolvers were something new to me, but Dan could sense that I possessed plenty of natural talent with guns, so he gave me some personal instruction. I caught on fast, and I was especially good firing in the squat and quick draw combat positions.

There was a yearly competition within the department, and even some of the FBI agents would participate. Over the years I collected a room full of trophies for my shooting skills. Each year I would win the combat event, which included the quick draw. Three years I won first place shooting from the standard positions.

The sheriff's department seemed like a family, the deputies like brothers. Ron Parsons was a new sheriff. He had been a deputy for Arnold Pigorish, who was reelected sheriff many times. Ron loved motorcycles. He probably had been as close as you could be to being a Hells Angel type, without the tattoos, beard and long hair. Apparently sometime back in his youth, he was chased through a couple of states by the law until they caught him. For what, I don't remember. Ron was a good sheriff, but he did put his foot down and make one thing clear to us. The police around the State had lost too many officers when the military reserves were called up for the Detroit riots. If we weren't required to be members, with a sign-up commitment, we would have to choose between the military reserves and working for the county. I had joined up with my old Marine reserve unit right after discharge, so I had to give that up for awhile. Nowadays they couldn't force you to do that, but then they got away with it.

Even though there was overlapping authority, we had an excellent relationship with the Grand Rapids City Police, the State Police and the neighboring county sheriff's departments. Many times our backup would come from one of these other agencies or we would be helping them.

My first high speed chase was in aiding the Grand Rapids City Police. On my radio, I heard they were in high speed pursuit of some guy. When I happened to look in my rearview mirror, I could see them all coming up behind me. This guy was in the lead, with several city police cars in pursuit; all their lights and sirens going. I turned on my overhead lights and wouldn't allow the guy to pass me. When he tried to change lanes and go around, I just moved to the other lane and blocked him. As I started to slow him down the other police cars caught up. Before long we had him boxed in and forced him to stop.

The Grand Rapids officers immediately jumped out of their cars and with their night sticks broke every glass part of that guy's car. Windshield, headlights, side windows, mirrors and rear lights all were shattered. When they were through with the car, they hauled this guy out and literally beat the shit out of him. I don't know what the guy was supposed to have done, but the force they used seemed a bit excessive. These were not new officers either. One was a lieutenant and two were sergeants. That scene sure made a lasting impression on me.

One of my first encounters with the State Police happened over in Comstock Park. I had been called to the intersection of Four Mile and Main Street where a car was blocking traffic. This guy was slumped over his steering wheel and the car doors were all locked. The drivers window was down a couple inches, but not far enough for me to reach in and unlock the door. Through the partially open window I could easily smell alcohol. The man's car was still running. I thought if I woke him too abruptly he might try to drive off, and end up causing an accident, so I called for backup.

This State Trooper pulled up. He looked like he had been around for a few years. "What can I do for ya, rookie?"

I explained the situation and he said, "I'll take care of it."

He climbed up on the hood of the man's car and from there to the roof where he jumped up and down at least a dozen times. The roof eventually was crushed down enough to almost reach the guy's head. Finally he woke up and I hollered to the trooper, "He's awake. Stop jumping." The trooper hopped off of the car, brushed his hands together and asked, "Anything else you need?" With that, he climbed into his cruiser and left. The man unlocked his door and I arrested him for being drunk and disorderly. His car was impounded.

A week later I ran into the same guy in court. He said to me, "You know, I was so drunk I don't even remember getting into the car that day. Someone sure did beat up my roof in the parking lot, didn't they?" I had a little trouble keeping a straight face. "Yes, I guess they did," I replied.

It used to be that to pinch anyone for drunk driving you actually had to observe them driving while intoxicated. If they were passed out on the side of the road and it was obvious they had been driving, you still couldn't get a conviction.

It was in the middle of the night and I noticed smoke coming up from the opposite side of the road I was patrolling. Turning my car around I could see a vehicle down in the deep roadside ditch. When I approached the car I could tell by the engine noise that the accelerator was full down. There was lots of smoke and steam from the exhaust and spinning tires. I expected to find an injured driver, but instead here was this guy actually still trying to steer the car. He was in a drunk stupor with hair hanging over his face, snot running out his nose. It was summer and his window was wide open, but he didn't even notice I was outside his door.

Finally, I started running in place beside the car and at the same time began slapping the roof over his head. "Pull your car over mister, put on your brakes!" He turned his head, looked at me wild eyed and then abruptly turned the steering wheel to the right, taking his foot off of the accelerator. I said to myself, "Driving or not, this guy really needs to go to jail."

When somebody was arraigned the arresting officer had to personally escort him into court. I said to this guy, "Do you remember me? I'm the officer that pulled you over for drunk driving."

He said, "Yep", and pleaded guilty. Even though his breathalizer test showed him to be double drunk, he still wouldn't have been convicted if he had fought it, unless I had lied and said I'd seen him driving. I'm not sure I would have done that.

Grandpa said that he wished we hadn't bought the house in Kentwood, because he would have given us some land and we could have built right beside him. My folks had moved down just off Plain-field Ave. and were living in a trailer. The old prefab home I was brought up in hadn't been lived in for quite some time. It was pretty well rotting away.

Dee and I decided that we'd like to raise our kids out in the country, so we took Grandpa up on his offer. He said that eventually his land would be ours, he might just as well deed it over right now.

We tore the old house down and drew up some plans for a new one. With his carpentry experience Grandpa was a great asset. When he figured the supplies needed to build something, there would never be much left over. He always did have a way with numbers anyway. When I was a kid laboring over some tough math problem, with a slide rule and several sheets of work paper, he would simply glance at the problem and jot down the answer while I struggled. He could work it all in his head.

In less than a year, we finished a nice house only 150 feet from Grandpa's. We sold the house in Kentwood for about double what we paid for it. Now our kids could be brought up in the same area and attend the same schools that I had.

Bob Bond owned Kentree Stables where I broke horses. He was quite a famous polo player, specializing in training horses for the wealthy people that enjoyed polo. Even after we moved to Sparta, I continued breaking his horses. Some I rode in his indoor arena and some I trained in an outdoor breaking arena that I built next to our place in Sparta.

There was one horse that Bob was unhappy with. At first I think he believed that I hadn't broken him as I said I had, but when I climbed on him he rode perfectly. I even rode him around swinging a polo stick while Bob watched. Unfortunately no one else could ride him, so when I left the stable and we settled up what Bob owed me for my work, the horse belonged to me as part of the settlement. No one else could ever ride that horse. Occasionally I would offer $5 to some hot shot cowboy if he could just touch the saddle horn, but the minute the man would reach for the saddle horn, old Guy would pin back his ears and go crazy.

Grandpa still had a small barn where he used to keep his two

Belgian work horses. I remodeled it a little, making room for six tie in horse stalls. In addition to Guy, we acquired a registered black quarter horse. The horse had been used for polo by Dr. Chris Helmis, an ear and eye doctor from the local area. This horse's name was Candy.

Candy had been an excellent polo horse until she broke her leg. She was sent to MSU where they put a cast on her and the leg did heal up. When they used her in polo matches though, she would start limping. Dr. Helmis said he wished he had a nice German Shepard dog, so we made a trade. We bought him a registered German Shepard and we got Candy.

The dog was sold to us by a good friend who used to raise them for a blind leader dog school. We also had one of these fine dogs given to us. There were two left in the litter, one he gave to Spiro Agnew who was the Vice President, the other we got. There had been some break-ins around the area. I thought the dog might give Dee some peace of mind in regards to the prowlers.

We also started boarding some horses. Two girls stopped by after they noticed our horses. They wondered if we would consider boarding theirs. As there were four empty stalls we decided to do that and thought about expanding even more.

Since we moved back to Michigan, I had developed a much closer relationship with my stepfather. We went deer hunting together in the fall, usually ending up in Luther. That's where he owned property when I was a kid. Finis changed jobs regularly, but he now worked for Ron's Wrecker in Grand Rapids and really liked it.

When we were up at Luther, I would always stop over in Baldwin and introduce myself to the county sheriff. It was a courtesy call, just to let him know I was in the area in case he ever needed some assistance. I'd usually visit with him for awhile to find out what was going on up there.

The first sheriff I met was Bob Radden. I told him I started school up in that area when I was a kid. We spent time on different occasions making small talk about police stuff and developed a nice friendship. He was found dead one morning in his patrol car. The car was crashed up against a tree. There were no witnesses, though friends afterward said that it had been a murder and made to look like an accident. Bob knew too much about what was going on in Lake County. I wasn't aware of this at the time, but was sad to hear of his death.

Each month we would change patrol shifts. From night to day and then the swing shift. There would be about 20 deputies on a shift and to celebrate the end of one month's shift we would have a little party. Many of the get-togethers ended up in my basement, as I had a nice bar, with both pool and ping pong tables. The guys would bring their wives and girl friends over, and we would have a fine time until the wee hours of the morning. More than once there would be a bunch around to eat breakfast with us.

More and more I was enjoying my work as a patrol deputy. There was always something new and different happening, sometimes funny, sometimes sad. It seemed that 90 percent of our calls were for domestic violence and abuse. Normally, we'd arrive on a call to find that some guy had beaten up on his wife. When we tried to take the guy away the wife would change her mind, or drop the charges in court if he had been prosecuted.

Once, I was by myself on a call and found that this lady's husband had really beaten her up good. As I tried to haul him away she blocked my path. I was determined to take him anyway, for her safety. The next thing I know she's on my back stabbing me with a dinner fork.

On another domestic disturbance call, a lady responded to my knock and opened the door. She looked fine. "What's the problem here?" I asked.

"I didn't call, he did." She pointed to a man lying face down on the couch. As I came in the guy rolled over. Holy cow! She had beaten the shit out of him. There were bumps, welts and bruises all over his face and head. He started begging me to get him out of there. From what I could tell she had warned him about what would happen if he came home drunk one more time. She had followed through on her threat.

Terry Debold and I responded to what we figured was a family dispute in a trailer park on Plainfied. There had been a report of a single gunshot, so we approached the trailer carefully, he from the front and me towards the back door. Terry finally pounded on the door and this lady jerked it open and frantically said, "He's in here! Hurry! Hurry!"

"Who's in there?"

"It's my husband, he's been shot!"

We hustled into the room she was pointing to. There on the floor is this dude that is naked, a western belt and holster strapped to his waist. The holster is tied down by his knee which has a nice bullet hole just above. He's bleeding good and in a lot of pain. Terry and I look at each other and wonder if we should inquire as to how this happened.

"How did this happen sir?" I finally asked.

In a painful voice he responded, "Each night before bed I practice my quick draw. I'm really quite good, but I didn't realize the gun was loaded." As he attempted to draw the revolver, the front site caught on the holster and his reflexes just continued the rest of the practiced movements, hammer back and then the trigger. Despite the man's obvious pain we had trouble keeping a straight face.

We lived just south of Sparta, so the city police were good friends and occasionally I'd swing a patrol up though their area. Sparta is known for a nationally famous rodeo, put on there once a year. The rodeo grounds were not far from our farm. I was in Cedar Springs picking up uniforms from a dry cleaning shop, when a call came over the radio requesting help for a personal injury at the rodeo grounds. I had my wallet out to pay for the clothes when I faintly heard the radio call from the squad car outside. Dropping everything I rushed out the door and reached through the passenger window to pick up the mike. This appeared to be dispatch's second or third call, and when I answered they directed me immediately to the rodeo grounds. A young lady had fallen off a horse, was in convulsions and had stopped breathing. I just pushed myself the rest of the way through the passenger side window, caught the siren and lights and was on my way.

There was a crowd around the girl on the ground. They parted for me to get through. Her tongue was between her teeth and the convulsions had caused the teeth to grind away on it. With my thumbs I managed to pry open her jaws, and to keep the teeth off the tongue I inserted my pocketknife so the teeth clamped on it's plastic sides. This wasn't an easy process. Some of the girl's boozed up friends didn't like or understand what I was doing to her, so they pulled me off. Time was critical if this girl was going to live. I pulled my gun and backed those boys off, then I was back down cupping off the girl's nose and mouth so I could blow some air into her lungs. Just before the ambulance arrived I had her breathing again.

After the ambulance left for the hospital, I started back to my car and realized that I had left my wallet in the dry cleaning shop. It was Saturday. The shop was now closed. Through the window I could see my wallet on the counter by the cash register. It laid where the cashier wouldn't have noticed it.

I called the lady and she agreed to come down and open up for me. Once inside I thankfully picked up the wallet and glanced inside. All the bills were missing, about 40 dollars. In those days 40 bucks was like a hundred today. The lady said that after I left she had just one customer and she would give me his name and address. The last name was the same as that of the girl I had saved at the rodeo grounds. I checked out the name and found that the man had an extensive record for petty larceny and some other offences, plus he was the uncle of the injured girl.

It was the first of the next week, and after about the third knock the guy opened the door. At first he denied even being in the dry cleaning shop, but then I reminded him that the cashier had his name on a ticket and remembered him. Well, maybe he had been in there, but he denied taking money out of any wallet. I looked him straight in the eye and said, "There is no way that I can prove that you took my money, but I want you to know that while you were doing that, I was breathing life into your niece at the Sparta rodeo grounds." The color went out of his face and he looked like he was about to pass out. I just turned around and left.

The girl was without oxygen long enough to incur some brain damage, but she did survive. I was awarded the department's Medal of Valor for my work.

In the early 70's Kent County hired two black deputies. One of them was to become my partner. His name was Joe Davis.

I did not grow up around black people. There weren't any black children in the orphanage, and it wasn't until the Marines that I had any association with them. It seemed like they had a difficult time passing the swimming test, but other than that, we were all just a bunch of young men trying to get through Marine basic training.

During my tour with the Marines I don't recall any racial problems or tension. I was aware that prejudice existed and knew what that was like. I had experienced it first hand growing up. On more than one occasion it was pointed out that I was half Indian, and somewhat inferior. I did have some very good friendships with fellow Marines that were black. Many I still have contact with today.

When the first black deputies were hired into the Kent County Sheriff's Department, I noticed there was plenty of deep-seated prejudice evident there. I didn't feel that way. I looked forward to having Joe Davis as a partner. There were a lot of things I didn't know about blacks, and I figured that maybe Joe could satisfy my curiosity. From the first time we climbed into a patrol car, Joe immediately put me at ease. He was more than willing to answer my questions, and had a few questions of his own about white people. We developed a solid friendship. There was nothing put-on or phony about it, we genuinely liked each other. We did have minor disagreements, but they never left the patrol car and they were never racial. Usually they were about a court decision that let some bad guy loose. There might be some heated discussions about whether we could have done our job better, to make the case more convincing, or whether it was just the court's fault. I think Joe was more mature than I was and had the potential to be a sheriff himself.

When we were off duty our families got together either at his house or mine. Our wives liked each other, our kids played together too.

While Joe and I were partners, Dee and I made some major plans to expand our stable. We built more barns and a large indoor arena. I bought the 60 acres across the road that Grandpa used to rent, and worked out a deal to harvest the hay growing on the Sparta rodeo grounds. We started boarding more horses, purchased several pieces of farm machinery and now had a drawer full of bills. Everything was purchased through some sort of payment plan.

I worked a lot of nights. When I came home in the morning, I did chores and usually slept until noon. In the afternoon I would be baling hay or doing some other farm work, then chores again before starting the next night shift. Joe understood that I was trying to build a business to supplement my deputy's salary, so that I might have a better life for my family. More than once he drove the patrol car while I rested my head against the window and caught some sleep. If we got a call he'd wake me up early enough that I was ready. Sometimes, if he had been working at a project or out partying, I'd drive for him, but most of the time I was the tired partner.

In the winter there was less farm work so I was rested and drove on patrol more. Some of the deputies were always getting their squad cars stuck or sliding off the road, but I prided myself on my winter driving skills. When there was a storm and things were quiet on the radio, I'd drive us out in the country where there were some deep snow drifts across the road. Joe would say, "Oh, so I suppose we're going out poofing snow drifts tonight?"

I'd reply, "Ya, I guess we are." With a good head of steam I'd bust through the drifts and for seconds everything was white. On the other side we'd be able to see again and have firm footing on the highway. Sometimes I'd have to dig the snow out of the grill so the radiator wouldn't overheat.

I was on patrol one night when the Lieutenant called and asked if I knew how to operate an M-79 grenade launcher. I told him I had trained with it in the Marines.

"Come on in and pick up the weapon, then report over to Lowell. Gumby's got a hostage situation going there." Gumby was the nickname for Bob Grummet, one of the deputies.

When I arrived on the scene, a Lowell police officer, Steve Pace, met me. We carried the M-79 and ammunition over to where Bob Grummet was and asked about the situation. There were two men and a woman in the trailer. One man was holding the other two hostage. It was an ex-boyfriend or lover; some type of love triangle, and Bob wanted some tear gas in the trailer to get them out.

I loaded one tear gas round, took a spare and Steve carried the other ammunition as we crawled up to the blind side of the trailer. Moving quickly, I put the barrel right in the window and pulled the trigger. Shortly after the round went off, one man and a woman burst out the front door. Another round fired in still didn't bring the bad guy out, so I asked Steve for a third. It was dark and Steve didn't read the label before he loaded it in the M-79. When it went off in the trailer, it was obvious that we had loaded an incendiary round by mistake, and now the trailer was on fire. I said to Bob, "Now we've got to go in and get that guy or we're going to have a dead body on our hands."

I had forgotten to bring along any gas masks. Entering the trailer we were going to be exposed to the gas and the possibility of getting shot. I explained to Bob that from my Marine training I learned that you could breathe the gas if you just didn't fight it. We should stay low and keep our lungs as relaxed as possible.

I kicked in the back door of the trailer and Bob followed me in. The guy wasn't visible; however there was linen stuffed around another room door, which was probably barricaded. We forced that door open and could see the guy down on the floor by the bed. He was trying to stay away from the fumes. We both jumped on him and had one hand cuffed when he broke the arm free. He was lashing out with that hand, swinging the loose hand-cuff when Bob slugged him. The guy's head hit against the floor and he was unconscious. We locked the other hand behind him.

By the time we dragged the man out of the bedroom the tear gas was really getting to us. Reaching the front door, we literally threw the guy out to the snowy ground below. The news media was outside with the cameras rolling. The trailer burned down, the bad guy went to jail and on the 6 p.m. news, the cops were again portrayed as using excessive force.

It was in the late summer of 1974, when my stepfather was diagnosed with small cell carcinoma; terminal cancer. Since I joined the sheriff's department, we were much closer than we had been while I was growing up. He adopted me, but I was angry at being left in the orphanage. And, as a teenager I was out of control. I slept at home and ate at our table when I felt like it. Dad tried to talk to me, but I wouldn't listen. My Grandfather was the father role model.

While working for Ron's Wrecker, Dad was popular with the deputies. The police knew all the wrecker drivers on a first name basis, because they worked together so much on accidents and impounding cars. Dad was a favorite with them though, especially since he helped out one of the deputies, Ed Baker.

Ed was trying to arrest a drunk driver after an accident. There was a struggle and the guy got Ed's gun. Ed had hold of the guy's hand that held the gun, but the guy kept trying to point it at Ed, occasionally pulling the trigger, firing off a round. Dad came upon the scene, saw Ed and the guy rolling on the ground with the gun. Dad took a tire iron out of his truck and hit the guy over the head, knocking him unconscious. In his drunken state of mind, if the guy had ever gotten his hand loose he would have probably shot Ed. Dad was considered a hero by the deputies. They figured he saved Ed from being shot.

I knew that the deer hunt of 1974 would be our last together, and I wanted somehow for it to be special. Dad always hunted with an old single shot 12 gauge shotgun. I thought it would certainly be nice if he had a new gun, but I just didn't want to give one to him out of sympathy.

One day we were riding along in my truck and he was lighting up another one of those cigarettes that was killing him.

"Dad, why don't you give up smoking?"

"I guess I just never had a good reason," he replied.

"Would a new deer rifle be reason enough?" I said.

"Yep," and with that he tossed the cigarette, lighter and almost a full pack out the truck window.

The cancer had weakened Dad, but we did get our last hunt together up at Luther. He had the brand new Marlin lever action 30/30 I bought him. I was sure he might get his first deer, but he didn't.

We were positioned fairly close together when this nice buck went by. I was hoping Dad would spot him. When I saw him raise the rifle, I knew that he would probably soon be taking a shot. Dad hadn't hunted with a lever action before. I'm sure he was overly concentrating on how he would rack the second shell in, because instead of pulling the trigger and shooting, he just stood there moving the lever back and forth until the shells were all racked out. Before the buck got out of sight, I knocked him down. When the deer was dressed out and hanging up I took Dad's picture with him. We had a good laugh over it, I don't think he cared whether he shot him or not.

It wasn't too long after deer season that Dad's condition deteriorated. He was admitted to the hospital for a long and agonizing stay, while the cancer claimed him. The Kent County Sheriff's Department sent an honor guard and several patrol cars to his funeral.

One night that same fall of 1974, my partner, Joe Davis and I had just started our patrol. Joe seemed serious and quieter than he usually was, but I didn't think much about it. Finally he said, "Bob I've got a terrific backache. Why don't you drop me off at my house."

My response was, "Right, ha, ha. We just get started and you got a backache and want to go home."

"No, I'm not kidding. It hurts so bad I can hardly stand it. Please run me home and I'll call in to the desk sergeant and have him put me on sick leave."

I dropped Joe off at home and told him I hoped he felt better tomorrow. Joe wasn't one to complain. As I drove off I wondered what he had done to screw up his back.

Joe didn't show up at work again. I checked to see what he'd found out was causing his back problem, and was devastated by the answer. Joe had gone in for tests which confirmed he had cancer of the spine. He was soon confined to a hospital room real close to my father's.

The pain that Joe suffered was terrible. They gave him all the morphine they dared to, and it still didn't deaden it. Joe said, "Bob, if we ever were friends you will bring my service revolver in here so I can end this suffering."

"Joe, it's because we're friends that I can't do that. You have your family to consider for one thing. If you commit suicide there will be no insurance money to take care of them. If I bring you your gun, they will know how you got it and my career will be ruined."

Like my father, Joe died a slow agonizing death. In fact he passed away a week after my father. Dad was 53, Joe about 26. I not only lost a father and a partner, I had lost two real good friends.

It used to be that when there was an accident with extensive injury, the accepted method of saving life was to rush the victim in an ambulance to the nearest hospital emergency room. Other than limiting excessive blood loss and providing oxygen, there was little done to cope with the injury on the scene. Kent County decided to formulate an emergency response unit (E-Unit) within the sheriff's department. There was a call for three police officers to volunteer for paramedic training. Myself, Larry French, and Ed Wierda volunteered.

At this time, there was no formal state or college program to become a paramedic. In the police academy we were only taught first aid. Now, volunteer doctors gave us extensive training on procedures far beyond basic first aid. Under their direct supervision we were actually delivering babies, suturing up drunks that came into the emergency room after fights, starting IV's, even running EKGs. The way our training was conducted then would, in today's world, create lots of lawsuits for a hospital.

The three of us were the first police officers in the state to receive this training and become certified paramedics. I had the individual distinction of being the first police officer in Michigan to be issued and use a defibrillator and to start an intravenous injection at the scene of an emergency.

If there was an accident with serious injury or any other medical emergency, the E-Unit was dispatched to administer lifesaving procedures before the injured person was transported to the hospital. The intent of all this was to get more people to the hospital alive, and in a more stable condition for the emergency room crew to continue treatment. I thoroughly enjoyed this work, plus, I was good at it.

As you can imagine, there were plenty of gruesome and heart wrenching scenes to cope with. Of the automobile, train and bus wrecks, many produced amputated arms, legs or crushed bodies. In an accident involving a pregnant woman, in one instance the baby was ripped from the mother's womb.

Once, I arrived to find a young mother sitting alongside the road, cradling her young boy who had been run over by a truck. The boy was over picking and eating wild strawberries before he crossed back across the road into the path of the truck. His airway was plugged with strawberries. I had to suck these out before I could get air into his little lungs. I felt so sorry for the mother, and of course for the boy. The smell of strawberries still bring back that memory. I have little taste for them today.

The worst accident I remember involved a truck driver that fell asleep at the wheel, crossed the median and rammed into another semi going in the opposite direction. The truck was loaded with cases of canned food which was scattered all over the accident scene. It happened on US-131 by 32nd Street and motorists were stopping, doing their best to scavenge the freight from around the food truck. Both drivers were dead. One body was removed, the other wedged in the tractor that rested on its side. The impact had been so great that the driver's head had been torn off. It lay wedged between the other truck's front tire and the pavement. The cab door was jammed shut, but peering through the truck window I could see the rest of the body in the debris.

No one was volunteering, so I took off my gun belt, locked it in the patrol car and lowered myself through the broken passenger-side window. The fire department stood by with body bags, the medical examiner and mortician were there too. I could see that one leg was separated from the rest of the body.

I handed this out first. The rest of the body, the trunk, with the buttocks and other leg, was hard to maneuver, so another officer, Sergeant Eli Roberts, stood on the side to help. I grabbed the body by the belt to roll it over so I could push it through the hole. Eli also grabbed the belt, lifting as I pushed it up over my head. As what remained of the body started through the window the intestines and body fluids came out of the hole created from the missing limb. This all fell on my head with fluids all over my face, hair, eyes, even in the corners of my mouth. By this time, the fire fighters that were watching started throwing up. Eli got the body in a bag, plus the rest of the organs and parts that I handed him.

When I was outside of the truck the fire fighters held up blankets while I removed my clothes. They provided a fine water spray for me to wash the fluids off. I tried to rinse out my clothes as best I could, because I had nothing else to wear.

By the time I finished my night shift report, it was almost noon. Arriving home I was exhausted, smelling terrible. The uniform I dropped in the garbage can, then sprayed myself down with a hose as best I could. Dee put lots of baking soda in my bath water to kill the smell, but it still took plenty of scrubbing before it wasn't noticeable anymore.

There was another non-life threatening injury, and I was trying to break a young deputy in on some first aid procedures. The accident victim had a pretty bad leg laceration, with extensive bleeding. The young deputy stood over me as I examined the wound, "What do you want me to do?"

I put a compress on the wound and said, "Here, hold this firmly in place to keep the blood from gushing out." He took my place, but in a couple minutes laid passed out next to the victim. I broke an ammonia capsule under his nose and had him back holding the compress until he passed out the second time. I covered the wound with a blanket so he wouldn't see the blood, popped another ammonia capsule and this time he held the compress just fine. He drove the E-Unit for several years after that, with no problem. We did laugh about that incident for several years. With 30 years in, he is about to retire, so I'll spare him the notoriety and not mention his name.

One of the volunteer doctors that taught our paramedic training was the infamous Dr. Raymond Lang. He was, and maybe still is, the Kent County Medical Examiner. Dr. Lang additionally had all the powers of a deputy, including that of arrest authority. I remember that he used to drive a front-wheel drive Toronado, which was packed with medical gear, a couple guns and riot equipment. If he was free, it wasn't unusual to see Ray Lang responding to about any kind of police call. When a high speed chase ended up out in some field or down a two-track dirt road, Ray's Toronado, with the little portable flashing light up on top, was in hot pursuit. The Toronado received lots of damage during these chases, but Ray covered all these repairs out of his own pocket.

Dr. Ray Lang was also a very brave man. Once, an individual held another policeman, threatening to kill him with a shotgun. When Ray came on the scene, he was dressed in a suit and tie. To the man holding the shotgun, he said, "I'm Doctor Lang." Then, he simply moved between the officer and the shotgun saying, "If you shoot this officer, you'll have to shoot me first." There was a moment of hesitation and during that moment, Ray grabbed the shotgun. For that, Dr. Lang was recognized with a commendation for bravery. I guess, if the job would have paid enough, Ray would have been a full-time police officer, rather than a doctor.

Even though I was now a trained paramedic, I still pulled the routine patrol duties, arresting drunks, settling family disputes, writing tickets. What might seem routine at first though, could suddenly be threatening to a deputy's life. Though it's not a textbook answer, nor one the psychologists want to hear from a prospective police officer, it's probably the threat of the unknown that is the biggest motivator for people that want to be police officers. No one talks about it, but the threat of physical danger creates a rush of excitement; not knowing whether the speeding driver will poke a license or a gun barrel through the window, what will be behind the door you smash down or will the guy you're struggling to arrest have a knife to use on you.

I was on patrol with my E-Unit one night when the dispatcher gave me a Plainfield address to check out. He didn't tell me much about the call, but it seemed routine enough. It ended up being my closest brush with death while I was a Kent County deputy.

It started out when the dispatcher received this call from a person complaining about the loud noise or party going on in the neighbor's house. The dispatcher found the neighbor's telephone number, called the guy up, told him to cease and desist with the noise. The dispatcher exchanged words with the noisemaker, who had been drinking. Eventually they were both very angry. The guy said, "If you're brave enough, you'll come over and make me turn my stereo down!"

The dispatcher says, "I'm on my way over right now, to kick your ass!" The young dispatcher didn't realize that the whole conversation was being recorded and would later prove rather embarrassing to him.

Well, the dispatcher just told me to check out the address. After I knocked on the door it opened with the end of a 12 gauge shotgun looking right at me. The guy pulled the trigger, the hammer snapped down, it didn't fire. He slammed the door in my face and I figured it was time to call for backup.

When help arrived, we filled the guy's place with tear gas, then stormed in to arrest him. I found the shotgun, with the shell still in the barrel and a small dent in the primer.

My first impulse was to whip that dispatcher's ass, but with a certain level of maturity I instead explained to him that his actions could have gotten me or the other guy needlessly shot. Lenny Hamp was sorry for what he did and turned out to become an excellent dispatcher. I'm not sure whether he's retired now or not, but we became good friends and in time could laugh about what happened.

When I met with the guy that poked the shotgun in my face, with tears in his eyes he convinced me that he really feared for his life that night. He thought that he was going to get beat up by somebody, not realizing that he had talked to a police dispatcher. I asked the judge to dismiss the charges, which he did.

As a police officer, my experiences nurtured a new level of maturity. When I went to work as a jail guard, I figured that all prisoners were lowlifes, they deserved to be there. On patrol I had good control of my temper, but if I was challenged by a bad guy who didn't want to go to jail, I had no bones about rendering him semi-conscious for his trip to the cross-bar hotel. Now, I had more compassion for those that were arrested. They weren't all as bad as I first perceived them to be.

I felt that I was really close to the people I served too. It was not unusual to have a judge or dignitary riding along on patrol with me; I felt comfortable with that. In the poor neighborhoods, I was equally in touch with those folks.

We were always having somebody run into a deer and I made sure the meat didn't go to waste. Of course the driver of the damaged vehicle was first offered the dead deer, but they usually didn't want it. We were supposed to bring it in for the inmates, so they could have some meat. I didn't do that. From a list of poor families I'd choose a name and drive to their house with the deer on the trunk, tied from the back door handles. They never did turn down my offer. I'd simply tell dispatch that the driver wanted the deer. A pail of water washed the blood off the trunk and I drove back on patrol knowing a family would have some fresh meat.

With my Marine diving experience, I also served as co-captain of Kent County's underwater recovery team. There were many traumatic experiences, just like the blood and guts of the E-Unit duty. Fishing out the bodies of little kids; the loss of life seemed so needless. What shook me up as much was when I had to bring up a fellow police officer from the bottom of a gravel pit. He had been skinny dipping with his girl friend in about 100 feet of water.

Sometimes when you see so much, the E-Unit officers and medical examiners can develop kind of a morbid sense of humor. That sense of humor got me into real trouble once; it was in the recovery of another drowning victim.

We had been called to Grandville to assist in the recovery of a body. For two days we searched the river bottom with no luck. On the third day I found the body. It wasn't in very good condition, for it was bloated up and looked like turtles had been gnawing on it. At the surface it had a pretty raunchy smell too. According to Michigan Law, we couldn't take the body out of the water until the medical examiner offically pronounced it dead, so I stayed with it until he arrived. After all that time looking for the victim I was exuberant that I had found him. Holding up the arm that still had the wrist watch on it, I did my best rendition of John Cameron Swazee's Timex commercial, "It takes a licking, but keeps on ticking." Little did I know, the cameras were rolling. It didn't play too well on the evening news. During my ass chewing, I was threatened with two weeks off without pay. Guess what I said was cold hearted, though it seemed funny at the time.

Our farm was now called Sparta Stables. Things were really going well. We had expanded, buying an additional 40 acres, building more barns and constructing 54 stalls for boarding horses. There was a well-lighted indoor riding arena with bleachers. And, a club house with a pool table, table tennis and pinball machines. Now we had a western store, selling saddles, harnesses, and even horse trailers. The tax assessor was impressed. He said I was the only guy he knew with 40 acres under a roof.

The good thing about it was that our drawer full of payment books was about empty. We were boarding, training and raising our own horses, plus, we were growing our own hay for feed, which saved a lot of money. We were accumulating plenty; however work dominated our lives. I hired two full-time workers for the stable. Their combined wages were more than I was making with the sheriff's department E-Unit. I could now make a full-time wage from the stables without the county job, but I did love the E-Unit work and really didn't want to leave.

What changed my thinking was the new policies and procedures instituted by the Captain in charge of the E-Unit. He felt that it would be a good public relations thing to have registered nurses (RNs) ride as partners with the E-Unit officers. I'm sure he had practical reasons too, but I was totally against this. Nurses were probably more knowledgeable than we were on emergency medical care. They did not however, have any police training, would not be prepared for the violence associated with this work, and would jeopardize my safety because I would be trying to keep them out of harm's way.

An even bigger reason why I disagreed with the RN's as partners, was that I knew it would put great strain on the marriages of these officers. The RN's for the most part were single, pretty and would be spending eight hours with these guys each day or night. The officers would be spending more time with these women than they would their wives. They also would have plenty of work-related things in common with these women, while at home they would hear about sick kids and things that needed repair. This was dangerous territory, first an affair then a divorce. I couldn't imagine getting a divorce from Dee. I couldn't imagine putting my kids through what I had been through. The yelling, the screaming, the feeling inside that you would be responsible for the divorce.

It wasn't long before what I predicted happened. One of the E-Unit officers dumped his family for his partner. He announced they were in love and engaged to be married. I would do anything to avoid that. It's great to be strong willed in the face of temptation, but it's smarter to eliminate the temptation. Like my Granddad used to say, "Locks on the door keep the honest people out."

I had been staying around because I liked driving old E-63, and I loved my work. We had no union to fight these new policies, nor would it do any good for me to argue with the Captain. I did have an alternative. Sparta Stables was doing well financially; there was a good income for me there.

In 1976 I left my full-time position with Kent County. They allowed me to retain a commission as a Captain, and work with the Search and Rescue Mounted Division. It wasn't long after I left that the RN partner program went by the wayside, too many divorces. The Captain that arranged the program was demoted to Sergeant and spent the rest of his career as a second shift supervisor.

As I put additional time into the stable business, it prospered even more. We had an excellent reputation. Disney's Wonderful World of Horses made our stable a regular stopover with the Royal Lippizan Stallions. Every year that the Lippizans elegantly performed in the Grand Rapids Civic Auditorium, they contracted with us to use our barns and train in our indoor arena. They were beautiful, and we had numerous private showings. I enjoyed having my kids grow up around horses. We sponsored trail rides and took them with us when we trailered the horses up to a federal forest area for a camp out.

In addition to the stable, I formed a small excavating business. I had seven dump trucks, two loaders and a 450 Case dozer. This business was successful too, but now I had to run it, plus the stable, and make sure the inventory was up in the western store.

The same year that I quit being a full-time deputy, we had an opportunity to haul a horse to Florida for the entertainer Wayne Newton. We had been working so much that Dee and I thought this would be a great opportunity for a little vacation too. I found someone to do chores and Grandpa said he'd look after things. He was 83 then, still doing what he could to help. Usually that amounted to dragging a hose around to water the horses. The kids went to the grandparents and with another couple we were on our way to Florida.

With the horse unloaded we got some tickets for Disney World. I never was one for calling home much, even in the military. We weren't going to be gone long, I just figured everything would be fine.

After a few days we hooked the empty horse trailer back up and were on our way home. When we drove in, Mother was there with the kids to meet us. Something was wrong, I could tell. Mother was a basket case.

While we were in Florida, Grandpa had come down with congestive heart failure. He was admitted to the hospital in critical condition. Mother said he kept crying out for me, he wanted me by his bedside. No one could find us. They had the Florida Highway Patrol and local police try to find us, but we couldn't be located.

Grandpa died a week before we got home. Mother would make no funeral arrangements until I was there. The funeral director asked her, "Weren't you married to his son?"

"Yes," she said, "But his next of kin is his grandson." The funeral director had to put him in the freezer until we returned.

The grief didn't set in until the day we buried Grandpa. Afterwards, we walked back to the car to leave the cemetery, and as I started to open the car door I completely fell apart. All the things he had done for me, the times he had always been there for me, when he needed me, was asking for me, I wasn't there for him. The guilt was enormous. It didn't make me feel any better, but I remembered one of his sayings, "You come into this world alone and that's the way you go out." There's nothing much I could have done except be there and hold his hand while he was dying. Guess I still carry some guilt.

Grandpa only got angry at me once or twice. He never laid a hand on me, but I knew when I displeased him. With so much respect for him, just being in his disfavor was punishment enough. I do wish I had paid more attention to his stories and philosophy. Perhaps, if they would have echoed in my ears just one more time, I would have been able to avoid some of the future tragedy and sorrow in my life.

The year after Grandpa died, I rejoined my old Marine Company in Grand Rapids. Even though it was only one weekend a month, plus two weeks in the summer, it gave me a chance to reconnect with the Marines. It was a chance to regain some of the comradery I missed.

From 1976 until 1979 we worked very hard at all our businesses; the bills were finally all paid. We also accumulated a lot of adult toys, a mobile home, snowmobiles, motorcycles and nice cars. These were all things that we felt deprived of in earlier years. There was even money in the bank, but somehow all this wasn't enough. Our lives were so full of work, so busy accumulating things, that we hadn't really taken time to enjoy it. Dee and I started to take stock of our relationship, the time we spent together, the time we spent with the kids.

While I was working at the sheriff's department, running the stable and starting the excavating business, Dee was working full-time at the western store, raising the kids and maintaining a household. When I was working night shift in the summer, doing farm work during the day, we hardly ever got to sleep together at night. My kids were essentially growing up without a father. I was physically there, but I really wasn't there in any other respect. When I quit the Kent County job, things improved considerably, but I joined the Marine Reserve and just managed to pack a little more work in my days.

Running the stables just didn't give me the personal satisfaction that police work did. I really missed that part of my life. What little I did on a part-time basis with the Kent County mounted division didn't fill that void. Another law enforcement opportunity had presented itself in Lake County, so Dee and I decided to try for that, change our life style and sell the stables.

Lake County had a real problem trying to keep a sheriff in office. When I went up there hunting I always took a little time to visit with the sheriff. Bob Radden had been the first, but was killed in a car crash. Others weren't in office for a year and they were gone. I couldn't get a clear picture of why they were leaving until I talked to Sheriff Guy Lee. He said there was so much corruption and organized crime that anyone trying to change it would be killed. Other sheriffs had been so fearful of the death threats that they just resigned. Apparently Bob Radden also received threats before his fatal car crash. Sheriff Lee said he didn't want any part of it. He was leaving too.

It was hard to believe that Lake County could possibly have an organized crime problem. Well over half of the county was state or federal forest. There were 156 lakes, 46 trout streams, 1000 miles of snowmobile trails and probably not more than 8000 residents. Certainly it didn't appear to be much of a haven for crime. There was no doubt that the deer hunters and snowmobilers liked to cut loose, raise some hell when they were up there. Prostitution was pretty open too, but that didn't seem too sinister.

I'd never met Lonnie Deur, the sheriff they were trying to recall. I had been following things in the news though. Sheriff Deur just didn't seem like much of a sheriff if his own deputies were organizing a petition to get a recall ballot organized. The news media was not supporting him either.

Apparently Sheriff Deur had been shot for some reason. He claimed it was because he knew too much, that it had been an assassination attempt. Deur had recovered from his physical wounds; mentally he must have been paranoid or something, because he allegedly went into the jail and beat a prisoner with his flashlight.

Lake County Sergeant Bill Moore was on the evening news. He sat kinda tipped back in a chair as he answered the interviewer's questions. No, he really didn't want to be sheriff; he felt forced to run though, to fill the void created when the voters recalled Sheriff Deur.

Here was the opportunity I had been waiting for, a chance to get back into law enforcement. All Lake County seemed to need was a top notch professional to be sheriff; someone that could be fair and impartial. I felt I had all those qualities. Here was a job where I could finally settle down and pursue the quality of life. My kids were in their teenage years, needing a full-time father. Dee and I needed to rekindle our marriage, plus Lake County could offer an environment for me to unwind, for I loved hunting and the out-of-doors.

The Lake County voters did indeed recall Lonnie Deur. It was the first time a Michigan sheriff had ever been recalled. Now that the voters had removed their sheriff, they were required by law to hold a special election in the fall to fill the vacancy. I decided to run for sheriff on the Republican ticket.

Immediately I changed my residency to Lake County by purchasing 40 acres and settling my mobile home on it. I would now have the required six months of residency to be qualified to run, so I filed my petition to enter the primary.

By spring we had the stables up for sale, my campaign began in Lake County. I started knocking on doors, attending township hall meetings, explaining my views and qualifications. Lake County was one of the poorest counties in Michigan; people needed to know that I could relate to being poor; I had lived in a tent and started kindergarten here. I was not really an outsider for I had been coming up here in the summers and in the fall to hunt for years. My father had owned property outside of Luther.

Things went well in the primary. I was the Republican candidate against Sergeant Bill Moore, the Democrat. After that, things got rough. It became obvious that there were those that didn't want me to become sheriff and did their best to discourage my candidacy. Twice my tires were slashed; even my windshield was smashed out.

The local paper printed everything that Bill Moore said as gospel. He questioned my qualifications and shed doubt as to whether or not I was ever a paramedic, even how much actual police experience I had.

The newspaper was the Lake County *Star*. Several people assured me not to worry about what the *Star* printed, that its best use was for wrapping up fish or lining your garbage pail. I didn't respond to Bill Moore's accusations, but I did keep a list of them for future use. I figured he was digging his own hole. Also I reserved and paid for a two page center section of the Lake County *Star* for the week before the October 3rd election.

I pressed on with my campaign and a week before the election I filled my two page newspaper ad with copies of all my certificates, my Marine discharge, letters of commendation, all the things that Bill Moore said I didn't have. An artist drew a cartoon of a deputy sheriff with an axe and a hood over his head. The caption in old english letters was, "Let the axe man chop away at this." Bill Moore was caught in all his lies. I was elected sheriff by a two to one margin.

Top: 1970 Pistol Champs — from left, Art Tannis, Jim Porter, Larry Flynn, Chuck McArthur, Bob, George Peck

Bottom: 1976 Mounted Division State Champs — from left standing: Charles Crowe, Harold Christrup, Earnie Graham; kneeling: Bob, Del Walker

Kent County Sheriff's Underwater Search and Rescue Team — from left: Bob, Sheriff Parsons and Ed Baker

Top: Bob trimming Guy's feet

Bottom: Lippizan training at Sparta Stables

Son Buck and daughter Karen at home in Sparta

Sheriff Lonnie Deur

As Lonnie Deur put it, Lake County is just a real bad memory for him. Like a soldier back from combat, it is Lonnie's desire to keep the bad memories buried. He has a deep fear that somehow the past will rear its ugly head and destroy the life he has worked so hard to create for himself and his family.

I spent an evening visiting with Lonnie, and think I now have a better appreciation for what he went through. Honoring his wishes however, most of our conversation is not part of this story. There is one exception, and quite frankly only he could tell that as it happened.

It is necessary to review what happened to Lonnie Deur, because it will provide a better understanding of what the situation was when Bob Blevins became sheriff and the people he would encounter. More important though, it offers some vindication for Lonnie. He was an honorable, courageous and dedicated lawman who should be remembered for more than just being Michigan's first recalled sheriff.

During Lonnie Deur's term as sheriff the weekly newspaper, the *Star*, gave a good account of the events that led to his downfall and why Lake County is such a bad memory for him. I have used that news coverage to reconstruct those years.

By the time that Bob Blevins had announced his candidacy for Lake County Sheriff, Lonnie had completely divorced himself from the place. In addition to carrying bitterness over being recalled as sheriff, he was also physically and mentally stressed out. Lonnie was a fighter, and had done his best to turn law enforcement around in Lake County, but he just hadn't made enough of an effort to campaign against the recall vote. Instead, he believed that people would remember and support him for what he had accomplished.

The undersheriff, Bill Moore, led the recall effort and I'm sure J.T. Nelson and many of the commissioners did their utmost to make sure Lonnie would join the ranks of former Lake County sheriffs. The news media seemed more than happy to gobble up the untruths they were fed, so if a Lake County resident based his vote on what he read or watched on TV, you couldn't really blame them for what happened.

In his fight against crime in Lake County, Lonnie had butted heads with the board of commissioners, the prosecutor and the county judges. I'm sure people wondered, "Could the whole legal and governmental system be screwed up, or is the sheriff the problem?"

When Bob Blevins was elected sheriff, Lonnie knew that there was a lot of valuable information he could offer, but he really didn't know whether Blevins would even want to talk to him. Lonnie's reputation was so scarred that it might be a big political liability for Blevins to be associated with him. Lonnie decided to stay away, so unfortunately Bob had to start from square one.

Before Lonnie was involved in Lake County law enforcement, he ran an ambulance service in Baldwin, Marion and Reed City. The business was quite successful and he had about 38 employees. Unfortunately he went through a divorce and the business was split with his ex-wife. She continued to run the service in a couple cities and he took the rest.

Law enforcement was something that had always interested Lonnie, so eventually he decided to sell the ambulance service and go to work for Bob Johnson, the sheriff. He started as a dispatcher and corrections officer. For a year he worked in that capacity and then in 1973, Bob sent him to recruit school for police training. When Lonnie returned, he was made a full-fledged deputy and pulled regular road patrol. This was a real small sheriff's department, so Lonnie was fortunate enough to get a wide variety of law enforcement experience in a rather short period of time.

Lonnie was not a stranger in Lake County. Through the ambulance business he had made a lot of friends and also some enemies. His determination to run a first-rate service with safe equipment had caused a few rifts between himself and the county commissioners. It was usually a battle to get the funds he needed to maintain the service.

J.T. Nelson was a commissioner from Webber Township who was on the opposite side of more than one issue with him.

After a year or so Sheriff Johnson moved his undersheriff, Ron Voss, into a detective position and appointed Lonnie his undersheriff. This caused some hard feelings within the department, for there were several deputies more experienced and who had been in the department longer than Lonnie had. The sheriff felt he needed more administrative experience in his undersheriff and decided to shuffle the cards. What Lonnie lacked in law enforcement expertise he more than made up with his proven administrative ability gained from running the ambulance service.

In late 1974 a major political rift developed between the sheriff and the commissioners. Shortly after that Sheriff Johnson took some time off and went to Florida. Lonnie Deur was left to run the department.

When the sheriff returned from Florida he had decided to leave Lake County and planned to take a position with the St. Petersburg Police Department when his term ended here. He liked Florida and had been conducting a little job search over the winter. Both Ron Voss and Lonnie decided to enter the race for sheriff, however Bob Johnson gave his support to Lonnie. Voss was defeated in the primary and Lonnie won again in the general election.

Ron Voss continued to work for the sheriff's department for awhile, then quit and took a job in Newaygo. Eventually he was elected sheriff there. He was a fine individual and an excellent investigator, but things were just not the same after the primary election.

When Lonnie started as sheriff things seemed to run real smooth with the board of commissioners. They provided funds to improve the physical appearance of the sheriff's department and to buy new uniforms and acquire more and newer patrol cars. Lonnie acquired a grant for some new police radios and even managed to come up with some money for additional training.

With the facilities improved and the personnel looking better, Lonnie now secured contracts to house state prisoners. This brought a lot of additional money into the county coffers. Before the board of commissioners, he explained that this additional money should benefit all county officers, not just the sheriff's department. The only thing he needed was a couple of extra correctional officers to make sure the jail was properly staffed. He had already secured funds from a special grant to facilitate regular road patrol.

Breaking and enterings (B&Es) in Lake County really kept the sheriff's department occupied leaving little time to pursue investigations in other areas. When Bob Johnson was sheriff there were some investigations being conducted that pointed to misuse of county funds within the courthouse and also established a connection between illegal activities and J.T. Nelson.

Lonnie's big fallout with J.T. Nelson and the Board of Commissioners seemed to start in the spring of 1977. One of Lonnie's young deputies pulled over J.T. Nelson for what he believed was a clear case of drunk driving. The officer didn't know anything about J.T.'s importance in the community and treated him like any other motorist under the influence. When he asked J.T. to get out of the car, J.T. said, "No!" and tried to run him over. Lonnie and another deputy responded to the officer's call for backup and arrested J.T. for felonious assault with a motor vehicle and operating under the influence of alcohol.

J.T. was more than unhappy about being in jail. He used all his influence in the community to put pressure on Sheriff Deur to release him.

Despite the pressure, Sheriff Deur would not give in and J.T. was not about to sit in jail any longer. He developed heart attack symptoms and was taken up to the Reed City Hospital. Eventually he sued the sheriff and county for monetary compensation for the stress brought on him and the cost of his medical bills. The insurance company decided it would be better to settle the claim than go to court.

The local judge and prosecutor dismissed themselves from J.T.'s case and a special prosecutor was brought in. Nothing ever came of the charges against J.T. That certainly was no surprise, but at least a record had been established on him.

When Lonnie took office as sheriff, Lake County was the third highest crime county per capita in Michigan. Before he left office he would have the county down around eleventh place. Also, at an American Medical Association convention in Milwaukee, Lake

County was the first department in the nation to receive a jail medical accreditation from that organization.

Contracts to house inmates from out of the county were bringing in close to $150,000, but since the battle with J.T. was out in the open, the sheriff's department now would be in a constant fight with the county board of commissioners to keep its budget from being slashed drastically. The board was more than willing to spend all the money brought in from housing state prisoners, but supposedly now there weren't funds to continue the sheriff's budget as it was.

The first thing the board tried to take away was the detective position that was funded through a grant. In an interview with the Lake County *Star*, one of the commissioners, Harrison Wilson, stated that there really wasn't any need for a detective anyway. For awhile they got away with the elimination of that position, but later they were forced to reinstate it because the money for the detective was provided from outside of the county.

Next, the effort was made to slash the secondary road patrol. This would mean cutting the sheriff's department's personnel by about 33%. J.T. was using all his power to bring the sheriff's department to its knees. It served the dual purpose of getting even for throwing him in jail and taking the pressure off his extracurricular money-making activities.

Commissioner Harrison Wilson now took legal action. He filed suit to block any grant money from being used by the sheriff until the sheriff's department hired some blacks. That didn't work because Lonnie had already been working with the local NAACP to help recruit qualified blacks.

After all the hard work and success in improving the department and reducing crime, Lonnie wasn't about to see it all hacked up without a fight. He started assembling some resistance within the community. From the Lions Club, Jaycees, Lakefront, and Riverfront associations, he gathered support and asked them to come to the commission's budget hearing and each provide one person to represent that group in speaking out against the proposed cuts in the budget. Even the local NAACP chapter lent its support.

Normally four or five people would attend the budget meeting, so when a couple hundred people marched into the commissioners' meeting, they were caught completely by surprise.

J.T. Nelson was an immaculate dresser, but on this particular evening he had not planned on any spectators attending, so he had worn his bib overalls. J.T. and the board were completely humbled as one speaker after another addressed the board asking that funding for the sheriff's department not be cut. Even residents not enlisted showed up and voiced their concerns. The sheriff was present, but said nothing.

The board was so frustrated that they willingly agreed to sit down with the sheriff and totally reorganize the budget. Some little old lady who was not a spokesman for any group raised her hand. "Excuse me Mr. Nelson. Do you mean to tell me you are going to let the sheriff organize the budget for the clerk and prosecutor's office too? That's not what we're asking for."

J.T. answered, "No, no, we will talk to each department head."

Well Lonnie got his road patrol back, an additional corrections officer and an offer for a secretary. He said that he didn't need a secretary.

"I'm not here to get things I don't need, only things I need."

That evening Lonnie won the battle, but now he was really J.T.'s target, and J.T. was very powerful. He had been on the county's board either as a supervisor or commissioner for a long time. He had been president of Michigan's Association of Counties, on a national steering committee and a member of numerous state and local organizations. J.T. had lots of political friends in Washington, Lansing and within the Lake County government and judicial system.

It was October, and a lady by the name of Ruby Nelson seemed to be developing some serious mental problems. It started at the school where she strangled a teacher to the point of unconsciousness. Next she threatened and hit the postmaster. She was escorted up to the Traverse City State Mental Hospital a couple of times, and after a short period, each time she was returned home with their assurance that she was of a normal mental state.

Towards the last of October, Ruby Nelson went to a local home and reportedly beat the resident's dog, then yelled at the resident, "I'll kill you, I'll kill you." She laid a canvas on that person's porch, poured gasoline on it and lit it on fire. That supposedly was what precipitated the order for Ruby to again be committed to the Traverse City mental hospital.

Rick Cooper was the probate judge, but he was out of town. Judge Wickens was filling in for Cooper and on a Monday the 1st of November he called and informed the sheriff that he had issued a court order to serve on Ruby Nelson to again commit her to Traverse City. Wickens was insistent that the order be served immediately. This didn't sound like a good idea to Lonnie and he tried his best to explain to the Judge why it was not a good time to serve the order. The children were home from school. With the seasonal change of the clock it would also be dark when they arrived and that wouldn't be good. It would be much better to put her house under surveillance for the night and then in the morning, after the kids go to school, move in and pick her up.

Judge Wickens wanted Ruby picked up immediately whether Lonnie thought it was a good idea or not. When Lonnie again balked, Wickens said, "If you don't serve that order I'll hold you in contempt and throw you in your own jail."

****

The following statement is Lonnie Deur's account of what transpired on November 1, 1977 when he attempted to carry out Judge Wicken's order and pick up Ruby Nelson:

I didn't like what I was being forced to do. It would be dark soon and if I had to do this I would have to plan it quickly. It would be better to have all the help I needed to make this pickup, rather than going in and taking a chance on something going wrong. As Ruby was black, it would be important to have some black representation there, so I asked the jail matron Gwen Warren to accompany us. Roger Dean, a representative of the Lake County Mental Health Clinic, would also go. He and Warren would ride in one car. Deputy Hemmerich and another officer would take a patrol car. Myself and undersheriff Piper would be in the unmarked car. As I left my office I felt a little naked. My bulletproof vest hung in the kitchen, still wet. I had washed it that afternoon.

Ruby and her three children lived in a small house out in a fairly remote wooded area. The marked patrol car was parked out of sight. Gwen Warren and Roger Dean approached the house and tried their best to talk Ruby out. They didn't have any luck at all. I felt that we needed to get her under custody before dark. If we couldn't do that, I'd wait till morning. It was decided that we would force open the porch door, then try to get at least part of the way in the main door to talk to her.

The porch door was kicked open. We tried the main house door and it was locked. In a loud voice I asked Ruby to please cooperate and come out. No response. Hemmerich then kicked open the main house door. Piper yelled, "Here comes the hot water!" Quickly we moved to avoid the boiling hot water thrown through the door opening and exited the porch. The door slammed shut and we could hear it being barricaded. We again stepped up on the porch and tried to reason with Ruby.

Hemmerich was married and his wife was pregnant. There wasn't any reason why I should let him be the one exposed to the dangers here. We would try once more. I would be the one to kick open the door, and if possible get inside and talk to Ruby. If this didn't work we'd wait until morning. Again the three of us stepped up on the porch and I forced open the house door and stepped inside. I could not see Ruby, but I'll never forget how the children looked. They were lined up according to age. The 15-year-old boy had a baseball bat raised above his head, Ruby's ten-year-old girl held a pan of steaming water and next to her the other boy who was 9 held a chunk of firewood. The girl said, "You want to burn Mister?" I decided I'd take my chances with the kids.

"If you want to hurt me, ok. But before you do, please listen to me. We're not going to hurt your mom, we just want to take her to a hospital and make sure she's ok to come home and be a good mom for you."

The older boy lowered the baseball bat. The girl seemed to act less threatening with the pan of hot water. I thought to myself, "If only I can get these kids to help, maybe we can get Ruby safely on the way to Traverse City and no one will get hurt."

To the right, not more than 10 feet away was the bedroom door.

Ruby stepped out without saying a word. She held the pistol in both hands as she fired. I could see the puff of smoke from the barrel and instantly felt a combination of burning and freezing sensations through my whole body. I screamed, "The bitch shot me!"

I knew I had been hit, but I wasn't sure where. My right leg was numb and I couldn't move it. Things seemed to be happening in slow motion. I found it difficult to step back out through the door. Another shot to the body would probably finish me so I kinda held out my left hand and moved it up and down while trying to get my gun out of the holster with the right one. I needed to get out of that doorway and into the shadows of the porch. Ruby held the gun steady with more puffs of smoke coming from the barrel. The bullets were aimed at my moving hand. My gun was in my right hand, but I was having trouble getting my finger inside of the trigger guard. I did have a right to shoot Ruby. But what if I pull my aim and accidentally hit one of the kids. Then shame on me. Imagine what the press would do with that. The best scenario would be that a white police officer was shot by a black, with the black taken into custody without further incident or injury.

Somehow I staggered back out the door. Three more bullets went over and under my waving arm, as I fell to the porch floor. My gun was ready now, in case she came at me through the door opening. The door slammed shut.

My deuputies had disappeared. The left leg seemed to work a little. The right didn't work at all. With my elbows and the good leg to push forward, I crawled off the porch and through Ruby's picket fence to my car and the radio. The deputies had taken cover when the shooting started. They began to emerge from the darkness.

Someone suggested shooting into the house. "NO!" I said, "There are kids in there. Get on the corners of the house, out of the line of fire. Shoot only if someone comes out shooting. I'm still in charge here, until I can transfer my command to the State Police."

Gwen Warren and Dean Piper got me into the car and we were headed for the Reed City Hospital. We were only about a mile or so down the road when it seemed like I was out of my body looking down inside of the patrol car. Gwen Warren said, "He's dead!"

I said to myself, "I must go back and last until the hospital." As we bumped over the railroad tracks I was back inside.

At Reed City Hospital there were attempts to stabilize me, then I was placed in an ambulance and we started towards Grand Rapids' Butterworth Hospital. There was a police car ahead and behind us and one on the side. The sirens and flashers were all going. A police car blocked every entering road until we got to Grand Rapids. You would have thought the U.S. President was coming, but it was only a policeman being looked after by his fellow officers.

Ruby had only hit me once, however it was with a 32 cal. hollow point. The bullet missed my heart by one hundredth of an inch, my liver and spleen by a thirty second of an inch and traveled through my spinal cord to end up near my sciatic nerve. It also tore off the corner of my cigarette pack. I kidded that this really angered me.

I asked the doctor how my injury looked. He said I realistically had only a slim chance of surviving the surgery and I might come out a vegetable.

I survived. The bad news was that now he said I would be paralyzed from the waist down with limited use of my right side. I told him if he didn't have any more good news that the ski season would be starting in a couple of weeks and I planned on skiing.

I was slated for a two-year stay at Butterworth and probably two-and-a-half at Mary Free Bed. My improvement and recovery were far beyond anyone's expectations. This made everyone happy, especially at Mary Free Bed. I was not a model patient, and I think they were anxiously anticipating my departure. When I didn't need the wheelchair anymore I shoved it in the hallway. Shortly thereafter, I threw the walker out, quickly followed by the crutches. What I requested was a cane.

I had been in Mary Free Bed for two weeks, and would be released after one more week. As was customary, I was allowed to go home for the weekend prior to my last week there. Instead of going home I made a beeline straight to Boyne Mountain. I walked into the ski shop with my cane. The guy behind the counter acted as if he knew me. He gave me the once over and said, "I suppose you want to ski too?"

"Yes, but I don't have my equipment."

"If you think you can ski, I'll give you your equipment and a weekend pass free."

Even though it was rather painful, I skied all the runs at Boyne. After so much time laying around the hospital, it was absolutely great being alive and being outdoors on the hills. It was the kind of mental therapy I needed.

When I arrived back on Monday morning, the doctor was soon there to check on my progress.

"How'd your weekend go?"

"Good."

"How were things at the police department?" he asked.

"I wouldn't know. I didn't go there."

"How are your folks?"

"I don't know, I didn't see them."

"What did you do with your weekend then?"

"I went to Boyne Mountain."

"What did you expect to do there?"

"Same as everybody else, ski."

For a few minutes my doctor lost his cool. "You can walk. Will you please admit you have a problem and let me determine how you walk?"

"Sure, but I'm in a hurry. I have a lot more skiing to do."

I really looked forward to getting back to work at the sheriff's department. The cane was a necessary prop though, and I would be driving back and forth to Grand Rapids for therapy for some time, so it would be awhile before I could completely resume my duties.

While I was in the hospital some reporter asked me how I felt about Ruby Nelson shooting me. I stated that I didn't hold any grudges. It was a bad situation to begin with and she probably felt she was just defending her property and her children.

****

While Lonnie was in the hospital Ruby was sent down to the Forensic Center in Ann Arbor to determine if she was fit to stand trial. They sent her back and said she was. There was a preliminary examination, formal charges determined and by December 1st she was released on bond. J.T. Nelson started a legal fund-raising effort on her behalf. Her court-appointed attorney was dismissed and Ernest Goodman from Detroit took over her case.

The community treated Lonnie very well as he returned to resume his duties. There must have been a different atmosphere in the department though. The undersheriff, Dean Piper, had started a recall petition. It was apparent that there were others in the community supporting his efforts. The prosecutor, Ed Duckworth, was quoted in the *Star* as saying that if the sheriff didn't agree to psychiatric exams he would call for his resignation. That seemed to be an obvious attempt to smear Lonnie's mental capability. Lonnie then informed the *Star* that he had been receiving both physical therapy and mental counseling in Grand Rapids. Before he completely resumed his duties as sheriff he would solicit a clean bill of health validation. He also pointed out that the counseling had already been set up and underway long before Ed Duckworth decided to make it a public issue.

I'm not sure how or why the undersheriff got involved in the recall effort. I would guess either someone put pressure on him or there was some kind of deal offered. Whatever it was they failed to get enough signatures for the recall petition to be valid, so it failed.

By the end of March Lonnie received a clean bill of health from his doctors and resumed all his duties as sheriff. He demoted Piper from undersheriff to desk duty. The demotion was to encourage Piper to leave the department. He resigned on May 11th.

Ruby waived her right to a jury trial, so Judge Wickens acted as a one-man jury. The effort to secure funds for Ruby's defense paid off very well. J.T. had contacts all over the state and a local civil rights activist named Robert Williams also solicited funds for her. She ended up with a first-rate defense team.

On May 17th Ruby went to trial with a not guilty plea being entered because of temporary insanity. Her attorneys stated that Ruby was unaware of what she was doing when she pulled the gun and shot Lonnie. They said that she had a long history of epilepsy and tests had proven severe mental retardation. She was just acting in an instinctive manner to protect her children. They went out of their way to portray the handling of the situation as being clumsy.

During one of the earlier court sessions Ruby had an epileptic sei-
zure. Now, according to her attorneys, she had proper medication to
control her epilepsy. She had turned her life around. Ruby had even
gotten married. By the time her defense team finished even the prosecutor agreed she should be found not guilty. Judge Wickens declared
Ruby was not guilty by reason of insanity, but did require that she
spend some more time down at the Forensic Center. In less than sixty
days she was back living in the community.

The local Rotary named Lonnie their Rotarian of the year. But,
there were those in the community that took exception to his receiving
the award and in letters to the *Star*, they were quite critical.

The effort against Lonnie now seemed to be spear-headed by Joyce
Radden. She thought that it was terrible that he had demoted Dean
Piper, forcing him to resign. He just wasn't a very good sheriff since
he had been shot. Obviously that's why the recall petition was started
before, and now Joyce felt compelled to initiate a new one.

Joyce Radden was the daughter of a former sheriff that had been
killed in an auto accident. Judging from her public statements it was
obvious that she was part of Piper's recall efforts initially. Now she
was leading the drive herself.

In the meantime, Lonnie hired a new undersheriff. His name was
Bill Moore. He was from the southern part of the state, a township
chief of police.

The recall effort launched by Joyce Radden didn't seem to pick up a
lot of steam until the jail incident. That happened in the fall of '78.

A man by the name of Winters was brought in drunk. He was making a terrible disturbance in the jail and had been warned several times
to be quiet. Lonnie was in the security garage, ready to leave for the
night. When he could still hear the guy yelling from his cell, he
returned to the cell block, and again tried to quiet him down. He told
the prisoner that his outbursts would certainly not do much for getting
him out of there. As Lonnie started to walk away from the cell, the
prisoner grabbed Lonnie's jacket through the food slot. Fearing that
he could be thrown against the cell, Lonnie swung his flashlight down
in a single blow to the man's hand, causing him to release his coat.

When Lonnie returned to work the next day, he learned that the inmate had been transported to the Reed City Hospital to receive medical care for the hand that he had hit with the flashlight. It didn't seem right that a single blow could cause an injury requiring medical care. It soon made sense.

In less than a week the undersheriff Bill Moore and another deputy, Vic Bunce, are testifying at an executive session of the board of commissioners. Bill tells the board that originally he supported the sheriff 100%, but that the sheriff has completely changed. Several incidents in addition to the inmate beating have caused him to come forward as a concerned citizen. Vic Bunce was a close friend of Moore. Bill had convinced Lonnie to hire him shortly after he took over as undersheriff. Bunce said that he witnessed the beating of the prisoner. He also stated that in an intoxicated frame of mind the sheriff had also made advances toward his wife.

Now the prosecutor, Ed Duckworth, is quoted in the *Star* as seeing the need to call in a special prosecutor to handle the assault and battery charges that will be filed against the sheriff. He reiterated all the other accusations that Moore and Bunce had alleged.

Bill Moore decides that out of citizen concern he will relaunch a recall position to get the sheriff out of office. He was quoted as saying that he was tired of trying to make the sheriff look good. Not satisfied with the lies he had already given birth to, Bill tells how the sheriff had bragged after he beat the prisoner, saying, "I guess I showed him who was boss." Bill also insinuated that Lonnie was against hiring blacks in the department.

Due to his adversarial relationship with the sheriff, Duckworth dismissed himself from the prosecution. A special prosecutor was brought in and, based on Moore and Bunce's testimony, decided that Lonnie Deur should be charged with assault and battery. An all-day jury trial took place in May and it only took 35 minutes for the jury to find Lonnie innocent.

Soon Moore was being quoted in the *Star* again. The special prosecutor brought in just did a good job covering things up. The charge should have been one of felonious assault. He reiterated all the other allegations against Lonnie, including not spending enough time on the job.

Lonnie finally broke his silence and told the *Star* editor, Denise Smith, that Bill Moore was on a personal vendetta against him. He had even been threatening sheriff's department employees that if they didn't sign his petition, they wouldn't have a job once Lonnie was ousted.

Apparently Moore had been coercing the inmates too. The *Star* published a letter from one state inmate who had been threatened by Moore to be sent back to Jackson prison if he didn't cooperate. The county jail was a picnic for those prisoners compared to Jackson. They could be in the county jail on a rotational basis for 90 days, and that period could be extended if they behaved well.

The news media was having a field day with all the happenings in Lake County. First the sheriff is shot by a black lady through a bungled pickup. Next he beats a prisoner. The prosecutor is against him. His undersheriff and all his deputies except one are heading a recall effort. The board of commissioners have asked that he resign, which he refuses to do.

It didn't seem to make any difference that Lonnie had been found innocent of the assault charges, or that he'd been shot in the line of duty in a situation that had been forced upon him by the local judge. In the *Star* Lonnie also stated that, through his own admission, Bill Moore had been approached by some commissioners in the late fall and asked to assist in Lonnie's removal from office.

The crowning blow came when the *Star* editor Denise Smith called for Lonnie's recall. By the end of May the recall petitions had been completed and turned in. A recall vote was called for August 7th. By the first of August Moore was already politicking for the office of sheriff. He said, "Sure, I'll run for sheriff. But that's not the reason for the recall."

The election was held on schedule and Lonnie lost by 74 votes. Bill Moore was unanimously selected as the Democratic candidate for sheriff. One of the commissioners, Harrison Wilson, wrote a glowing endorsement for Moore's candidacy. The sheriff's department employees signed Bill Moore's campaign ad in the *Star*. All except for one, Mike Seelhoff.

Mike Seelhoff

Reed City was my home town, and that's where I first met Lonnie Deur. Lonnie owned an ambulance service there. After college and my Army National Guard basic training I started driving an ambulance for him. When he won the contract for the service in Lake County, Lonnie asked me to go over to Baldwin and help him start up the service.

The ambulance service had previously been run by a federally-sponsored health center located in Baldwin. We went over to pick up the ambulance and found that the equipment was dirty and poorly maintained. People told us that the service previously provided was very marginal at best. The drivers only had a basic first-aid card and weren't highly motivated. The standard joke around town was that once a week the drivers would all pile in the ambulance, turn on the siren and drive down to the local Dairy Queen for ice cream.

There wasn't much public confidence in the health center either. The Vietnam War was going on and several of the doctors assigned were conscientious objectors putting their time in here.

Lonnie and I had no place to stay in Baldwin. Guy Lee was the sheriff then. He didn't live in the house attached to the jail, so he said if we wanted we could sleep in the basement. There were a couple of beds there. He also informed us that we could get gas for the ambulance from the sheriff's department.

Lonnie and I were both bachelors at the time, so we worked seven days a week, thirty days a month, to get the service up and running properly. Guy Lee's dispatcher was a cute young thing. Her presence made the job tolerable, but after a year I got tired of the pace, and went to work for the ambulance service back in my home town. Phil

Rathburn now had the contract. Supposedly I was to manage Phil's operation. But, his wife also had that intention so eventually we clashed and I returned to Baldwin.

Things had changed quite a bit since I left. Guy Lee had been sufficiently scared by the lawlessness in Lake County. He was ready to leave. Whether he had been personally threatened or what, I don't know, but Guy didn't make any secret out of the fact that the job wasn't worth getting killed for. He was a welder by trade, so he left the sheriff's job before his term had ended and went to work on the construction of the nuclear power plant near Midland.

In a special election Ron Voss finished out Guy Lee's term and then in the next general election Bob Johnson became sheriff and Ron worked as a detective for Bob. Bob had a position open for dispatcher so I quit the ambulance service in Reed City and went to work for the sheriff's department in Lake County.

The sheriff's department had very little to work with. The patrol cars, uniforms and equipment were all old and worn out, as was the jail itself. There was no overnight police patrol. I was the only one working on dispatch at night, and at the same time I had the responsibility for jail security. Every hour I would go back and do a cell check. I left the back door ajar so I didn't accidentally lock myself in.

It was about my fourth night on duty when the burglar alarm went off downtown. Bob and his wife Georgia lived in the sheriff's residence attached to the jail. I called over and Bob said he would be right over. It was only a few minutes later when Georgia strolled in. She had her bathrobe on, her hair in curlers and you could see her fuzzy slippers sticking out below the robe. Georgia said, "Move over. I'll take over here." Bob was right behind her. "Here take this." Bob hands me a shotgun.

Now you gotta understand, I had never had any police training, and had only hunted a few times with a shotgun. To say that I was a little apprehensive would certainly be an understatement.

Bob drove us down to the drug store where the alarm had sounded. He left me by the front door with these instructions. "Stand here. If anybody comes out, shoot 'em." He took the patrol car around to the back of the store.

I remember specifically that my attire that night was plaid pants and a yellow sweatshirt. Looking through the front window I could see two guys crawling around on the floor behind the counter throwing prescription drugs into a garbage bag. Finally I didn't see them anymore, but I could hear a commotion developing in the back of the store. Apparently they had run into Bob. Next I heard what sounded like several shots from the rear. There was an old Thompson machine gun in the department from the days when Bob Radden was sheriff. I wondered if that's what Bob was using.

About this time the State Police came racing in from Reed City. Apparently Georgia had alerted them. There I am in civilian clothes holding a shotgun. I've only worked for the department for a few days and they don't even know me. I say to myself, "Oh God, I'm gonna die."

Apparently Bob had got a hold of one of the robbers and in the struggle the guy broke loose and started to run. Bob fired and hit the man. The other burglar ran away. Though it sounded like Bob had used a machine gun, it was only the echo in the alley from his service revolver.

The next morning someone recognized the out-of-town car from the night before, and they caught the other burglar in the coffee shop across the street. Guess he just decided to have a cup of coffee and see what was gonna happen. They were from Muskegon. Both were wanted for a string of drug store B&Es. Apparently they had a little drug habit to support.

Lonnie Deur sold his ambulance service and by the time I went to work for Bob, Lonnie was working as a corrections and marine control officer. Bob then sent Lonnie to police recruit school and when he came back he regularly pulled road patrol.

Bob Johnson didn't present the image of a real professional police officer. He was big and overweight. It wasn't unusual to see him with just his uniform pants on, a T-shirt riding above his belly button and a candy bar sticking out the corner of his mouth. Despite the image and the poor facilities Bob Johnson and Ron Voss were both tremendous investigators with excellent minds.

Ron Voss and I also became good friends. In fact, he and his wife stood up with us when Mary and I got married.

Mary's father had operated the Government Lake Lodge for some time. When we were first married I worked part-time as a bartender for him and he quickly filled me in on all the local politics. He told me about J.T. Nelson, the supervisor from Webber Township. Other than his supervisor's position, J.T. didn't work, yet he always seemed to have plenty of money, drove nice cars, and was an immaculate dresser. In cold weather he wore an expensive leather coat topped off with a nice black cowboy hat.

J.T. was rumored to be the man in charge of fencing all the stolen property from the multitude of B&Es committed there. He also ran a little protection racket. My father-in-law said that J.T. had frequented his bar several times, suggesting to him that protection could be provided for his business for a fee. J.T. was usually drunk and my father-in-law wasn't one to back down from anybody, so each time J.T. was usually just evicted from the bar. Finally one of J.T.'s enforcers fire-bombed the bar. The fire department arrived in time to get the blaze out, and J.T. didn't try anything again. J.T. made no bones about bragging about all the people he had in his pocket and how he could have anybody hit. You could easily perceive him as a big blowhard, but he really did have power and people did fear him, it wasn't just talk.

There was plenty of crime in the county and occasionally Bob would be shorthanded and assign me some duty other than dispatch. One day he told me to go take an early lunch, as he would have some all-day duty for me guarding a crime scene until the State Police crime lab arrived. I told Bob that I had some sandwiches and could go anytime. Come to find out, I was to be guarding a crime scene and a body.

Dr. Nelson, the County Medical Examiner, finally arrived on the scene. He walked up to the body, kinda kicked it a little bit with his foot and said, "Yep, he's dead." Doc always had a pipe in his mouth. He stepped back from the body, cocked one of his legs up and rapped his pipe against his shoe to clean out the bowl. I didn't think much about it until the State Police crime lab arrived. They were quite excited about the tobacco they found on the ground until I explained where it came from. Bob Johnson got a little egg on his face because one of his men had let the crime scene become contaminated, but eventually Ron Voss solved the crime so everything was ok.

From some fine detective work, Ron had learned that the white victim liked to mess with the black prostitutes down at Rosanne's Place in Idlewild. Apparently this big fellow told the guy that he would take him out and introduce him to some fine lady, so they got in a car and left the bar together. Well, the supposed pimp directed the guy out to this remote area and when they got there he told the guy that he was the girl. This guy had no intention of making love to a queer so a struggle took place and the big queer hit the guy so hard in the back of the neck that it severed his spinal cord.

The area where the murder took place was where the county buried road kill deer. Some kind of acid was used on the deer carcasses and Ron traced the acid to the bottom of the queer's shoes, plus the guy had scratches and skin of the dead man under his fingernails. There was some plea-bargaining and finally the guy ended up spending seven years in jail. Of course, he probably was in seventh heaven there as the queen of the jail.

Doctor Nelson was quite a character. He was up in years and seemed to be well-respected. Although he was supposed to be wealthy you certainly wouldn't know it from his office facilities. I remember going to him once when I had a real bad earache. It was necessary to walk around all sorts of clutter and stacked magazines to reach his examining room. Doc couldn't find his instrument to look into my ear, so he took an old funnel and a dim flashlight to look in. "Yep, looks infected to me," he said. He wrote me a prescription and I was on my way. If you went in with a bad cold and said, "Doc, I need some penicillin for my cold," he'd just take your word for it and write you a prescription for the penicillin. I don't think J.T. and Doc Nelson saw eye to eye, but they co-existed. Doc's wife owned plenty of property in Lake County. She owned a restaurant and some other businesses, but it seemed like the most of her holdings were low-income housing units. In those days the government assigned rental payments directly to the landlord. Doc had a daughter who came to live in Lake County too. She was also a doctor and left a practice in Chicago to join her parents.

Bob Johnson eventually left the department and moved to Florida.

There were a lot of rumors floating around as to why he left. Some folks said he had been threatened and was scared for his family. Others said he liked Florida and just wanted to move there. Being Lake County sheriff wasn't a real bad deal for him. His house was furnished. His wife worked as jail matron and dispatcher, which brought in more income. Plus, the jail kitchen must have brought in a little extra money, because they sure didn't spend a lot feeding the inmates. It was corn flakes and a cup of coffee for breakfast, a baloney sandwich for lunch, coffee and a TV dinner for supper. The trustees would do almost anything for some cookies or an apple. When I worked night dispatch I parked my car in the security garage. The trustees would be out cleaning or waxing my car just to get extra food.

Bob joined the St. Petersburg Police Department in Florida and Lonnie Deur was elected sheriff. Ron Voss was disgruntled because he had lost to Lonnie in the election. Ron had previously been replaced as undersheriff by Lonnie and Lonnie becoming sheriff was the last straw. He came to work, but didn't do much. Finally he left for a deputy position in Newaygo. For one term he was sheriff there. Eventually Ron got a tracking dog and several years ago he died of a terrible cancer. Ron was a big healthy man who didn't smoke or drink but the disease got him. At the funeral they let his dog into the room and the dog obediently laid by the casket during the service. Several hundred tough policemen had a rough time keeping tears back.

I continued as night dispatcher. Also I worked part-time in the Idlewild Post Office. In the summertime Idlewild was a real popular resort for rich black people. While I worked there I had an opportunity to meet a lot of the older residents and made some very good friends. When they came in I would routinely help them fill out money orders, address their correspondence, whatever I could to help. They seemed real appreciative. The post mistress didn't believe in providing more than the basic services. "You don't have to do all that," she would tell me. Hell, I didn't have anything else to do. I liked meeting people and it was a pleasure to help.

Bob Johnson told me that I would make a really good cop. Along with dispatch he had allowed me to do some weekend marine patrol and wanted to send me to police recruit school. It was a tough deci-

sion, but I was married then and there was just too much political baggage connected to law enforcement in Lake County. I decided I was
also fed up with the low pay I was receiving and had been studying on
the side to become an insurance agent. Finally I went to Lansing,
passed the test and became a licensed insurance agent. We moved up
to East Tawas and I went to work with my brother-in-law who sold
insurance for Prudential. That lasted six months. Selling insurance
was not going to be my thing so we moved back to Baldwin.

I hadn't been back to Baldwin for more than a day when Dean
Piper, the undersheriff, pulled me over. I wondered what I'd done
wrong. Didn't think that I'd been speeding.

"Mike, what ya doin' these days?"

"Nothing right now."

"We need a night dispatch, you interested?"

"Well, I guess so. At least till I get something else. When do you
need me?"

"Tonight at midnight."

Lonnie Deur finally convinced me that I should go to police recruit
school. Lonnie was doing a lot to improve the image of the police
department. We had newer patrol cars, new uniforms which included
new shoes and gun belts, plus there was some money available for
training. Lonnie was also gaining some notoriety for lowering crime
in the community. The only problem was, that like the other sheriffs
before him who were interested in fighting crime, he would eventually run head-to-head with J.T. Nelson and all the people that he had
influence over.

I wasn't there when Lonnie was shot, but I heard about it over the
scanner. Lonnie made a miraculous recovery only to return and learn
that Joyce Radden and his undersheriff, Dean Piper, had teamed up to
get a recall petition to force him out of office.

Nineteen Seventy Eight was a real rough year for Lonnie. Not only
was he trying to go through rehabilitation for his injury, he was also
trying to fight the budget cuts that the commissioners were trying to
force on us and regain control of his police force.

Dean Piper finally resigned. Lonnie brought Bill Moore on as the
new undersheriff, but Joyce Radden was again trying to get the recall

effort started against Lonnie. Several members of the board of commissioners were also doing their best to give him bad press too, and
someone finally got Bill Moore to help out. Lonnie was accused by
Bill Moore and another deputy of beating a prisoner with a flashlight.
I had been around Lonnie Deur since the ambulance service days
and I knew that beating people was just not Lonnie Deur's style.
Well, they all went after Lonnie in force. J.T. Nelson, the prosecutor,
several of the commissioners and now Bill Moore felt it was his duty
to lead this effort. He got everyone in the sheriff's department to
sign Lonnie's recall petition, except me. I told him I wasn't playing
politics.

After Lonnie was recalled, Bill came around again and asked me to
endorse his campaign for sheriff ad. I refused. This became a big
thing around town, but I just didn't want to get involved in politics or
with Bill Moore. Bill was running the show until the election, so to
punish me for not endorsing his candidacy I was taken off of road
patrol and put on night dispatch again. That was just fine with me.
Bill's buddies in the department wouldn't talk to me anyway.

Two weeks later Bill came around again. If I didn't endorse his
campaign ad he would file some charge against me. I signed his damn
ad with a felt tip marker and it really looked strange when it came out
in the *Star*. My signature stood out from all the rest of the department
employees and actually worked against old Bill. When anybody
around town asked me about it, I told them that I was threatened with
my job if I didn't sign Moore's ad. Bill found out what I was saying
and was livid. He said that when he was elected sheriff he would get
even with me. I was mad too, because I didn't want to get pushed into
supporting him.

Bob Blevins was the Republican candidate. I mentioned to someone that I would like to meet the guy, to see what he was like. Shortly
after that, Bob called and came to my house. We visited for a couple
hours. Bob explained what his goals for the department were and
asked if I would support him. I said that I certainly wasn't about to
support Bill Moore, so I guessed I was in his corner. Bob never used
my name on his campaign literature, even though I was backing him.
That, I appreciated.

All the good things that Lonnie Deur had accomplished went down the drain after Bill Moore took over. The young deputies enjoyed partying and the control over their activities was very loose. Bill Moore really wanted to be sheriff and had done what he thought was necessary to gain support from those in power. The people had other plans, they handily elected Bob Blevins sheriff.

New Sheriff

In the county clerk's office, I raised my hand and was officially sworn in as the new Lake County Sheriff. From my experience as a patrolman in Kent County, I was very familiar with the criminal and traffic codes. But, what I wasn't knowledgeable on was the administration of a county sheriff's department. The Michigan Sheriff's Association provided a school for newly elected sheriffs; however that was every four years, after an election. As I was filling the office of a recalled sheriff, I was the only sheriff elected this year. They said it wasn't practical to provide the school for just one sheriff, but they did give me some starting pointers and suggested I solicit some advice from a neighboring sheriff.

My opponent in the election was Bill Moore. He had been acting sheriff since Lonnie Deur had been recalled. It was time to officially take over the reins of the sheriff's office from him. I was mentally prepared to leave the slashed tires, broken windshield, all the other election dirt behind and get on with the job.

Bill Moore was standing in the office as I entered. I started the conversation in as pleasant a manner as I could. "Bill, I have been in contact with the Michigan Sheriff's Association. They have given me a list of things I should get from you. The first is an inventory of all the property in the sheriff's department, guns, ammunition, patrol cars, everything."

He tossed a wad of keys on the desk top. "This is all I got when they recalled Lonnie, this is all I'll give you." With that he turned around and walked out the door.

I thought I should have someone to verify a property inventory, so that sometime in the future I wouldn't be held responsible for property that wasn't there to begin with. A member of the board of commissioners spent two days with me, listing all the equipment, and then a copy of that list was filed in the county clerk's office. I found that there had been a list when Moore took over, but it had disappeared.

Before I had an opportunity to contact another county sheriff for advice, one stopped into my office to say hello. It was Art Speers, the sheriff in Osceola County. Art congratulated me on the election victory, then handed over a copy of the 1963 Michigan Constitution, plus the state statutes that clearly defined the county sheriff's powers. He seemed genuinely friendly, offering to fill me in on all the administrative requirements of the sheriff's department. Art also invited me up to Reed City to walk through his jail. I was eager for such a visit, but first I needed to evaluate the work force and establish my authority here.

Dale Gibbs had been a part-time deputy sheriff in Kent County. We had spent hundreds of hours together in a patrol car there, and he volunteered to come up north to support my election effort. Dale was not only a good friend, he was someone I could trust as my undersheriff. It was now time for us to interview the present employees, reviewing with them the policy and procedures we expected to follow in running the department. Dale and I had put together a small guidebook that simply outlined everything from haircuts, firearms policy, off duty work, plus specific responses to certain situations a deputy might encounter.

During the election campaign many citizens came forth with unfavorable comments about the present sheriff's department. During the door to door conversations or the town hall meetings, people were giving me names of specific officers and incidents. A typical comment would be, "Do you realize what some of those deputies have been doing? That one guy is always at the Government Lake Bar in uniform, drinking beer, with the patrol car in the parking lot."

There were numerous comments on the deputies' long hair, sloppy uniforms and the tennis shoes a lot of them wore with the uniform. I kept notes on all the things I'd been told, so when we brought each employee in for an individual interview, we had some good background information to start with.

Approximately 15 people were employed by the sheriff's department when I took over. That included road patrol deputies, jail guards, a dispatcher and one man for marine patrol in the summer. Before we called them into the office for the interview, we allowed each employee some time to review our policy guidebook.

Dale Gibbs ushered each employee into my office. I would stand, introduce myself as the newly-elected sheriff, then introduce undersheriff Dale Gibbs, who would be their immediate supervisor. I would ask if they had reviewed our policy guidebook and then say, "I don't care how much you supported Bill Moore's campaign against me, that doesn't matter. What I need to know is can you follow these procedures and give me an honest day's work for your pay?" If they said they could, then I would bring up specific incidents that had been brought to my attention during the campaign.

"I have been told that you occasionally go down to Government Lake and have a beer in uniform. Is that true?" Some would admit to it, but say that Bill Moore didn't have a problem with drinking in uniform. Others would deny the charge, but the denial didn't last long when I threatened to bring the witness in for verification.

"I have promised the people of this community a professional department. You are my representative to them. Regardless of the past, if you can follow these rules, look and act like a deputy should, then you will have a job. That means no drugs, no drinking in uniform on duty or in the patrol car."

Four of the six road patrol deputies were very negative. It was either a wishy-washy answer, or simply "I don't like you, I don't think I can work for you." I let them go. Losing the job didn't seem to bother them much. Probably they just figured the union would get the job back.

Unknowingly, by firing these deputies I had just gotten myself and the county into some future hot water. The Michigan Statutes clearly stated that the county sheriff was the chief law enforcement officer for the county, with the authority to hire and fire at will. What the statute didn't say, nor did my neighboring sheriff, Art Speers, happen to mention, was that there were several court decisions, attorney general opinions, along with a union agreement that the sheriff must honor. There must be a record or progressive discipline and punishment for cause to lead to a dismissal. Firing those dudes was my first bad mistake. This lack of labor law knowledge could have broken the county's back financially, but all this wouldn't come to light for awhile.

Mike Seelhoff was one of the few road deputies that I retained. I had visited with him before the election and was impressed with the man. He was also the only employee in the sheriff's department that had not buckled under to Bill Moore's threats and endorsed Lonnie Deur's recall petition. That certainly said a lot for him.

Shortly after I took office a black prisoner by the name of John Philpot hung himself in my jail. Philpot was charged with murder and conspiracy to commit murder in the death of Phyllis Whittler.

Phyllis Whittler was a Baldwin resident who had been killed a year earlier in a gruesome contract killing. Only recently two other men and two women had been implicated in the slaying.

Somewhere in Detroit this guy by the name of Robbins was telling an undercover police officer how our soldiers in Vietnam would collect ears off Viet Cong they had killed, but that he collected fingers. When the undercover officer showed some curiosity and interest in the guy's story, Robbins produced a jar of vinegar with fingers in it. He pointed to some on the top, "Those belong to Mrs. Whittler from Baldwin." Shortly thereafter there were other police officers on the scene and soon the man was on his way to Baldwin to stand trial.

Robbins decided to cooperate and testify against the other participants in Whittler's murder when he heard street talk to the effect that there was a contract out for him. George Pratt, a Reed City State Police detective, spent a lot of time taping Robbins' statements. Eventually the tapes were used in court proceedings and the *Star* carried all the details for the public.

John Philpot left a suicide note before hanging himself. He was implicated as the go-between person for the contract. A man by the name of Alfredo Robinson was the man actually contracted to kill Whittler and for $50 he had hired Robbins to help.

Mrs. Whittler was director of the Lake County Child Guidance Center. She was living in a trailer and had been separated from her husband, Jim Whittler, for a short period of time. Jim owned a mens clothing store in Baldwin and Mrs. Whittler was about to testify that a recent fire in Jim's store was not accidental, that it was arson. Someone didn't want her to testify and put the contract out.

On the 12th of October the murderers stopped at Mrs. Whittlers to ask for directions to a store. The next day they again pulled up close to her residence. This time they raised the hood of their car and sent Robinson's girl friend in to supposedly call for help to fix their car. When Mrs. Whittler opened her door the men hiding in the bushes outside burst in. At gun point she was directed down to a bedroom and raped. Robinson told Robbins to go into the kitchen and get a knife, then stab Whittler. Robbins got the knife and stabbed Whittler in the chest area. One of the men again raped Whittler. He moved the knife around to cause spasms, creating more sexual pleasure. The knife wound was not killing Mrs. Whittler so they took the pillow off her face and Robinson put his gun to her head and pulled the trigger twice. The knife was pulled out and used to cut two fingers off her left hand to prove the contract had been fulfilled. Later the medical examiner concluded that, from the profuse bleeding evident, the heart was still pumping when the fingers were cut off.

Jim Whittler was a prominent Baldwin citizen and never was implicated in the crime. Occasionally he would run for local office and that December he was reelected president of the local NAACP. It was about a year or so later that Jim disappeared and his store was sold. One of the local state troopers told me that they found parts of Jim Whittler's body in several different plastic bags around Detroit. I'm not sure whether Jim pissed off someone in Detroit's organized crime or what, but they finally got him. It was interesting to note that for lack of evidence the women involved were released. One of them turned out to own the house where the murder contract was discussed. Records showed her to be Jim Whittler's first wife, who at the time of Mrs. Whittler's murder still owned part of Jim's store, plus several other parcels of land around the Baldwin area.

When I first walked into the Lake County sheriff's dispatch room, I saw the radio and mike, but no computer. "Where's your computer? How do you get into the LEIN (law enforcement information network)?"

"We don't have a computer, never have," someone answered.

"If we need information we call one of the Michigan State

Police posts."

I got on the phone with the Michigan Sheriff's Association. "Is it possible that in this day and age Lake County has never had a LEIN machine?"

"Well, some of the smaller northern police departments don't have one, but as far as we know, out of 83 counties, Lake is the only one that has never had one," was the answer. Eventually I would convince the commissioners of our need for this capability and get the department tied in with the rest of the world.

My first priority was to get a 24 hour road patrol set up. As it was, there was no road patrol after midnight. The last deputy on duty would just call the Reed City State Police Post and let them know the sheriff's department was closed down for the night. Probably there would be only one state police cruiser out for two counties, so crime could have free rein. I started to hire replacement deputies for those I had fired, eventually having the capability to schedule at least one deputy on the road during the early morning hours. That wasn't a lot, but one on the road was more than we had before. Also, there now was some additional money available from the state for secondary road patrol. If I got the grant paper work out I could quite possibly hire two additional deputies.

The sheriff's school provided some excellent instruction on how to maintain and run a jail, state jail inspections, prisoner nutrition, plus the guidelines to receive state grants. It would be a year before the school would be available, so I'd have to wing it until then. I was spending a lot of time with the Osceola sheriff, Art Speers. Art seemed genuinely interested in showing me the ropes. He had been a sheriff for almost a full term, so I viewed him as the expert. Whenever I went up to Reed City to visit, Art would insist that we go out for a drink and dinner. Invariably we'd end up at the Osceola Inn, where they had excellent food. Art liked to drink. He usually downed several drinks before we ate. I had done my share of drinking before joining the Kent County Sheriff's Department, but now I stuck to soft drinks.

In public, Art seemed to be a bit of a showoff. With a few drinks he became a little more boisterous; at times I was embarrassed at the way he conducted himself. Still he was really helping me; I could put up with it. One other thing made me uneasy. After drinks or a meal I would reach for my wallet. With a head shake and hand signal he would refuse my offer to pay, we would get up and leave. I figured that Art was certainly generous. He must have a tab with the place, as he made no offer to pay when we left.

Art's wife, Lucinda, was also becoming a great help to me. Known by most as Cindy, she was the county clerk. The county clerk was required to be a Lake County resident, so when she and Art were married, in December of 1977, she just kept her maiden name of Keefer and used her mother's Baldwin address as hers. She actually lived with Art in the house Osceola provided for their sheriff. Her residency caused a lot of controversy for her over the next few years, but she still managed to remain the county clerk.

Cindy went out of her way to help me get started in the sheriff's office, keeping me abreast on the board of commissioners' activities as they related to my position as sheriff. If she knew ahead of time that the board would be asking me about something in particular at their next meeting, she would give me a call so I could be prepared. Much of the paper work associated with the sheriff's office was filed or forwarded through the county clerk's office. Cindy gave it all her personal attention, that it might be handled properly.

Once each quarter the sheriffs from Michigan's 83 counties would get together someplace for a conference weekend. I was very interested in the conference discussions, listening intently, even taking notes. Some of these sheriffs had been in office for quite some time. Many would come over and introduce themselves to me and I could have some one-on-one discussions with them. I was the new kid on the block, needing to tap into their expertise.

Smoke Meyers was the sheriff from Roscommon County. We struck up a friendship rather quickly and he gave me some good advice on obtaining state grants for my department.

On some of these sheriff's conference weekends, there would be some social activities, with the wives invited. Occasionally we went with Cindy and Art, even shared a hotel room with them.

A couple of the sheriffs noted my close friendship with Art Speers, cautioning me that he might not be my best role model. For some reason they didn't figure he'd ever get reelected.

I did feel some loyalty to Art, for the help and friendship he and Cindy provided as I was getting started. Still, there were a lot of things about Art that bothered me, especially his personality when he had been drinking. I decided to start distancing myself from him and I did it gradually enough that he didn't seem to notice. He probably just thought I was becoming more knowledgeable in my job, needing less of his help.

Lake County probably didn't have over 8000 full-time residents when I took over. With all the 156 lakes and 46 trout streams though, there were many additional seasonal homes. When the vacationers left in the fall, that was the signal for the thieves to start cleaning the valuables out of these residences. A lot of the places were more than just a summer cottage. Some were very elegant homes filled with antiques. Previous law enforcement had been so lax that the burglars were getting sloppy, making lots of mistakes. They would clean a place out in the middle of the day, then drive down through main street with their loot piled in the back seat or protruding from the car trunk. Motorcycles, refrigerators, lawn mowers, antiques, TVs, anything that looked valuable was removed.

We started catching these amateur thieves in large numbers. The jail was filling up. The prosecutor was busy bringing charges and we were trying to contact owners of the burglarized homes to claim their belongings. The larger counties, with more jail space, would have held these guys longer than we did. But, at least the word was out that they would have to be more careful or risk being in jail again for a more extended sentence.

I started some new programs in the county, not only to cut down theft, but to counter violence and make the community a safer place to live. The thrust of the new programs was to get the people more involved in looking out for each other. In various meetings around the county I explained that law enforcement was reactive. That we didn't have enough deputies to keep crimes from being committed, we could only react after the incident. They must care about each other's children and property. Neighborhood and child watch programs had proved effective in other areas, so I started these programs in Lake

County. Also, the department purchased eight electric engraving tools that people could check out to put an identifying number on their valuables. This would make it easier for us to spot stolen property being sold, and for the rightful owners to reclaim it.

After being on the job a few months, in one of the board of commissioner meetings, I asked about the possibility of the county buying a patrol car for my use. Most sheriffs did have their own patrol car, I explained, and it would allow me the capability to respond directly to incidents from my home in something other than my private vehicle. They said they would consider my request. Eventually, J.T. Nelson, the board chairman, visited my office. He was very friendly. "Sheriff, we think you are doing a great job. You live in the outskirts of the county and winter is coming, how would you like a 4-wheel drive vehicle?" I gave J.T. a positive, enthusiastic response to that idea and he laid several brochures on my desk. I could decide which I wanted, a Chevy Blazer, Ford Bronco or a Dodge Ram. I ended up picking the Dodge Ram. When it was delivered to me, it was all gussied up with extras and a nice black and white police paint job. It was certainly more than I expected. Everyone that saw the vehicle knew that the sheriff was around.

I had heard some bad things about J.T. Nelson, but I kinda liked the guy. So far, he had been quite helpful. He was intelligent, friendly, a fine dresser and made things happen in the community. I didn't like to judge a person based on someone else's opinion anyway. It would be best to take some time for my own evaluation.

Even though I was busy with my new job, I still was having more quality time at home with the family; the new house would be started in the spring, the kids were settled in school and life was easier for Dee. Unfortunately, there were some other things that were not going as well.

When I decided to go back into law enforcement, we wanted to simplify our lives so we put all the farmland and the riding stable up for sale. We were in a bit of a hurry to sell, ending up with a buyer that only had enough up-front money to cover the realtor's commission. Although the man didn't have the cash for a proper down payment, he did seem like a responsible individual that would succeed in our business. We wrote up a 30 year mortgage and left it all in his hands. It was an established, prosperous, very reputable business that we had struggled for many years to completely pay off. There was plenty of money in our savings account at the time, so we really could survive with my sheriff's salary and the monthly mortgage payments. But, after a few months the new owner got several payments behind.

When I drove down to talk about the missed payments, I could see that the operation was really slipping. He said that he had lost much of his horse boarding business, and I could see why. The boarding stalls were dirty, with a collection of manure. It wouldn't be long before the stable reputation would be completely ruined and this guy would be out of business. I started foreclosure proceedings.

Included in the sale of the farmland and stable were two homes. It was late one evening when we received a call that the home where the owner lived was on fire. By the time we arrived, the fire department had put out the blaze, but there was extensive damage to our old home. The family was huddled out in the yard. The man explained that the three-year-old had been playing with a candle that caught the curtains on fire. It soon was beyond their control.

There was another call the next morning. The fire had rekindled, burning the rest of the house to the ground. The man assured me the house was fully insured and that he wanted to rebuild, and to continue the business. Later he even mailed me a couple of the missed payments. I felt sorry for the guy; the foreclosure actions were put on hold.

The drawings for the replacement dwelling looked good. On my next visit there were several carpenters busy and it seemed as if the new house would be framed in shortly. A Wicks lumber truck was also backed up with a load of plywood.

The insurance check was made out to both of us. He said that funds were now needed to start settling with the lumber company; would I mind endorsing the check. Well I did, and what a hell of a mistake that was. The next day the guy disappeared with the $70,000 insurance money. Nothing was paid for. We returned as much of the lumber as we could, but still had to dip into our savings to pay for the rest.

We put the property up for sale again and accepted the first offer we received. When I tried to transfer the property title to the new buyer, I learned that the State of Michigan had put a lien on it. The other owner had previously owned a furnace business, and never paid any state or federal taxes. I had to come up with $12,000 to settle this debt for clear title.

The offer we accepted this time gave us approximately half what the property was worth. This deal was also on a land contract, so now we were getting strapped for cash to finish our new house, plus we were weary of the whole ordeal. I decided to take a 20% discount on the contract, cashing it out. What the hell. I'd already lost my shirt, what's another 20%.

I had to chalk the whole thing off as just a bad experience in life. Some kind of lesson, I guess. It did seem a shame to see all our security for the future go down the drain. We had worked day and night for ten years to build and pay for it. It was all gone so fast.

Now was the time to dwell on the things that were going right. Dee and I had a solid marriage. For all my misadventures, mistakes, and the suffering I had put her through, she was still with me. When we were running the stables, it seemed that the family got very little of my time. I was just too busy. I had more time with them now, plus plans for our new house were coming together. The 40 acre lot covered half the frontage on Stowell Lake. I was building the house into a hillside, with a 40 foot waterfall cascading down to the lake past our bedroom window. The trailer living was inconvenient, but soon the house would be started.

My next goal to improve county law enforcement was to have a deputy assigned to the village of Luther and another to Idlewild. The villages previously employed a police officer of their own, but I explained to the community leaders that it would be a great advantage to them to instead have a county deputy assigned. They would still have to pay the officer's wages and provide for a uniform and patrol car, but he would be tied into the county sheriff's office by radio if he needed backup. The county prosecutor could then process their warrants and prosecute their cases rather than their having to hire a city attorney to do this. They would have control of their officer, but he would work as a deputy with the county sheriff's department, wearing the same uniform, driving the same color patrol car. They could hire who they wanted for the job, but I must approve of the officer. This arrangement would not cost them any more than they had already been paying, probably even less. They accepted my offer.

From what I know, Idlewild is Michigan's first all-black community. I felt that I needed some solid law enforcement representation there, not only because it seemed to be presently lacking, but because I wanted to cement good relations with the large black population in Lake County. An old friend of mine from the Kent County days agreed to come up and work for me. Idlewild seemed the perfect place for his services and I thought that Les Fisher would be acceptable to the local officials.

Les was nicknamed "The Count". Many years ago he had a large band and followed Count Basie from city to city on tour. Count Basie and Fisher were close friends and someone nicknamed Les the Count also. He was a drummer, but had plenty of talent with the piano too. Les's wife was of Latin descent and she was an accomplished singer with a beautiful voice.

When I first met Count Fisher I was still working in booking and receiving for the Kent County Sheriff's Department. He was working for the Friend of the Court's Office. Count would come in late at night or in the wee hours of the morning and we would have coffee together.

Count was a big man. His wrists were so large that a set of handcuffs would not go around them. He could also really be a bad ass if he wanted to be. He'd go into some of the worst areas of Grand Rapids with a pocket full of warrants, and come out with a back seat full of prisoners. More than once the sheriff grumbled about that big black bull from the Friend of the Court's Office filling his jail up with prisoners.

The Count was also a martial arts expert of the Korean style. I had studied some of the Okinawan moves in the Marine Corps so we talked about the different styles and when things were quiet at night we got into a little hand-to-hand combat. Occasionally, when I was lucky, I could take the Count down with a move he wasn't expecting. That move would never work the second time. The receiving room didn't have any furniture, so we had plenty of room to maneuver without hitting anything. The marble floor was a little hard though, causing a skinned knee once in awhile.

We also had our share of fun together. When there was a new deputy around the Count would come in for coffee and I'd throw some racial remark at him. He'd answer, "Come out from behind that desk white boy and I'll kick your ass!"

I'd counter with something else then jump over the receiving desk and we'd put on our little make-believe martial arts fight while the poor deputy stood there wide-eyed. He'd usually try to get us to stop then run down to get the shift Lieutenant who had seen us do this before. The Lieutenant would always go along adding some choice words to make the whole scene a little more believable.

The Count also liked to put on a little show in public. One morning I was sitting in the Red Lion Restaurant waiting for some breakfast, when he comes in. He spots me at the table and in his meanest voice says, "What you doin' in here honky?"

"Don't give me any shit you black S.O.B.!" I countered.

I got up from the table and we moved towards each other in a threatening manner. Meanwhile the patrons were looking for the most convenient exits. As we reached an arm's distance from each other we rushed forward and embraced. When our arms were around each other we broke into grins from ear to ear. Needless to say, there was a mass sigh of relief in that restaurant.

Once I started on patrol duty the Count would spend lots of hours riding along with me. I'd even take him to visit some fathers that weren't paying support. One way or another the Count got them back on track. Both he and his wife were wonderful people and we became the best of friends.

Les Fisher did help me provide good law enforcement in Idlewild, plus he was a strong positive influence in the community. He also provided good backup for the rest of my deputies if trouble cropped up.

One humorous incident involving the Count happened late one night when one of my young deputies pulled over a truck and trailer rig. The truck driver was trying to intimidate the deputy. Sensing a bad situation developing, the deputy called in a secret code over the radio which meant that he was in a volatile situation and needed some help. Count Fisher responded and with his lights out parked some distance behind the deputy's patrol car. The truck driver was trying his best to threaten my young deputy. He sat in the passenger side of the patrol car. "Sonny, why don't you lay your gun and badge on the dashboard. Let's step outside and settle this like men."

The deputy said that all of a sudden the truck driver was busy looking out the open side window. His head was moving up and down surveying something. Then he turned away from the window, looked straight ahead and said nothing. Count Fisher stuck his face in the window. "I know, I know. I'm the biggest fuckin' nigger you ever saw. Right?" The trucker didn't respond, but when the deputy finished writing the ticket he just got in his rig and quietly drove off.

Information came to me detailing local storage areas for stolen property. If a person talked to the right folks, he could purchase at very reasonable prices anything from a boat to a lawn mower. Bill Burden, an old partner from my Kent County days, now worked in the detective bureau there. The time seemed right for a little sting operation in Lake County. I called Bill to see if he could spare a couple of his undercover boys to set up a buy of some stolen property. Bill said the undercover men could be made available, but it would be my responsibility to come up with a couple thousand dollars of marked money to facilitate the buys.

The county commissioners were informed of my planned sting operation, and they agreed to provide the money for it. Bill sent up some scruffy looking characters to make contact with the folks that might have some stolen property to sell.

Word somehow leaked out. There was no stolen property to be bought. The locations where it was supposed to be stored were empty. Someone had gotten the word of warning out. I was left with a little egg on my face. There would have to be a different arrangement if I were to catch any thieves.

Before the board again, I explained that it would be much more convenient if we could keep some sting money in the prosecutor's safe. That way, if I needed the money on short notice, at night or on a weekend, it would be there and I wouldn't have to bother the commissioners for it. There was no mention of a suspected information leak.

Ed Duckworth had been the prosecutor for three years before I became sheriff. He seemed like a good enough man, was proficient in processing the bad guys and getting me warrants when I needed them. We put about $4000 of marked sting money in his safe, then let things settle down for a few months before we tried it again.

Some time later, I learned of the availability of a large quantity of stolen goods. Arrangements were made to schedule the undercover officers again to pose as potential buyers. At the prosecutor's office I told Duckworth that I would need to check out four thousand dollars for a sting operation. He seemed receptive but said there was a problem.

"What's the problem?"

"I used the money."

"Used the money?"

"Well, I needed some cash and really thought I would have it paid back by now."

I knew that I couldn't go back to the commissioners for more cash, otherwise the word would be out again. Neither could I afford to use my own money and have it tied up as evidence for several months. Maybe the bank would loan it. Bob Smith, the bank president, was a friend of mine. I asked what he would charge for interest if I borrowed the money for a few months. He agreed to a no interest, signature loan.

Everything was put in place for our operation. Some bearded shady-looking undercover cops started flashing some money around at the local bars and pretty soon, bingo, we hit the jackpot. We had more stolen property than you could imagine. The basement in the jail was quite large, but we now had it jam packed with every valuable household item you could imagine, from toasters to deer rifles and bicycles. The stuff had been stored all over the area in garages, barns and resident houses. We got the people selling it along with those providing the storage. This would put a sweet dent into Lake County's stolen merchandise market. The sting was so successful that the bank's money would be tied up for quite some time as evidence. There might be interest to pay, so I brought this possibility before the board. Their response was, "We gave you the money; why should there be interest to pay?"

"Because the prosecutor spent the money."

Eventually the prosecutor was charged with embezzling public funds. He pleaded guilty to a lesser charge, paid back the $4,000, was fined $1,000 and had his law license taken away for five years.

Usually in the spring of the year, when the seasonal home owners started returning, the sheriff's department was flooded with burglary reports. During the wintertime there was no road access to a lot of these homes, and not many neighbors around to watch things. The thieves could arrive on a snowmobile, break in and clean out a residence. It would be several months before anyone would discover the loss. By that time, the trail would be cold. I approached the county board of commissioners with the idea of getting some snowmobiles for winter patrol. I was reminded that Lake was a small poor county. They had already gone over budget to purchase my new sheriff's vehicle. Snowmobiles were out of the question.

Dave Maney was president of Kawasaki Mid-west, in Grand Rapids. He was a friend of mine and I approached him with the idea of me helping design a prototype snow vehicle for law enforcement. One large enough for a radio, first-aid kit, along with flashing red and blue lights, and siren. It would be light and rugged enough to go over loose snow on ungroomed trails. He was receptive to my idea. In the meantime, as a public relations venture, he was willing to provide four top-of-the-line, Kawasaki 440 snowmobiles to get the winter patrol started. Not only did he provide the snowmobiles, he included suits, helmets, all the equipment needed for each machine's rider. Our winter patrol was in business.

One of my priorities was to reactivate the county sheriff's posse. In addition to providing men and horses for its traditional role of off road search, it provided a man to ride with my deputies on weekend patrols, when the law enforcement force really needs to be visible. Now, with the new snowmobiles available, I was getting even more volunteers to ride around checking out all the seasonal homes and cottages. It wasn't long before some fresh burglaries were discovered, with nice tracks leading to where the thieves stashed the goods. A search warrant got us into the buildings to seize the stolen items and with that evidence, the property owners became willing to cooperate, exposing the culprits.

One of the early visitors to my office was a man that I had heard about, but never met. He strolls in one day, lays this scrapbook on my desk and says, "I'm Robert Williams. You've probably heard about me already. Thought I'd stop around and confirm what you've heard so there's no doubt in your mind. As proud as you are of being a Marine, I am that proud of being a communist." Opening his scrapbook, he slowly turned the pages. There he was with Fidel Castro, Communist China's Chou En-lai, Mao Tse Tung, all the prominent communists of the day.

"I am a civil rights activist. I don't profess violence as I did in my younger days, but if I'm pushed ..."

Weighing well over 200 pounds, he stood there in that leisure suit and looked very intimidating. Even though he was probably in his fifties, his height and full beard made him look pretty tough.

Robert Williams lived over on Paradise Lake by the black community of Idlewild. Supposedly, his home was at one time a summer retreat for Chicago's famous criminal kingpin, Al Capone.

Williams routinely addressed small community groups and touted himself as one of the first Black leaders to advocate self defense in the South against violent attacks by the Ku Klux Klan. He often bragged that he along with Malcomb X, and others, founded the Black Panthers organization.

In 1961 Robert Williams was accused of kidnapping a white couple. He fled the United States to Canada, then eventually made his way to Cuba where he lived in close harmony with Fidel Castro for five years. From Cuba, Williams traveled with his family to The Peoples Republic of China, where he was Mao Tse Tung's guest for four years. He was there during the cultural revolution and on a 1967 National Day celebration Williams stood beside Mao Tse Tung and addressed two and a half million people. He was also on speaking terms with not only Mao, but Chou En-lai the Chinese foreign minister, Ho Chi Minh the North Vietnamese leader and Che Guevara, Castro's guerrilla representative to our southern hemisphere.

The Nixon administration wanted to improve ties with Communist China and appraise the situation following the cultural revolution.

Williams was supposedly one of their first contacts. He stated that he also collaborated with the State Department for Nixon's China visit.

When Robert Williams finally returned to the United States he was slapped with a federal fugitive warrant, but they were never able to convince Governor Milliken to sign the extradition order sending him back to North Carolina. Eventually during the Carter administration all charges were dropped and Williams successfully retrieved the many volumes of files the FBI had accumulated on him. It was interesting to note in the files that the Chinese had sent a petition with over 12,000 signatures to help aid him in his six year extradition fight.

Since Williams resided in Lake County, he seemed to live very well even though he didn't have any visible job. He did make a yearly visit to China, perhaps that's where his income came from. His wife occasionally worked at 5-Cap and ran for some political posts in Pleasant Plains Township.

From time to time Robert Williams returned to the jail on a fact-finding visit. If a black inmate fostered a complaint, Williams was there to check things out. Despite all I'd heard about the guy, I tried to give him the benefit of the doubt. Maybe there were legitimate concerns to be dealt with.

Williams wasn't just involved in our area, he seemed to be present wherever there might be a racial incident. In Kent County when a school board tried to fire a black superintendent for not being properly qualified, the news media picked right up on it as a racial incident, covering it night after night. In the background of many pictures you could see the tall bearded black man, Robert Williams.

It soon became obvious to me that Robert Williams wasn't merely looking after black civil rights, more so, he was doing his best to convert any event he could into a racial incident. It seemed he never missed the opportunity to over-dramatize something, gaining press and TV attention for an incident he knew wasn't racial at all.

Once I was convinced of his real goals and motives, when he visited the jail I let another deputy hear his complaints. I simply ignored him.

One morning the people of Baldwin woke up to see three big Ks painted on the side of their water tower. The local folks were pretty well convinced it was just a kid's prank, but Robert Williams didn't feel that way. The next thing we knew there was a Detroit Free Press reporter in Baldwin asking questions.

On the 5th of May, Robert Williams was pictured in front of our town water tower in the Detroit Free Press. Below the picture Williams was referred to as a national civil rights leader. The reporter, Eric Sharp, did a pretty good job on the article though. I don't think it was sensationalized nearly as much as Williams hoped for. Sharp interviewed me and I told him it was just a prank and that we had the best race relations of anywhere in the state. Williams agreed with me saying that we had good race relations, but that the whites just weren't willing to listen to him and admit this was the work of the Ku Klux Klan.

J.T. Nelson was also interviewed. He said one third of the citizens were black and that even though race relations were pretty good here, there were a lot of white people moving in from Chicago and Detroit. He felt that it wouldn't take much to turn the local whites against the blacks.

The reporter even interviewed the town maintenance man, a guy by the name of Radar. Radar said he didn't like heights and he didn't want to climb up there and paint over the Ks. The tower was supposed to be repainted this summer anyway. Finally Williams threatened to bring a lawsuit if the Ks weren't painted over immediately. Radar was forced to make the climb and cover the letters with spray paint. He said he sure didn't like that job.

It wasn't long after all our press coverage that a little investigation revealed that two fifteen-year-old high school students were responsible for the three Ks on the water tower. They admitted that it was all done on a dare. One of the kids was black and his initials were K.K.K. They just decided to paint his initials up there for fun, but it certainly caused a fuss. Robert Williams and the news media had no further comments on the subject.

On a regular basis, Lake County boarded state prisoners and prisoners from other counties. This brought needed revenue into the county; however, there had been long term discrepancies with the jail. If these weren't corrected soon, the state could cancel our contract. There was a requirement for a prisoner visiting area, with tables and chairs so the prisoners could converse with relatives that came to see them. Also, regulations dictated that there be an exercise area. Repeated warnings had been issued by the state, because none existed.

I still had the backhoe and bulldozer from my excavating business, so I again appeared before the board of commissioners with a proposition. If the board would purchase the gates, poles, fence and barbed wire, I would use my equipment and trustee labor to construct an exercise area. The board was reluctant, but I explained that the requirement was from the Department of Corrections, who could cancel the state contract with us if we didn't soon get one built.

The board finally agreed to buy the materials I had asked for. I brought in my backhoe and dozer, and with the help of a half dozen trustees, a month later we had ourselves a state-of-the-art exercise area, complete with a basketball hoop and picnic table. Now, with a guard present, a prisoner could get sunlight and exercise. The added benefit of course, was that if a prisoner was unruly or failed to keep himself out of trouble, the exercise privileges could be denied.

The sheriff's department was really becoming something I could be proud of, especially my deputies. They were young, many just out of the academy, but they were eager like I had been and ready to write lots of tickets and catch all the bad guys. Lake County didn't pay very much. I knew they would only be here two to four years, then move on to a larger department where they could double or triple their pay. It didn't disappoint me when these young people left for greener pastures. I took pride in what they learned under my supervision, and the fact that I didn't hold them back. Because we were small, the deputies were allowed to do things they couldn't have done in a larger department. We had no detectives, so I even let them do investigative work that would normally be assigned to a detective.

There were a few deputies that had to be pushed some, to do something besides just drive around in the patrol car. Most however, I had to throttle back. I explained that Lake County was so small that if they issued a ticket every time someone went 5 mph over the speed limit, they would soon piss off a lot of good people. It wouldn't take long in such a small community to put a large percentage of the population in that state of mind. In a larger city you could get away with giving lots of tickets, and not create any repercussions.

Another point I tried to impress on my young patrolmen was that people were motivated to obey the law from fear or respect. After pulling a motorist over, they had only a few seconds to decide what type of driver they had stopped. If they felt the driver had respect for the law, a warning might be appropriate. On the other hand if the driver was motivated through fear of the law, then they must issue a ticket, to maintain that fear.

The board of commissioners certainly had been supportive in my first year as sheriff. There was just that one year of Lonnie Deur's term to finish out, then I would run for reelection in the coming fall elections. Bill Moore had already announced that he would run against me again. He said that in the last election he had been too busy with police work to campaign. This year he would campaign door to door and the results would be different.

J.T. Nelson came to my office one day, opened his wallet and laid three one hundred dollar bills out on my desk.

"I know with the election coming up you'll probably need that. I'd like to contribute to your campaign."

I shoved the bills back towards him, saying, "Please pick up your money, I don't want to offend you, but I promised the voters I would not take money from groups or individuals that I might then owe payback to."

I could see that my reaction worried him. Maybe he just felt insulted that I had rejected his money. But, if I read him correctly, I would say that he knew he had misread me, and that my influence probably couldn't be bought with money or material things.

When I was young, we had very little. As I became successful in businesses, I had the money to buy lots of adult toys to satisfy what was lacking earlier in life. If this had not been the case, I might have taken J.T. Nelson's money and climbed on the bandwagon for a lot more. Even though I lost plenty on the sale of the stables, I could still afford to spend what was required on my election campaign. I didn't need outside help.

For as much as the commissioners had done to help my efforts, the county was certainly getting its money's worth from me. The ambulance service was again quite unreliable. When the director quit, I offered to take over the post for free. With my experience as a paramedic I didn't feel it would be that difficult to provide a reliable service. J.T. Nelson questioned why I should do this, and along with two other commissioners voted against my proposal. The other commissioners gave their approval. It wasn't long before the Lake County *Star* was receiving letters to the editor complimenting the ambulance service. It also saved the county from paying the $12,000 director's salary.

At the turn of the century the town of Luther once had been heavily involved in the timber trade. These days, the town had less than 1000 people. But most of the people still were employed as wood cutters, loggers or some phase of the timber industry. It was an all-white population that worked hard and played just as hard. Some might classify a good percentage of the people as Red Necks.

There were two bars in Luther. When there was no 24 hour road patrol, they fell into the habit of violating the state's closing hours. Sometimes they would keep the bar open all night and have breakfast specials for those still there.

One of my young deputies stopped by one night and was unsuccessful in trying to close them down. Some of the drunks grabbed his hat, tossing it back and forth in a game of catch. This humiliated the deputy. He just stormed out the door and climbed in his patrol car. There were lots of jeers and laughter as he left. I decided it was time to get these boys back on track.

It was Saturday, a big night for the bar crowd. With two deputies, I waited until after 2:30, the time that the place was legally supposed to be closed. Then, with the two men behind me, I walked in, and from behind the bar took down their liquor license. I told the owner the bar was closed until further notice, that his liquor license was suspended for violating state closing hours. The customers started getting a little hostile, making comments to the effect that we might not have enough deputies to close the place. It looked like a fight was eminent. I called dispatch on my portable radio, asking for backup. For backup I had coordinated with the Luther Fire Department to be on the scene. It was a common riot control technique to cool down the hot-heads.

When the locals came out, they were face to face with their own fire department, who had the hoses out and ready for a wash down. With no further protest the bar was closed. The liquor control commission fined the owner and eventually he got his license back, but from that time on there were no more problems closing the Luther bars.

Right across from the old Totem Pole Bar in Luther was George's Cafe. George had been there as long as I could remember. One day I stopped by for a cup of coffee and a donut. "George, do you remember me or the name Blevins?"

"No, can't say as I do."

"Do you remember the 14-year-old kid that cleaned out the bar with the chain saw?"

"Well I sure do, so does everybody else. Was that you?"

"Yep."

We had a nice visit and a good laugh over my role in breaking up the brawl that night. My Dad sure would have been in rough shape if I hadn't done something. There certainly wasn't any law enforcement nearby.

On September 3rd, 1980, Governor Milliken hosted a crime symposium at Ferris State College for all the county sheriffs and police chiefs. The Governor talked about his newly organized Crime Task Force, and also explained that he was providing counties six million dollars in new funding for secondary road patrol. The funds would be distributed as grants according to the population of the county. But, even the smallest counties such as Lake would now be able to hire additional deputies. The grant would cover their training, salary and additionally provide the extra patrol cars. Larger counties might be able to hire as many as 20 more deputies. This new funding would create a small war in Lansing, for the six million was coming out of the Michigan State Police budget.

As the symposium was winding down, Governor Milliken was at the speaker's podium again. He stated that while the crime in most areas of Michigan had been on the rise, there was one county where the sheriff had actually lowered the crime index by 14%. I looked around and kind of wondered who that sheriff might be.

"I'd like to recognize that sheriff's efforts at reducing crime and have him say a few words. Would Sheriff Bob Blevins please come forward."

I was stunned. The State Police compiled the crime index figures and I had no idea that I was the sheriff he was talking about. Governor Milliken presented me with a commendation and they took our picture together. Wow! was I proud.

Even though I hadn't been sheriff for much over a year, I had a solid record to stand on, one I was ready to brag about. Our crime reduction success was now a matter of public record. With volunteers we had increased our police force by 15 officers at no cost to the taxpayer. I had actually cut over $26,000 dollars from our budget to help get the county out of debt. We had generated a host of new programs to get the community involved in crime reduction and they were working. Congressman Guy Vander Jagt gave me his personal endorsement for reelection.

Bill Moore won the democratic primary and came back to run against me in the general election, but I defeated him with over 90% of the vote. Things were going great, all that I had hoped for. I was walking on clouds. But it was time to get back to business.

During deer hunting season the prostitution business really came out in the open. It was considered to be a normal part of deer hunting to see the mobile homes and trailers from Flint, Detroit and Toledo set up in the Idlewild bar parking lots. It was not at all unusual to see out-of-town hunters lined up outside of a mobile home, waiting to step inside to relieve their stress with some professional lady.

In cooperation with the Michigan State Police we made a concerted effort to reduce this activity. Female state police officers would enter the bars, posing as prostitutes. When a hunter made an offer to acquire the lady's services, she would take him around the corner where one of my men or a state police officer was waiting to arrest him. Also Channel 13 would videotape the fella for his wife and family viewing back home.

We ran the same sting operation for each hunting season; however, after that first big one, open prostitution became less and less, almost disappearing in a couple years.

The prisoners that we boarded for the state all were on the last few years of their sentence. Most of them had been trustees at the Jackson Prison before they came to us. In the beginning they were sent to Jackson for committing some horrendous crimes. When we received them, they were about as close as they could be to rehabilitation. And they didn't have their mind set on causing trouble or trying to escape. They wanted out, not to go back to Jackson.

When I started in the jail at Kent County, I viewed all the prisoners as guilty scum of the earth. Now, after a year as sheriff, I had changed my attitude towards them. I found that they did in fact have personalities and I came to believe that some of them needed to be given a second or third chance. I do consider myself a victim rights advocate, and feel that many times we look out more for the person committing the crime than the victim. The criminal gets a free ride to a warm cell and three square meals. The victim may not have that luxury. Still, I believe that unless we want to keep building more and more jails we must try to give prisoners a chance at rehabilitation, even if the success rate is not high. I do believe in the death penalty and that some criminals also need to be put in prison for good. There are others who could lead productive lives. Just warehousing them away does very little good for anyone, plus it is expensive.

Several of the trustees worked with me when the outside exercise area was constructed. They were hard workers and I felt that many were making a real effort to change their lives. Reading and studying, they were completing their high school through GED, or trying to learn a trade rather than just being proficient at breaking the law. Some turned to religion and became Christians.

Thanksgiving was coming up soon. Dee and I were discussing what the prisoners should be served for that meal. There are those that would say, "Hell, feed 'em peanut butter sandwiches." Probably I would agree with that to a point. Jail wasn't meant to be a pleasant experience, yet there did seem to be a time when compassion was appropriate. We decided to provide them a traditional Thanksgiving dinner with all the trimmings. Dee volunteered to help. Guess we had about 35 boarders that year.

As the food was being passed from the carts into the cells, one of the prisoners said, "You had your dinner yet Sheriff? Why don't you come in and eat with us?" I surprised them when I replied, "Ok." Setting with them at the steel tables, we made some small talk; I heard some gripes, answered some questions and then I took some more of my meal in another cell, and dessert in another. Some prisoners were standoffish. They didn't want anything to do with me. I felt likewise with them. Others were impressed that I would spend a good part of my Thanksgiving afternoon with them.

After that Thanksgiving, and for the rest of my years as sheriff, I would spend more time talking with the prisoners; asking about their ambitions, dreams, trying to understand feelings of hopelessness, whatever was on their minds. I didn't let them waste my time with piddley complaints. Jail was not meant to be a cakewalk. Yet, if there seemed to be a legitimate concern, I listened and tried to be firm, but fair.

With the Christmas season of 1980 I started a practice that would continue each year, turning out to be one of my best kept secrets. About 10 days prior to the actual holiday, I would start contacting the immediate families of my trustees. Because their terms were short, some inmates would have already moved their families to the Baldwin area. Others lived much farther away. I told them that I would deliver their husband on Christmas eve and pick him up again on Christmas day. They could bring a tree, gifts, whatever, and spend Christmas eve and the next morning together. I drove some inmates to homes as far away as Evart.

The understood bottom line was that they would be ready to return at a specific time. Also, if they decided to walk away they would completely ruin that opportunity for other Christmases, and I would make sure the other inmates knew who the person was that made the program end.

Most of the trustees that enjoyed my special Christmas gift were those that had spent many years in Jackson. I felt that they had earned this gift, for they worked hard for me and didn't cause any problems in the jail. I have no regrets that I did this. No one ever found out. I didn't tell and the inmates didn't tell. You can't believe the loyalty I had from them after that, and the problems that I didn't have. More than once one of them would leave a note to tip me off about something that was going on in the jail. They felt a closer kinship or bond to me than the inmate to inmate connection.

All in all my first term as sheriff had been as good as I could have hoped for. With the new programs established, I had connected with the community and it was starting to pay off with crime reduction.

State Senator Phil Arthurhultz presented a special tribute to our department to recognize "the remarkable manner in which these individuals have safeguarded the property and lives of the people of this area of western Michigan." The tribute cited the Lake County Sheriff's Department as "one of a kind in preventing crime."

In January I would start my first four-year term. There was plenty still to be done, and like my attitude in the Marines, I planned to be the best sheriff I could possibly be.

Lake County Sheriff Bob Blevins

Dee Blevins: Lake County Jail Matron

Top: Governor Milliken presenting Bob with a commendation for reducing crime in Lake County (1979)

Bottom: Lake County Snow Mobile Patrol — Bob far right; Undersheriff Dale Gibbs next to him

Top: Lake County Courthouse

Bottom: Sheriff's Department

Count Fisher

It was Bob Blevins on the phone. I hadn't heard from him since the Kent County days.

"Count, I got a job for you."

"Where you at?" I asked.

"I'm Sheriff up in Lake County."

"Oh, from one frying pan to another, huh?" I laughed, but he sounded serious.

"I can use ya."

"What you got?"

"Just come up and talk to me, if the pretty lady says you can."

"What you want to do?"

"Just come up and talk, ok?"

I was running a business on Franklin Street in Grand Rapids at the time. When I got home I told my wife that Bob had called.

"Well if Bob is up there and he needs you, go see what he wants." She knew that Bob loves me and I feel the same about him. I would give my life for Bob Blevins, that's just the way I feel about him. Of course it would take a big sucker to take my life. Several have tried and failed.

My wife says, "Just go see Bob. I know he's your friend."

"He's more than my friend, he's my brother."

When I got up to Baldwin, I met Mike Seelhoff's wife. She was working dispatch in the sheriff's department. Bob came out of his office and I said, "What ya got for me?"

"Count, I don't want to talk business, let's ride."

Old Bob always had a way with me. And he wasn't about to tip his hand just yet.

We drove out to the island. That's the way the locals referred to Idlewild. It sat between two lakes, and it was where I had played as a kid. My grandmother and uncle are buried in Baldwin. Idlewild was a famous summer resort for black people. The boxer, Joe Louis, had his training camp here, and many other celebrities owned cottages on the lakes. There was Sugar Ray Robinson, and I remember a dance team that performed for the King and Queen of England stayed here. The richest blacks in Chicago were the Jones brothers, and they frequently came too. The rich people and celebrities brought their bodyguards, chauffeurs and maids, so there were lots of people in the summer. I remember that I was about twelve or so when I came to visit my grandmother. The streets weren't paved then and more than once I walked from Baldwin to Idlewild to help Benny Caruse set up his drums in the Paradise Club. Jones's Ice Cream was still in Baldwin. A visit wasn't complete unless you stopped there for their great ice cream.

Bob led the way as we walked along the sandy beach. A flood of memories came back as Bob Blevins knew they would.

"How do you like it Count?"

"It's just like I remember it as a kid."

"Count, how would you like to run this and be one of my officers?" Before I could answer he continued on. "I need ya Count. There are a lot of things going on up here and a lot of people I don't trust. We can work together. You're my brother, and with you behind me I'll be all right. Really Count, it's made to order for you."

"How long do you need me Bob?"

"As long as you want to stay, I'll be right here to back you up."

"You know that I'll need to check with my wife."

From Bob's office I placed the call home to my wife. "Honey, Bob wants me to come work for him for awhile." Her answer was no surprise, that's the kind of fine woman she was.

"Ok Bob, I'll be with ya."

Well Bob and I got in the squad car and drove right down to the town hall in Idlewild.

"Mr. Collins, I've got a chief of police for you."

"Oh you have?" Mr. Collins seemed a little surprised.

"I want you to swear him in right now."

There was some reluctance in Mr. Collins. "But, what are we going to do? ..."

"Don't worry, just swear him in."

From the town hall we drove down to Rosanne's Bar. The present chief of police was passed out with his head on the bar. Bob reached over and removed the star from the man's coat and pinned it on my chest. From there we drove over to Bob's office and I raised my hand as he swore me in as a county deputy too.

"There you black S.O.B., you're my man and now we'll clean up this town."

"Where do I live?"

"Get your stuff and be back up here on Monday. I'll have a place for you to stay."

Every town has a bully and Idlewild's was a big black boy named Rags. It was one of my first days on duty. I sat parked in my squad car downtown. I'm pretty particular about my dress and I'll admit I looked sharp. My uniform was all pressed, my brass shiny. Also, I was feeling good about being home after all these years. It was hot out and there were several girls standing around not too far away. Rags came sauntering down the street, spotted me and in a loud voice said, "What do we have here?" He ambled over towards my car, reached in and cuffed my badge with the back of his hand. So all could hear he said, "I see we got another one of these punks." Rags gives off with a disgusting laugh, "I guess I'll have to educate this new one too. What ya call yourself boy?"

Quietly I responded, "I'm your new chief of police."

"Ha, Ha, Ha, the new chief of police." Rags looked around for approval from the girls watching. They had a few tee hees for him, and one said "Oh Rags."

As I opened the door and stepped out of the squad car Rags asked, "What are we going to do to educate you Buddy?"

"I don't know."

"Well I'm Rags!"

"No shit," I said.

Rags laughed again then came around with a punch to my midsection. The next thing Rags knew he was spread against that car with the barrel of my 357 sticking in one of his nostrils.

"Mr. Rags, I'm not the punk you think I am. I don't drink or smoke dope and I'm not afraid. I could blow your head off." The girls couldn't hear what I said to Rags, but I knew Rags could because I was so close he could smell my breath.

"You will get to know me Rags. Believe me I could kill you, but right now you have the right to remain silent. Now boy, I know I don't need to put the cuffs on you, just get into the back of my car." As Rags started to get in it was evident that he had peed his pants.

I was so angry that I did something I wouldn't normally do. My foot pushed the accelerator to the floor and I spun the squad car around in the gravel and turned all the lights and siren on. I picked up the radio mike, "I'm bringin' Rags in!"

Well Rags had a real bad reputation. No one had been able to arrest him before, so when I pulled into the jail driveway the deputies were all out waiting. In the security garage I opened the back door for Rags. "Get out and get over against that wall Rags."

Removing my gun belt, I said, "If you're such a bad ass you can whip me right now. You do, and I'll be your flunky from now on, or anything else. If you don't whip me, then you're going to be good Rags, or I'll break you again."

From Rags there was just a scared "Un huh."

"You see how bad you are when you're off your home ground and your friends aren't around. There aren't going to be any bad people around here unless they are in jail or in the ground. You mess with me and I'll give you a highway to hell. As long as you listen, you'll be ok."

"By the way, the next time you see me just politely call me chief."

"Bob keep him locked up for awhile."

"You got it Count." Rags was not a threat after that.

Another young fellow had a habit of driving fast through town, so I turned on my lights and pulled him over. When I walked up to his car he started to say something smart, but before he could get a word out I said, "Shut up. I'm the new chief in town and I'm gonna do you a favor. No ticket today, I'm just going to take your license number, but next time I don't know what might happen to your car. We're going to change things in Idlewild and I'm the man that's going to do it. If I see ya doin' this again you're in trouble."

Well it didn't take long for word to get around that there was some new law and order in Idlewild. Some people didn't particularly like my methods though. They were starting to get pretty stirred up in this one town meeting, saying things like, "What's your new chief trying to do here, play God?" Another says, "What kind of law and order is he trying to bring in here?"

Mr. Collins tried to settle them down. "Let's hear your comments, one at a time."

Several people started grumbling about things I was doing wrong and then this lady stands up and tells them to all shut up. She looks over at me and says, "Hello Chief. You're doing a good job." With that she starts tearing into my accusers.

"You wanted law and order; you wanted to sleep at night; you complain about people breaking into your house. This man put his hand on two or three of those bad boys of ours and you all start raking him over the coals. What the hell do you want?"

Nobody said anything.

The lady started up again. "I'm telling you I back him if he does things right and so far he's doing things just fine. If he keeps this kind of law and order up I'm with him. Chief, keep on doing what you're doing and I'll back you. I don't give a damn what these people say."

That was the end of the discussion on that subject, and the agenda swung to money matters. When the meeting was over the lady that had spoken on my behalf was standing by her car. "I'm Muriel Castallante, your second boss. Keep on doing what you're doing and I'll back you Chief Fisher."

For two weeks I stayed in a trailer, and didn't like it a bit. "Mr. Collins, I'm bringing my wife up here. We've got to do something different about my housing. Collins' wife was a nice lady. She said that she thought she could help me out. Arrangements were made to furnish me a nice house. The house used to belong to the Jones brothers' mother. She didn't use it so Mrs. Collins said it would be mine as long as I was the township chief of police. Joe Louis had even been a guest in the house at one time. They even put a new roof on it for me, so now I had a nice place for my wife and I to stay.

Bob Blevins expected a lot of his deputies, but in return he gave them his respect. You could see that the way he worked with them on the shooting range. He demanded excellence, because our ability to shoot straight might mean saving his life as well as our own. No one could shoot as well as Bob. Boy could he shoot.

Bob had a hell of a job on his hands, but he was all man. He didn't believe in giving ground and wouldn't allow anybody to run over his deputies. I remember once when he came back from the courthouse where he had been talking to some judge. He was so mad and fired up he was just plain out of breath. "Count I don't know what I'm going to do."

"Bob, Bob. You'll never change. We're doing all right."

He glared at me. All the impatience and frustration had reached a peak.

"Get out of here!"

Bob treated me as good as you could treat a man. But he could be moody and short. He was under so much pressure, it's a wonder he didn't crack. I think he was made of 100% granite.

Although I was the chief of police, I also was a county deputy and would provide backup when they needed it. I usually participated when they had a big bust.

There were so many B&Es in Lake County you could hardly keep up. Once we thought we had a solid lead and would catch some culprits in a house down by Rosanne's Bar. Bob brought in a Mexican who was an undercover man from Grand Rapids and we all sat down and planned our attack. The houses behind Rosanne's were used like a hotel by the girls to turn tricks. This is where we expected to catch the bad guys. Bob was never hesitant about leading a raid. He had no qualms about being first into a dangerous spot.

We parked our cars by Rosanne's then approached our building. When we broke in the door it was beautiful, Bob went high, I went low and the Mexican came in from a side door. It was a thrill the way it all worked, producing a load of adrenalin in our bodies as we stood in a combat stance with our guns drawn. It was just a pretty sight. Unfortunately the building was empty, the whole effort a waste. Boy was

Bob pissed. It had either been a poor tip, or more likely someone had tipped off the bad guys. By this time there were people gathering around wondering what all the fuss was. We walked back to the cars, put our stuff in and Bob said, "Come on, I'll take you to supper."

Bob led our little procession up to a little restaurant in Mesick. "You guys did a hell of a job. Next time we'll get them."

We'd just kind of settled down after a nice meal when this joker comes over to our table. Looking right at me he says, "Damn they let anything wear a uniform nowadays." Boy did he make a mistake in timing. I won't go into detail about what happened other than to say I think this man was totally convinced that what he said was unacceptable.

We left the restaurant wide open for Baldwin. Bob's Bronco was between the two squads. When we pulled up by the jail this cat was runnin' in back by the fence. Bob says, "Look at that S.O.B. Let's get him!" From different directions we moved in and cornered him. Shucks he was only a kid. There was a jewelry box in his hands.

"Do you know who we are?"

"Yes sir, you're policemen. You got uniforms on."

"Where'd you get that chest?"

The kid pointed to a house not far from the jail. We escorted him to the house and knocked on the door. An older lady opened it.

"Lady, is this your stuff?"

"Oh yes officer. We're so glad to get it back. We didn't know what to do."

Back outside the jail Bob gives this kid a little clunk on the noodle with his riot stick and says, "Get your ass out of here. I don't feel like writing you up tonight, do you Count?"

"Hell no, let him go."

As hard as he tried to bring in the criminals, when they were in his jail, if they minded the rules, he treated them as men. If there was a diabetic he would give him shots and make sure they were all treated fairly. He even had a cart where they could get candy and cigarettes.

It just wasn't B&Es that we had to contend with, for at first there were a lot of murders. This one guy shot another at a dance in the Webber hall. We had him in jail, but he escaped. There were sources around that I could tap and eventually we had an informant who gave us the guy's location near a motel east of town. It was me, Mike Seelhoff and Randy Gerke at the motel door. I said this guy's a murderer. We need to break the door in rather than just knock on it. Mike backed up and gave it a hard kick. The door didn't budge so he tried it once again harder. We certainly gave the guy notice we were out there.

"Let me try it," I said.

With one good kick the door flew off the hinges and over on the guy's bed. There was no fight in this guy. Looking into the barrels of our guns he was pretty scared.

Another time I was out driving around and saw this guy laying on the side of the road. I thought he was drunk. Standing right beside him I said, "Come on, get up buddy." He didn't move cause he was dead. I called Seelhoff and the medical examiner.

It turns out the guy was an attorney and a politician. He'd been coming up early every year to get his hunting club's cabin ready. At the same time he was enlisting the services of a prostitute in Idlewild. Guess he just flashed too much money too many times. When they tried to roll him he objected, and they shot him dead. Bob got some good information as to who shot him, but by that time they had left town. Through some good investigative work by Undersheriff Finch and Sergeant Seelhoff we finally caught up with them. It sure was a feather in Bob's cap.

One of Bob's biggest problems was J.T. Nelson, the commissioner from Webber township.

J.T. Nelson was a piano player from Chicago who came to Lake County during the depression. From what I heard somehow he worked himself up to a supervisory position in the WPA. The WPA was one of President Roosevelt's work programs to give people jobs. From there J.T. progressed into local government and an influential position to guide some illegal money-making operations.

J.T. would like to have had Bob Blevins in his pocket like some of the rest around there, but he couldn't. I remember hearing Bob tell him, "I'm here to do a job. If you do wrong I'll get ya. If I do wrong you can get me. Just let me do my job."

When J.T. tried to talk against Bob to me I told him I didn't want to hear it, because Bob was my boss. J.T. didn't own Bob. Only the people owned Bob.

Robert Williams was a man that Bob Blevins didn't like, and Robert Williams didn't like Bob Blevins. It wasn't because they were on different sides of the law, it was because they believed in different forms of government. If Bob would have believed in Communism like Robert Williams, Williams would have loved Bob. Williams was a Communist and proud of it. Bob Blevins was a Marine and just as proud of his country the way it was.

Robert Williams was good to me. He invited me into his home and we had many interesting discussions. He had been in China during the cultural revolution and became close friends with the Chinese leader Mao Tse-Tung. In fact, according to an article in the *Star*, Williams persuaded Mao to make a statement in support of Black Americans. Wherever Robert went he was treated as a celebrity by the communists. He was completely convinced that under communism all black people would be treated as he was. Although Robert was very intelligent and well educated, I think he was confused and used by the communists. When he finally returned to the United States he was not such a celebrity. He was respected by a lot of the black community, however he did not have an opportunity to speak to the masses, just small church and school groups. Robert didn't like the way the system ran here, but he couldn't convince me to think his way. I told him that I was an American, and a Shriner and proud of it. This was the United States and the way the laws are. In his heart I think he thought he was doing things here to make the situation better for the black people. I think Bob Blevins was also out to make things better for the blacks in Lake County, but there were people trying to make his department look racist and that wasn't true.

Dr. Nelson was a friend of mine for awhile, but because of his old club house he wanted to shoot me and that was the end of our friendship.

They were very rich people, Dr. Nelson and his wife. I think they owned half of Idlewild. The doctor had cars in eight or nine different garages and although it hadn't been used in years he still owned the old club house. I was driving by one day and saw this little girl climbing up the steps and ready to look inside the old building. I stopped and told her, "Baby, don't go up there." Just touching that building would make it weave. I got a hold of her hand, and led her off the porch.

After I mentioned the incident to Muriel Castallante she said, "Chief, can you get that torn down?"

"Yes, if I get an order to do it."

"We've been trying to get it torn down for two or three years, but Doc Nelson owns it and he wouldn't let the other chief do anything to it."

"Listen, I just pulled a little girl out of there. Somebody is going to get hurt. You get the order, I'll get it torn down."

After Muriel gave me the written order, I approached two different local companies to get the building demolished. Both refused because they knew Doc wouldn't stand for it. Finally I hired an out-of-town company to do the work. Word started floating around the island that our Chief was gonna have them tear Doc's club house down. Word also filtered back to me that Doc said, "He better not come on the property or I'll kill him."

The day of destruction arrives and the out-of-town company equipment is rattling down the road toward Doc's club house. When I got there a small crowd had already arrived. Either they were here to see the old Paradise come down or for the showdown.

Someone said, "Don't do this Mr. Fisher. Doc's on the property with a gun. He said no tractor is going near his building and what he says goes."

"Here's the order. It's got to go."

Doc's wife drove up with him and their daughter. As the machine got closer Doc got out of the car. "Count Fisher I didn't think you'd do this to me. This is my property."

"I'm not doin' this to you, I'm doin' it for the town. Kids come up and play on it everyday. Which one do you want it to fall on? It should have been torn down years ago, but it's coming down today."

Everybody looks around as Doc starts to pull out his gun.

"Doc, don't do that. I've got too much respect for you. I like you, but this building is coming down. Don't do that." I flipped off the flap that held my gun in the holster.

Things were real tense for a few minutes until Doc's wife took him by the arm and said, "Let's go home Doc."

"Thank you, thank you very much." I breathed a sigh of relief.

Before Doc turned away, his parting words to me were, "Don't ever speak to me as long as you're here."

The third time the wrecking ball hit the old building it all just fell down. As I drove off I cried. Facing Doc was a hard thing to do. I did my job.

I figured Idlewild was pretty well cleaned up and the pay was so low I couldn't hang around any longer. Thanks to Bob Blevins and the support of Muriel Castallante and Mr. Collins I think Idlewild was a safer place to live. Had I known how bad it was going to get for my friend Bob, I would have stayed and backed him up as he had backed me. But I didn't know.

If the Walls Could Talk

I would have preferred to have my first full term start out on a better note. If I'd have been more perceptive, perhaps I could have seen it coming. For the last year I didn't have much contact with Art Speers except at the sheriffs' meetings. When Dee was around, Art went out of his way to flirt with her. I didn't feel threatened by this, in fact I felt rather proud of the fact that my wife was still attractive and could catch a man's eye. Art never was real smooth when he had been drinking anyway.

The predictions of the other sheriffs came true. Art was not reelected Osceola's Sheriff that fall, so on January 1st he would be out of a job and would have to find a new place to live. It was our last sheriff's meeting for the year and the women were invited. Art was really dipping into the booze. Earlier in the evening I had asked about his plans. He said to start out with, he'd just take some time off. It was evident he was feeling a little low. Once Art got a good buzz going he was back making eyes at Dee.

Later that evening I was sitting at a round table with a couple other guys. Art was there too, but he was reasonably quiet like he had something on his mind. Finally he leaned over and whispered in my ear, "Why don't we swap wives for the night?" It was like his suggestion set off a little bomb inside of me. I was ready to clean his clock and as I came out of my chair after him someone grabbed me. Slowly I regained my composure. It was just as well that I was restrained, for it would have created an incident that I might have regretted, and one that could have been embarrassing to the Michigan Sheriff's Association.

Not too long after it happened, Art called up apologizing for what he'd said. "Bob I drank a lot, was just kidding, would never have followed through. Please accept my apology. Let's let by-gones be by-gones." I accepted his apology, but I didn't believe him. Having laughed the flirting off before, this time there was no doubt in my mind what Art was after.

About two weeks into January Art Speers was standing on my doorstep begging for a job. If there had been a job opening, or even if there would have been someplace in the department I could have hid him, I probably would have made a dumb decision from the heart and hired Art. As it was, we only had a small department and there was no position open. I told Art I couldn't hire him. He stormed out of my office and I've never seen him since. I did feel sorry for Art. He had been living in the Osceola sheriff's residence, so now he had no home and no job. Cindy was still the county clerk and after Art left my office, his first stop must have been to give Cindy the word that I wouldn't hire him. Because, from that time on Cindy Keefer was very unfriendly.

Art certainly was a big help when I started my job as sheriff, but Cindy was even more important. She personally took care of documents and reports that were crucial to the department's funding. Now she wouldn't even speak to me. When I came in she would tell an aide, "Get the sheriff what he needs," then she would walk off. She never mentioned me not hiring Art, but I knew she was pissed and from that time on I got the cold shoulder. That also was the end of the calls she gave in regards to the board of commissioners' meetings. I thought she'd eventually get over it. Cindy never did though, and this was only the beginning of the trouble she would cause me.

I thought that the relationship between blacks and whites in Lake County was reasonably good, but there were others that didn't believe this. They felt that there needed to be changes. There was pressure from Robert Williams and a few others to have more blacks in the sheriff's department. In January there was a meeting to address these concerns. I explained that for the last job opening I had forty applications, but none from blacks. I had met with the local NAACP, the Executive Director of Michigan Department of Civil Rights and even consulted with State Police Posts to find ways to recruit more blacks.

At that time Lake County deputies were making around $12,000 annually. With such low wages being offered here, qualified blacks were going elsewhere for more money. Those present seemed satisfied with my efforts and attitude, plus shortly after that I promoted Gwen Warren.

Gwen Warren was a black lady who had been employed by the sheriff's department for almost 10 years. She'd worked in a variety of areas from dispatch to secretarial duties. She seemed to have a good feel for the community, was very intelligent, and knew how to make the office run smoothly. I promoted her to Lieutenant, assigning her the responsibility of running the entire jail. Gwen was responsible for the work schedules, booking and receiving and any inmate problems.

My good friend Count Fisher came up with the idea of holding a local inaugural ball to honor newly elected officials. Count and his wife, Renee, planned the whole celebration and decorated the Idlewild Lot Owners Association Building for the occasion. The music was provided by the Count Fisher Trio. Renee sang with the band and my undersheriff Bruce Finch even substituted with Count as the drummer. It was an absolutely beautiful evening. Blacks and whites seemed to blend into complete social harmony. I told Dee that I couldn't be happier. This was my Mayberry, my home. We were here to stay.

I was determined that my second term was not going to be just a business as usual operation. It was time to plunge into all the things I wanted to accomplish. The Department of Corrections people realized that handling and controlling prisoners became a much easier process when bad behavior was rewarded with denial of privileges. And, there had to be privileges established before you could take them away. When all you offered was three meals a day, plus a bed, there just wasn't much to take away to control behavior. Beating inmates into submission was just not an option of the day.

Under the jail was a huge basement that we used for storage of stolen property. With a little reorganization this could be turned into an excellent inside exercise area and weight room. There were no funds available for this project. Everything would have to be donated or come out of my pocket.

We separated off the boiler room, constructed a secure place for evidence and still had 75 percent of the area left for an exercise room. Peacock Furnace Company, on M-37 north of Baldwin, donated the eight-inch circles stamped of steel plate for the bar weights and drilled the holes for the bars. Earlier, they had provided the steel bars. Out of my own pocket I purchased some angle iron. Using my welder, we constructed padded weight benches, plus the other equipment that turned the basement into a fine exercise and weight room. A little paint job made the whole place look as good as any public training facility. I thought this exercise area would be very beneficial to relieve stress, improve the inmates' state of mind and outlook on life. Again, there are those who would say that prisoners don't deserve this and that jail time was supposed to be punishment. To them I would say that the privilege an inmate has to work out under supervision a couple nights a week, guarantees me inmates that keep clean cells, with no graffiti on the walls, give the guards no problems and keep peace among themselves. That alone makes it worthwhile. If one inmate gets out of line, the whole cell may lose its exercise privileges for awhile. In that case, the cellmates will administer a little attitude adjustment on their own, without me even having to lift a finger.

At one time the building next to the jail was the sheriff's residence. Some past sheriffs refused to live in it, so the county turned it into offices for the probate judge. I went to great lengths to convince the board of commissioners that the building should be returned to the sheriff's control. It could be used to satisfy some state requirements, in addition to relieving our cramped office space. They finally agreed.

With trustee labor we got busy remodeling and converting the old house into something we could use. I relocated my office into one of its old bedrooms, and another was used for the undersheriff. The kitchen was changed over for the department secretary, with filing space and a counter area for her to wait on people. In the basement we constructed a comfortable squad room for the deputies. It had lockers, coffee, a bulletin board and both a separate men and women's dress area. It provided an excellent place for the changing shifts to visit and pass on information to each other.

The jail area we moved out of provided more cell space for the trustees, plus a visiting area for inmates to receive family and guests, which was a state requirement.

In my second term Dee started working at the sheriff's department too. She went to school and was state certified as a corrections officer. Dee oversaw the jail kitchen, ordered food, planned the menus and worked with the cook and servers to ensure the food was cooked properly and delivered hot from the electric carts. Also, whenever we had a female inmate, she would be present when the inmate was transferred or searched. She did a great job.

Dan Rose was a long-haired hippie when I first met him. His sister boarded a horse at our place and for some reason long-haired Dan was there on his motorcycle. Well I visited with him and he seemed like a nice-enough kid so I asked him if he'd like to ride along in my patrol car that night. As we rode around in the patrol car with the radio chattering at us I could see some magic forming in Dan's eyes. The next night he was back again, this time with a haircut and shave. It wasn't long before he was asking about the police academy, and now he was a deputy in my department.

I looked out the office window one day and recognized a car that belonged to a fellow we had a warrant out for. Dan Rose was there. "Come on Dan let's get 'em."

We climbed into a patrol car and I called for backup as we sped after the two men in the car. They could see us coming full bore with our flashing lights on, so they pulled over at a gas station on the outskirts of town. Dan headed for the left side of the car, but before he could do much the driver was out of the car attacking him. For a second the guy got his hand on Dan's gun and in the struggle it was thrown about twenty feet away. To say the least, Dan had his hands full.

The fellow we had a warrant for burst out of the passenger side. With some martial arts experience, he braced himself on the car and door while he planted both his feet in my chest. That certainly took a few seconds out of my game plan. He and I were in a real wrestling match when a third man comes from the rear seat. I managed to get my friend in a headlock as the third bad guy heads out to pick up Dan's gun. Holding the headlock I reached down with the other hand and pull my second gun from the hideaway holster strapped to my leg. I holler at the third guy, pointing my gun in his direction. Fortunately he stops moving towards the revolver on the ground and raises his hands. Thank God, Count Fisher pulls up in the nick of time. He handcuffs my man while I go over and give Dan some help. Dan just about had the driver subdued.

All three men were arrested and hauled to jail thanks to Count Fisher's timely arrival. Dan and I were pretty well banged-up though. I had a ruptured sternum from where the guy had kicked me, but I guess we both were lucky to get out of the situation as well as we did.

It was April when the body was discovered. A hiker came in and reported that he noticed the blue box exposed below a newly fallen tree. He had made a little hole to see what was inside. First he put his hand in. Alarmed at what it felt like, he got down on his knees to peer inside. It looked like human skin, so he got the hell out of there.

People think they see all sorts of things, but you have to check them out nevertheless. We forced open the box and found a very small woman inside. She had been shot. Even though the skin was withered, the body was intact, including the facial features. We cordoned off the area and started a search for evidence while we waited for the medical examiner.

A check of our missing person files turned up a report that had been filed that summer. The husband was the individual that filed the report on his missing wife. The picture attached to the report looked like our victim.

I ordered an autopsy and we finally confirmed the identification through dental records. The husband was called in for questioning. He said that they had been having some marital problems, and that his wife had just taken off to go find herself. When she didn't return he became suspicious and filed a missing person report.

The blue foot lockers like the one she was found in were sold in the Big Rapids area. A search of the husband's property revealed two more such boxes in his garage, loaded with hunting gear. We suspected that he might be involved, and under intense questioning he admitted to shooting her. She had threatened to take away his kids and get the home too.

In June of 1980 Fred Corb came to me with the suggestion that if I hired him as a deputy, he would also provide the services of his German Shepherd tracking dog, Champ. Fred had several thousand dollars invested in Champ and the dog was a state certified tracking dog. He even offered to use his station wagon as a patrol vehicle to haul Champ, if the county would pay for his fuel. The board ok'd providing Fred his gas. I saw to it that his station wagon received a black and white police paint job, decals, plus a radio. Several individuals donated money specifically for the tracking dog program. This money went into our contingency account, then came out as needed to help Fred with the cost of shots and veterinarian bills. Later as a promotion, a dog food company donated a year's supply of dog food.

Champ proved that he was worth all this effort. In fact, he was the best tracking dog I've ever seen. Champ was instrumental in finding lost children, walkaways from nursing homes, in addition to locating those responsible for burglaries. He became quite famous and other counties, even the Michigan State Police, would call for his help on searches. Since arriving in Lake County, Champ was credited with finding 17 lost people and being responsible for 20 arrests.

Fred and Champ had been with us for almost a year when Fred announced he was resigning to take a position in Oceana County. I certainly hated to lose him and Champ, for I really appreciated the service they had provided. In an interview with the *Star*, Fred said, "It had been a pleasure to work for a sheriff who has done so much for professionalizing law enforcement in Lake County." That statement certainly pleased me.

In the spring of 1981, the Lake County Big Brothers and Sisters Organization gave me a certificate of appreciation for my dedication in helping to develop and maintain a program of friendship and guidance for the Lake County youth. This recognition really made me feel good. A law enforcement agency must be an integral part of the community, not a separate agency. I felt I was starting to achieve this strong tie to the community. At every possible opportunity I would speak at schools, and to community groups, enlisting volunteers in all phases of law enforcement. Recently I formed a county dive recovery team. With all the water recreational activity in the summer, this was certainly a needed asset. With a wealth of experience from the military and Kent County I was able to train an all-volunteer team. Here again, the county was getting plenty from their sheriff at no additional cost.

It took a couple of years before the union response began to take shape over my firing of the deputies. The union had been pushing, but not hard. We were now faced with the possibility of having to reinstate these men and pay them back wages. I had built a cohesive, effective road patrol and to bring these guys back would mean laying off the fine people I had hired. Paying back wages would really break our budget, plus nobody wanted these characters back. The present employees told the union that they didn't want these guys back. The union said that once the action was started they would have to continue to pursue it, as they had an obligation to the four men. The employees then fired the union, voted it out and joined another. This cut off all revenue to the union from our department. The union had a strong case but finally lost interest in representing the former deputies and the case fizzled out. I really lucked out having the employees behind me 100 percent.

I took great pride in the professional team I had assembled in the Sheriff's Department, but as in any other organization there would be changes for various reasons. Dale Gibbs was my first undersheriff. He was a good friend and a great help getting me started, but his wife still lived and worked in Grand Rapids. She didn't like the fact that Dale was hardly ever home, and she didn't want to move. Dale needed to do something or face a divorce. Bruce Finch worked for the Rockford Police Department. He wanted to move north, so a deal was worked out where they just switched positions. Bruce proved to be an excellent undersheriff and a tremendous investigator.

Todd Chaney was another real fine investigator. He was responsible for solving numerous cases and just seemed to keep improving in his ability. There was only one problem. Todd felt that on the cases he solved that he should be the one interviewing with the media instead of the sheriff. He wanted the limelight. I afforded him the opportunity to do this occasionally but most of the time I did the interviews. Never was I reluctant to give credit where it was due, but this was my department and I was the one to face reelection every four years. Todd didn't seem to grasp this. The more cases he solved, the better he got, the more resentful he became until he resigned.

In the summer of '81 there was a guard position open. A man by the name of Joe Kuhlman came by to apply for the job. The starting pay for jail guards wasn't much, only $3.79 per hour. Joe said that he was a retired master sergeant from the Marines. He was drawing his retirement pay and just wanted to settle in the quiet north country, make a few extra bucks and enjoy retirement. Joe was presently a jail guard in Ottawa County. Being a retired Marine told me plenty about Joe. Leadership, trust, loyalty were all things synonymous in my mind with a Marine. I did call the sheriff down in Ottawa to ask about Joe. The sheriff was Bud Grysen, who later became the Executive Director of the Michigan Sheriff's Association. When I asked Bud about what kind of a deputy Joe was, I was surprised by the answer.

"Joe has caused me all kinds of trouble. He's a drinker that can be sober for a long time, then go on binges. He's been in trouble for another problem here, enough to get suspended for 30 days. I'd fire him but this was his first offense and with the union I need to show progressive discipline for the problem. Nowadays you can't fire someone for alcohol abuse, that's considered a social problem." Bud didn't elaborate on what the problem was that Joe got suspended for, but did say, "I'd like to send him to you, but I sure wouldn't recommend hiring him."

I questioned Joe about the drinking thing. He said that he hadn't had a drink in two years, but yes, Vietnam flashbacks triggered some occasional heavy drinking in the past. As far as the suspension problem, he had been burned when someone else took an inmate's money after he was booked.

There was no indication of any alcohol problem when I met with Joe. Being in love with the fact that Joe was a Marine, I overlooked what Bud had told me, didn't look for any more applicants and hired him. It was a decision of the heart, not a rational mind. Not far in the future I would certainly regret that decision.

On the last day of their jail sentence, I had the long term (10-15 year) prisoners brought to me. I gave them a heavy duty stainless steel jail cup with a bullet in the bottom, then I told them, "For the rest of your life, I want you to have your coffee in this cup so you never forget about your experience in the prison system." If they didn't ask about the bullet, I didn't say anything. If they did ask, I'd say, "Just in case you ever think about coming back, use it." Several made it into a locket to wear on a chain around their neck, a symbol that they would never go back.

John was a state prisoner, finishing out his term as a trustee in our jail. He was the jail cook when I became sheriff and finished out his term in that capacity, though he preferred being referred to as a chef. John came in for his coffee cup with the bullet inside. I thanked him for the fine meals he had provided the inmates through the years, and I complimented him for being an excellent prisoner. John had something to say too. "There's something I want you to know, cause I don't know how to contact Lonnie Deur. Now that I'm leaving I can tell the real story about the prisoner beating."

"Lonnie Deur never beat that prisoner. They came in and talked to the prisoner. Cut a deal with him to conjure up the beating charges against Lonnie Deur in exchange for the prisoner's criminal charges being reduced to some rinky-dink charge, trespassing or something."

"He was in the next cell. I could easily hear the deal being made and from the reflection in the windows I could see this guy beating his hands on the bars to create marks and swelling. Since the sheriff survived being shot they decided to disgrace him, by accusing him of beating the prisoner's hands with a flashlight."

I thanked John for the information and after he left I thought, "Why would he tell that story if it wasn't true? What would he gain if it wasn't true, for he's already a free man?" I could understand why he hadn't said anything before; there surely would have been reprisals against him from the other prisoners and Bill Moore.

Later I happened to see Lonnie Deur and I told him what John had said. His response was for me to just let it go. He didn't plan to get back into politics; it was water under the bridge. Lonnie did thank me for telling him. At that time I should have considered the possibility that the same thing could happen to me, but I didn't.

As I said before Bruce Finch was a fine undersheriff. Unfortunately circumstances surrounding Bruce's marriage were affecting the sheriff's department in a serious way. This was a very uncomfortable situation for me but I felt it was necessary for Bruce to be aware of what I knew and that I was concerned. Following our conversation things happened that led to Bruce's resignation. Somehow I think he felt ill will towards me for bringing the problem out in the open. He was cold and distant from that time until his resignation.

Joe Kuhlman really had been a asset to my department. As a jail guard he did a fantastic job. He was pleasant and brought maturity and leadership to the jail; everything I would expect of a former Marine. There was no drinking problem evident. Joe had earned my trust.

"Joe, I've got an election coming up in a couple years. It doesn't look too good when I can't keep an undersheriff for more than a year. Now I'm looking for a third one. You've done an excellent job for me, would you consider being undersheriff number three?" There was no hesitation in Joe's acceptance. Within a matter of minutes Joe was moving into the undersheriff's office. He couldn't have been happier. I gave him an old-fashioned undersheriff badge that was mounted on a plaque. It was silver and very large. I wanted him to have it for his wall. Joe was to be my second in command, my executive, a marine buddy. I knew this would be a good relationship, things would prosper. I couldn't have been more wrong.

Early on, I learned from other sheriffs' experience how important it was to keep track of all money that cycled through the department. Quite often people would donate money and as time went on I would increasingly solicit funds for different worthwhile projects that the county couldn't afford or just wouldn't fund. In the beginning there was no set procedure for handling these funds so I had Todd Chaney and Bruce Finch set up a Sheriff's Contingency Fund. Their names were on the account and both needed to sign their names to any check disbursing funds from the account. There was money donated for Fred Corb's dog expenses and several officers had to pay for their own training so the contingency fund money helped to pay for some of their gas back and forth to school. Todd Chaney also used some of the funds to purchase some miscellaneous police equipment he felt we needed.

Once a year the Sheriff's Department had an auction of unclaimed stolen property. That money went back to the county's general fund, but we also auctioned off donated items and had a bake sale. That money went to the contingency fund.

I prided myself in being a good business manager. There were responsibilities I just didn't delegate either. I was the only one that spoke to my board of commissioners on department issues. I personally endorsed all revenue checks for the department and turned them over to the county treasurer. All department action was thoroughly documented, all monies thoroughly accounted for. There wasn't even a slush fund for coffee. I was paranoid about accountability and I was the sheriff 24 hours a day. No one was assigned power of attorney for me.

When Bruce Finch and Todd Chaney left, I had my new undersheriff Joe Kuhlman take over the contingency fund. Unlike the other department financial matters, I thought it was appropriate that someone other than myself continue to handle the contingency account. Joe did keep a ledger that I reviewed, plus I asked his wife Lois to balance the account.

Well into my second term, I couldn't have been more satisfied and proud of the sheriff's department's success. Just recently an investigation by Count Fisher and deputy Dave Fouke yielded $5,000 in stolen goods.

A lot of our success could be attributed to the community's involvement and support. That doesn't mean that everything always went well, because it didn't. There still was an occasional homicide and those can generate more public interest than several good programs.

The Republican chairman from a neighboring community was found shot to death on a Lake County road. The man was known to have flashed money in the Idlewild area, frequently using the services of a prostitute. Apparently someone decided to relieve him of his bankroll and in the process something went wrong and the guy got filled with 22 caliber bullet holes.

There was another shooting that infuriated me with our local judicial system, because I felt it could have been prevented. It all started back in April when I was called to Willie Scott's home for a domestic dispute, and Willie ended up pointing this shotgun in my face. Fortunately I was close enough to him to snatch the gun away. It was loaded and I had the distinct feeling that if the situation had run a few minutes longer I would have been shot. Scott had some serious mental problems and right after that incident he was committed to the Traverse City Regional Psychiatric Hospital.

When Willie returned from Traverse City he came over to the sheriff's department to claim his shotgun. I refused to give it back so he went over to David Hill, the prosecutor, for help. The prosecutor insisted since I had filed no charges after the April incident that we had no reason to hold his gun. I tried to explain to him that filing attempted murder or assault with a firearm against a mentally impaired person would just not stand up in court. Hill quoted me the law and said that we needed to give the gun back to the owner. The owner was Willie Scott's father, but I knew that Willie Scott was the dominant person in that family and that giving the gun back to his father was like giving it to Willie.

I went over to Judge Farabaugh's chambers and asked if there was anything I could do to prevent giving the gun back. He was no help either, so reluctantly the gun was returned.

It was on the 7th of July, about two weeks after the gun was returned, when Omar Mangram, a mental health caseworker, visited Willie Scott's home to see how Willie was doing. He and Willie talked outside the home for a few minutes and then Willie said he had to go in and use the bathroom. Omar had the feeling something was wrong and was in the process of leaving when Willie came out of the house with the shotgun. Willie chased Omar around a car parked in the driveway. From a distance of about 12 feet Willie fired, hitting Omar in the back of the head and neck. Omar feared he would get shot again so he got up and ran off in a wooded area where he collapsed. Willie's brother called the police. Fortunately the shotgun shells held #6 shot. Anything larger would probably have severed Omar's spinal cord or blown his head off.

Omar was rushed to Butterworth Hospital. I headed to Judge Farabaugh's courtroom with the shotgun. I was furious. Court was in session with lawyers, jurors and spectators present. I burst through a side door and walked to the front of the courtroom holding the shotgun up so all could see. To Judge Farabaugh I said, "You wanted me to give back this shotgun and I did what you asked. With this shotgun Willie Scott almost blew Omar Mangram's head off. Right now Omar is enroute to Butterworth Hospital and I don't know whether he's alive or dead. I'll tell you one thing, this shotgun will never be in Scott's possession again."

I stormed out of the courtroom and at the sheriff's department took the gun by the barrel and smashed it on the corner of the building. I also informed the media about the malfunction in our criminal justice system in Lake County and vowed I would never return a confiscated weapon to a violent person again.

Judge Cooper and Judge Farabaugh called a meeting with the media in my office a week later to defend their turf. Farabaugh charged that I should have filed an appeal to block the return of the weapon, that it was basically my fault. He called it a big publicity stunt; me hoodwinking the public in blaming the judicial system for returning the gun. He said that I was an inept lawman masquerading as a white knight. Well maybe I hadn't jumped through all the proper legal hoops as I should have, but I had made an earnest appeal to both Farabaugh and Hill for not returning the gun and it had fallen on deaf ears as far as I was concerned. It was hard for me to believe that a hearing or appeal would have changed their minds.

We also were starting to have some pretty major problems with the folks at the courthouse. My reports started disappearing. The secondary road patrol grant paper work was not being sent to the State on time, or not at all. Russ Webb from the office of Highway Safety and Planning called. I was the only sheriff that did not have his report in on time. I explained that my grant officer Mike Seelhoff had finished it early and I signed and submitted it through the county clerk's office a month ago. I should have been the sheriff with his report in first. I promised I would send a copy. Russ said he couldn't accept a copy, he needed a document with original signatures. Now I had to hand-carry the document, get the board chairman and county clerk's signature and run it to the post office.

This wasn't my only dilemma. Governor Milliken was ending the boarding of state prisoners in county jails due to insufficient state funds. In 1981 this brought over $100,000 into the Lake County treasury. This money was used to maintain the sheriff's budget and any leftover money went into the general fund. Now that these state funds were being withdrawn the commissioners were telling me that there would need to be huge cuts in my budget with layoffs because there was no money to fund it. Apparently they forgot that it was the original responsibility of the taxpayers and general fund to support the sheriff's department before we brought all this extra money in. At most every board of commissioners meeting there was some kind of skirmish between myself and the commissioners over the budget.

When I wasn't there to defend my department they would take even more action. For example in July while I was on vacation they cut another $38,000 from my budget. Roger Beilfuss was the only commissioner that spoke against the cut. In the *Star* he was quoted as saying, "This cut is grossly unfair... other departments are not cut... you just told him to cut $33,000 ... now you take out another $38,000." Of course Cindy Keefer was all for the cut and Commissioner Chairman Bartoletti backed her. The others just fell into place behind them.

I did show up at the next meeting and pointed out while they were cutting my budget they had increased their own budget for conferences. Also from the state prisoner money we had brought in the county clerk had received a nice fat annual pay increase and so had others. I also again outlined the things I had done to cut expenses and make the department more efficient. That didn't seem to matter.

There was one state prisoner program left that brought us some money. A few state prisoners were still housed in county jails if their life was in danger in the state institutions. The commissioners tried their best to pass a resolution to end this program too, but I finally salvaged it in the next meeting. It seemed to me that there was a concerted effort afoot to destroy my ability to produce income to support law enforcement in this county.

Normally when I appeared before the county commissioners at their meeting, I was only there for the time it took to cover the sheriff's department business. When the meeting notes came out I would review what happened in the rest of the meeting. The county medical examiner would also periodically appear before the board and leave death certificates to be filed. He received a fee for each certificate and the dead persons' names were listed in the meeting notes. One day I was reviewing the board meeting notes with another deputy and I remarked that one of the names listed by the medical examiner was that of a person who had died several months ago. That seemed strange. A little investigation revealed that all the people listed had died several months before; some had been dead over six months before their death certificates finally were turned in.

Dr. Nelson was quite elderly. He had been a physician and the medical examiner in Lake County for quite some time. We weren't sure why Dr. Nelson was waiting so long to file the certificates, but at first it didn't seem that big of a deal. After all he was getting up in years.

There were a couple of bright spots to my routine. The Michigan Crime Prevention Association was an organization of Michigan law enforcement people that met to plot strategy and exchange information on combating crime. They had a special meeting in Canada that I couldn't attend; however at that meeting the members recognized my efforts with an Outstanding Achievement Award and named me the Crime Practitioner of the year. That award meant a lot to me because it was presented to me by my peers.

The other bright spot that spring was a gal by the name of Sherri Holliman. Sherri was working as a bouncer for Club 37, a very nice eating and nightclub establishment. Club 37 was noted for their Walleye fish and prime rib dinners. Sherri wanted to be a policewoman and go to the police academy. She was a single parent raising a young son. Sherri had a very pleasant personality. She also wanted to be a policewoman for the right reasons. I hired and sent her to the academy where she graduated as one of the top two or three in her class. She would be the first female patrol deputy for Lake County, and a good one.

As I mentioned before, the village of Luther had a fair amount of law violators. The infamous locals were colorful and just plain didn't like police or any authority figure.

Sherri couldn't wait to patrol in the Luther area. It probably was not the best place to have a single deputy, let alone a woman, but I was also reluctant to hold her enthusiasm back. On her second or third night out she got into a good scrap. When she returned to the jail she was pretty well beat up. There was a cut lip, bloody nose and her shirt was nearly torn off. Sherri did have a back seat full of handcuffed prisoners, and in one night she had gained the respect of Luther, especially the frequenters of the Totem Pole Bar.

J.T. Nelson was responsible for supervising a government funded program (CETA) to employ county youth on worthwhile projects. One project was to refurbish several black cemeteries. Supposedly he put together a crew for this purpose and the summer project was well underway. Bob Smith, the bank president, said that J.T. Nelson would come into the bank with all these payroll checks for his work force. The workers would endorse their checks, J.T. would countersign underneath their signatures, cash the checks and leave with all the money. J.T.'s explanation for this arrangement was that there was a food vendor that visited the work site, providing meals on an I.O.U. basis. J.T. wanted to make sure that the food vendor was paid for his services in case some worker decided to leave with his pay before settling his account.

This certainly sounded like a fishy arrangement to me. We visited the black cemeteries of the county. They were in sad disarray. There were fallen trees, weeds as high as your armpits, even tombstones tipped over. No restoration activity was evident in any cemetery; in fact, in some of them, it looked as if there had been no activity since the last person was buried.

There were local township elections in the fall, with some Webber and Yates township residents complaining of the election results. Investigations of the voting records revealed that several deceased residents had been filing absentee ballots, voting in all the elections. Some had been dead for several years. If these folks were still voting, chances are that the families were still receiving and illegally cashing social security checks, long after the resident's death. I provided some of this information to the local paper. Channel 13 TV picked up on this news clipping and sent a reporter to do some investigation. At least a hundred names came up of dead people still voting.

Dr. Nelson was a very elderly, well-liked gentleman. No one really wanted to believe he was purposely withholding death certificates for any reason, but now it was public knowledge that there was some real illegal activity here, and I knew that it was just the tip of the iceberg. I went to the prosecutor with everything I had. Shortly after that, Dr. Nelson gave up his job as medical examiner, turned in all remaining death certificates, and the dead people were taken off the voting roles. That didn't address all the social security checks being illegally sent out, though it did stop them. I hoped that the State Attorney General might get involved and pursue some charges, but that didn't happen.

It wasn't too long after all this that Dr. Nelson passed away. No one will ever know if this whole scheme of withholding death certificates was an oversight, or master-minded by someone else that wanted to have control of an election voting block. One obvious outcome of all this was my fall in popularity among some of the black residents. Extended families had been relying on these illegal social security checks for years, and now they were gone. I was to blame for this.

With the county medical examiner's position vacant an arrangement was made to use the medical examiner from Newaygo County until the Board could fill the vacancy. Because of my paramedic experience, the Newaygo examiner appointed me his assistant so in cases where the cause of death was obvious, the body could be removed without the Newaygo examiner having to travel all the way to Lake County. If someone died in a nursing home I would get that person's doctor to sign the death certificate, and if the death resulted from an auto accident I could fill out the certificate. If there was any apparent foul play, or the death was questionable, I would immediately call in another medical examiner so I wouldn't be filling that role and that of an investigator.

Count Fisher came up to Lake County as a favor to me. He left a business in Grand Rapids to do this. Being a township police chief and county deputy didn't pay much, so it wasn't a big surprise when he came to me with his resignation. The Count, with the solid backing of Muriel Castallante, had cleaned up Idlewild. He also was a real effective county deputy, but most important of all, he had smoothed relations between blacks and whites in the community. The Count was one man the radicals didn't mess with.

If I'd asked him to stay I think he would have. Things seemed to be going pretty good then, so I didn't question his decision. I certainly would miss him. The Count was a true friend I could trust. Had either he or I foreseen the trouble on the horizon, the Count would have never considered leaving. I'm sure his presence by my side would have made a world of difference in how things turned out. Many times both Dee and I have wished that I would have offered him the county undersheriff position. That is all water under the bridge now though.

In the fall Robert Williams returned from his sixth trip to China and reported miraculous advances in all fields since the great Cultural Revolution. He was openly critical of the Reagan administration for not pursuing a one China policy. He said the Chinese were very upset. It was no surprise to me that Williams was one of their favorites.

Nineteen Eighty Two ended with a shooting that put the department back in the news. The young man involved had been in and out of trouble a good share of his life and since escaping from a state training facility he took up residence in a cottage out by Joe Kuhlman. One night a couple of days before Christmas this guy walked over to a nearby house, put a shotgun next to the glass in the front door and shot a woman in the back who was standing inside the house. There was no reason for the attack other than he was in the mental state of envisioning himself as Satan. After the shooting he was on his way back to his cottage when Joe Kuhlman came over to see what the shooting was about. The guy was yelling that he was Satan and was going to kill everybody. When he saw Joe he even took a couple of shots at him. Joe called in for help and myself, Mike Seelhoff and Jeff Chamberlin responded.

When we arrived the guy was standing on the cottage porch acting crazy, waving his shotgun around saying, "I'm Satan. I've got 500 rounds of ammunition and I'm going to kill everybody." It was dark out. I told Mike and Jeff that I would crawl through the weeds and get as close to the porch as I could, then turn my eight-cell flashlight in the guy's face and tell him to drop his gun. That just might startle him enough so that we could get control of the situation. Mike and Jeff were to come down on opposite sides of the cabin and be ready when I turned the light in his face. Joe would try to talk to him, keeping his attention.

On my belly I crawled to within ten feet of the guy. I had my gun in one hand and with the other I flashed the light in his face and yelled, "Drop the gun!" It worked. He dropped the shotgun to the porch floor and Mike and Jeff quickly moved up on each side of me with guns drawn. I holstered my revolver and was reaching for handcuffs when this nut jumps off the porch right on top of us. Mike's gun discharges and the bullet goes through Jeff's forearm, hits the guy in the stomach and exits through his back. Jeff's wound wasn't too serious, but this fella was on the ground moaning that he's gonna die and will never see his girl again. We called an ambulance and in the process of trying to load him he started to put up a good struggle. I think he must have been full of drugs.

My mother had just had surgery to strip some veins from her legs. When I went down to see her during recovery, guess who they had in the bed next to her? It was our self-proclaimed Satan from the night before. I asked the nurse if they could please move either him or my mother to another location. He survived and was sentenced to life in prison. The woman he shot was in terrible shape, but did make a miraculous recovery.

Top: Police Chief and County Deputy Count Fisher

Mary and Mike Seelhoff
at the inaugural ball

Count and Renee

Top: Bob, Fred Corb and his tracking dog Champ; the Science Diet Company furnished a year's supply of food

Bottom: Bob's son Buck became a diver and deputy with the Sheriff's Department

Too Many Battles

The beginning of 1983 was the high point of J.T. Nelson's career. J.T. had been involved in Lake County government for 28 years and now he was Chairman of the Board of Commissioners. J.T. was well known not only in the state, but also at the national level where along with over 5000 other county administrators he attended each National Association of Counties conference. At these conferences J.T. specialized in attending seminars on criminal justice and law enforcement. He also was considered an authority on financial management, CETA funds, and was member of the national steering committee on employment. In Michigan he was the past president of the Michigan Association of Counties and had recently been reelected chairman of the West Central Michigan Employment and Training Consortium Administrator Board. That body was charged with administrating funds under the Comprehensive Employment Training Act (CETA). J.T. was a former member and officer of Five-Cap (food distribution program) and a supervisor of summer youth programs. J.T. was a church member, Mason, and a life member of the V.F.W. and D.A.V. He certainly appeared to be one of Michigan's most respected citizens, but in reality he was a crook.

My investigation of J.T. Nelson was taking a fair amount of my time. Pete Hector, a local reporter, was offering a great deal of assistance in looking into J.T.'s activities. Pete had been covering Lake County for some time and took a bold interest in the crime there. Along with Pete's help was Channel 13 TV who was extremely supportive. There were plenty of tips from within the community too. We were beginning to learn that J.T.'s CETA Program was just a minor part of his extracurricular activities. He was also collecting welfare rent from unoccupied houses, falling down houses, even vacant lots. The recipient's check was just sent to a post office box. Some people were reluctant to expose J.T. because it was his policy to only take a percentage of the illegal funds he generated. Exposing J.T. meant exposing themselves and ending a source of income.

There was no detective in my department, so my investigative resources were very limited. Early in 1983, information on illegal activities involving another prominent citizen began filtering into my office and things got even busier.

There was a State program which basically was in existence to distribute food to the poor. The program involved several surrounding counties, each county having a director. The headquarters was in Baldwin, the county seat of my county.

A lady came into my office and stated that she had some information I might be interested in. Apparently she had been in attendance, or had talked to someone who had heard my impromptu speech after Governor Milliken presented the award for lowering crime in Lake County. She said she trusted me, and believed that I was really out to fight crime.

This lady was a director for one of the counties. She stated that she had personally been directed by the program director to deliver truckloads of food commodities to private homes and black-owned restaurants. She was given a list of addresses and explained that these places were providing free storage until the food items could be delivered to the poor. The lady said that over a lengthy period of time she was directed to make many deliveries to these places, but there never were any pickups to distribute the commodities to the poor. She knew there was illegal activity going on here, and didn't want to be involved anymore. The lady had resigned her position and wanted to expose what was going on and was willing to make an official statement, furnishing names, dates, addresses, everything she knew.

This lady's story was very believable, but before I continued I wanted to substantiate her credibility. I asked if she would mind taking a lie detector test. There was no hesitation in her response, so I scheduled a polygraph test at the Rockford State Police Post. When the test was over the examiner called me, and said, "That lady could stand right along next to the Pope at Sunday service."

I took a lengthy deposition from my informant and thanked her for being brave enough to come forward with this information. I assured her that we would do our best to continue an investigation into her allegations.

Harry Norman was a black state trooper assigned to the Reed City Post. Two years ago he had received state recognition for rescuing a nine-year-old boy from a burning auto. The boy's foot was lodged in the crashed vehicle and Harry received burns while trying to free the boy. Also on another occasion in Detroit while off duty, Harry walked in on an armed robbery. He was shot twice, but still apprehended the bad guys. After he recovered he was transferred to Reed City. Harry was a very likeable fellow. He was a big man who was in excellent physical condition.

There was a warrant out for this individual that was temporarily living in Idlewild. He made no secret of the fact that he was there. Being a martial arts expert, he bragged to folks that it would take a policeman of greater ability, or excessive force, to ever arrest him. We repeatedly sent deputies in, but were never even able to locate him. In February Harry Norman found out where this fugitive was staying. With a warrant in hand, Harry located the man and stated that he was under arrest. Now this guy starts into his martial arts repertoire to prevent the arrest. Harry was not into martial arts, so he went back to the patrol car and got his riot stick from the trunk. Harry and his riot stick didn't discourage this man's attitude. The fight was on. Harry and his stick against the karate expert.

When Harry drove up to the jail he looked like he had been in a fight; a little blood here and there, his uniform was messed up too. The karate expert was in the back seat. He looked much worse off.

With kind of a grin on my face, I said, "Harry, instead of bringing this fellow over to me, I think you better take him to the hospital. Sure looks like he needs some medical attention."

After they got the guy bandaged up and we had him tucked away in a cell, Robert Williams showed up on my doorstep. He was investigating allegations that excessive force had been used to apprehend a black citizen. I knew now that his only intention was to stir up trouble and create a racial incident. Williams appointed himself in charge of a committee to file complaints on the fugitive's behalf.

Williams filed complaints with the Civil Rights Commission against the Colonel in charge of the Michigan State Police, the Reed City Commander, Lt. Gary McGhee and Harry Norman. He wanted the Lake County Board of Commissioners to climb on his bandwagon, so we all appeared at one of their meetings for a hearing. I testified that I had known Harry Norman for quite some time and that I had never seen him angry or lose control. Also, I was aware that the fugitive was a martial arts expert with the ability to kill with those skills. Harry had tried to facilitate a peaceful arrest, but was forced to use force in his own protection while apprehending the man. I did not think the force was excessive.

The board listened to both sides, then without public comment went on with their other business. This certainly wasn't the community support that Williams wanted. His goal was to have Harry and the post commander fired. Halfway through the meeting Williams stormed out. Later he said he was washing his hands to the whole situation because of the defensive attitude of police officials. Delmoris White, president of the Lake NAACP, was more conciliatory. "I think the main accomplishment is that we came away with an understanding of what we would have liked accomplished."

Norman was transferred out of the area, but neither he nor the post commander were reprimanded. Harry is now stationed at the Lakeview State Police Post and we are still friends. My support of Harry also cemented a solid relationship with the Reed City Post and its Commander, Gary McGhee. It is interesting to note that, as much publicity as Harry Norman got in the Lake County *Star*, they never mentioned his name or made reference to his race. If a resident didn't know he was black it would have been easy to consider this another racial problem involving a white police officer and a black victim.

The State Police Post was not the only organization feeling the heat from Robert Williams' efforts to stir up some racial unrest. Williams was also alleging civil rights violations against the County Sheriff's Department too. In a board of commissioners' meeting I explained the circumstances surrounding the incidents Williams had brought up. I had detailed official reports for each incident. I also charged Williams with trying to fan the flames, saying, "I don't believe the black community is as interested in this as you would try to make us believe." I went on to say that the big problem in Lake County was drugs, not racial strife.

My cup was running over with information on illegal activities. I needed some outside help. Dan Bidwell, my police academy firearms instructor, was a local FBI resident agent and had become a close personal friend. We were both interested in guns and shooting, as well as horses. When I had the riding stable in Sparta, Dan's family visited periodically, eventually purchasing a horse from us. I confided in Dan the things that were going on in my county. Could the FBI get involved? Dan said that until the Michigan State Police called the FBI in, or there was evidence that would warrant federal involvement, the FBI would have to observe from the sidelines.

Next I called the Michigan Sheriff's Association and explained that I needed some outside help in investigating organized crime in my area. They suggested I get a hold of Vince Persante. I remembered that at the Ferris College meeting in 1980, Governor Milliken had talked about appointing Vince Persante as the head of a Task Force against Michigan organized crime. Apparently this task force was still in operation under Governor Blanchard.

Vince Persante came to my office with an assistant. I gave him all the hard-core evidence and depositions that I had compiled and I shared with him my speculation on what was going on in the county. My resources were limited; I explained that his help was needed. Vince and his assistant stayed for a week. Before Vince left he assured me that, although there needed to be additional investigation in several areas, there certainly were charges that could be brought against individuals. He said that I should work with the local prosecutor and he would do what he could to help.

My relations with the Michigan State Police were very good. We regularly exchanged information and worked together on projects. I hadn't asked them to get involved with further investigation of the director of the state program, instead I focused on J.T. Nelson and welcomed any help I could get. Since Vince Persante became involved, I had the feeling that the State Police were running their own investigation on J.T. Nelson. They were regularly asking about anything new that I might have, but didn't volunteer information on what they had.

David Hill and I had differences in the past and so it gave me a measure of satisfaction when he came to my office asking for my help. He felt that word had leaked out of his office somehow about the warrants he would be seeking on J.T. Nelson. There had been death threats and he was nervous. I asked him if he had a gun. David said that he had never fired a weapon before. We went out to the pistol range and I had him practice with an automatic. I told him to take the gun with him until things cooled down.

Apparently the death threats didn't stop because at the end of March David Hill resigned from the prosecutor's position and joined another law firm downstate. There was a brief article about his resignation in the *Star*, but the death threats weren't mentioned. I had to make a trip to the courthouse to retrieve the pistol I loaned him.

On the 5th of April nine persons involved in Webber township government were served arrest warrants by the Michigan State Police. The charges varied from tax fraud to embezzlement and J.T. Nelson's name headed the list. The *Star* reported that the arrests stem from a year-long investigation by the State Police.

Joe Kuhlman had been my undersheriff for nearly a year. I was bragging to myself about what a great choice I had made in hiring Joe. We had cultivated an excellent relationship. Joe did everything I asked of him without hesitation; he had my trust. We were both about the same age, and as two Marines, I thought we made a great team. Boy, was I disillusioned.

It was on March 30th of 1983 that the meeting took place at the Oasis Cafe in Branch, Michigan. I didn't know anything about the meeting until November when I happened to run across the tape in the top drawer of Joe's desk. The tape was the kind used in those small hidden recorders. Out of curiosity I played the tape and then asked the secretary to transcribe its contents. The recorded conversation helped put together some of the things that happened between the meeting and when I found the tape. It also raised other questions. Who arranged the meeting and why? Why was the conversation recorded by Joe? Even now, I can only speculate on those whys.

The conversation at the meeting was between Joe, my former undersheriff Bruce Finch and another former deputy, Todd Chaney. Finch seemed to be a real neutral in the conversation. Chaney had a definite axe to grind with me. When he left the department he had asked to be paid for comp time accumulated. I explained to him that we had no provision to pay for accumulated overtime, but that if he stayed I could give him that time off with pay. That didn't satisfy him so he went out in a huff. As I mentioned before, he also never felt he got enough credit for the cases he solved.

Joe said he felt the need to give me his loyalty, but both he and Chaney felt that I had exploited them. He went on to relate the fact that he had been offered fifteen thousand dollars to get information that would discredit me. His exact description of the scene went like this.

"Two guys approached me sitting out by the Baldwin Auto Parts, here, about three weeks ago, waiting for the state inmate to come out with the car parts. Two guys, a white and a black, walked out of the bank and just before that, Harrison Wilson (county commissioner) had walked out with his aging mother or whatever... er, maybe it been his wife, I don't know, she was old, older black... and waved and come over... and said how ya doing? Well, right behind him comes these two. He walked up to the car and said, you're the undersheriff ... Yeah ... How'd you like to make ... What do you make a year? It's a matter of public record $14,500. How'd you like to make that in one lump sum, clear. ... Hell's Bells, Yeah! All you gotta do is give us some information. I thought it was a joke."

Joe went on to say that he had told me about the offer, but that I had just brushed it off. That, of course, was a lie; the first time I heard about it was on the tape.

The indication was that Joe was considering the offer. "Fifteen thousand would be a lot of frickin' money for me. I could pay off my bills and live very comfortable on my money until the retirement check."

Joe tells about how decent J.T. Nelson, Robert Williams and others talked to him and how they were really out to get me.

Finch had doubts about the offer. "They're not gonna pay that kind of money just to get him out of there ..."

Chaney agreed. "They can do that other ways ... If they really wanted him out that bad, they'd just have to kill him and they could do that a lot cheaper than fifteen thousand dollars. Believe me. They could make one phone call ... There's probably a dozen people that I know of in Idlewild right now..."

Chaney talked about how nice it was before I was elected sheriff. "When we first started here, we didn't get involved in anything. And, we had good times ..."

It was obvious from the conversation that Joe would like to be sheriff; even Chaney had ideas in that direction.

Speaking to Joe, Chaney stated, "You're an idiot compared to the things I know about administration and the things you know about administration and running a department."

Joe agreed, "Absolutely."

Chaney summed up the bottom line for Joe. "Your alternative is to work to get him out of office and ... ah, or your other alternative is to leave. If you like living up here, which really is why we're here ..."

As I said before, until the fall of 1983 I considered Joe the model undersheriff. In fact everything seemed to be running pretty smooth in the department. The grant money was available each quarter. I personally made sure the paper work got sent in on time. Cindy Keefer wasn't any more accommodating, but I resigned myself to the fact that she probably wouldn't change.

Not too long after Herman Smith came on board as a deputy, I fired him. That probably stirred a little buzz in the community. Here Blevins has been complaining about not being able to recruit a black deputy and when he gets one he fires him right away. At least that's what people thought. I had him removed from the county payroll; didn't even tell the union. Really, Herman had agreed to go undercover for me to see if we could find out those involved in buying and selling stolen goods. We hit pay dirt and netted about $15,000 worth of stolen goods and made some arrests. Herman did a good job so when the operation was completed I got his picture and a nice write-up in the *Star* for him. I was pictured shaking his hand with a table full of stolen guns displayed. I think the community could see then that I didn't have any hang-up about hiring black people.

At the present time Robert Williams was more concerned about world affairs than he was about Lake County anyway. He had just returned from Libya and was busy singing the praises for Libya's new leader Muammar Qaddafi and the new form of revolutionary government being practiced there. In an interview with the *Star*, Williams said that he was very much surprised by Qaddafi's deep concern for all oppressed peoples. He said he found Col. Qaddafi to be just the opposite of what western politicians and media portrayed him to be. It must have been a terrible blow to Williams' ego when he returned from his visits with the world's top communist leaders, only to be asked to address the small town civic functions that he did. For instance, when he returned from Libya he addressed a Baldwin gathering that was raising funds for a playground. There he was pictured in the May 31st issue of the *Star* wearing his Chinese officer's coat telling the local folks about his great travels to those grand communist nations. It was true that those nations treated him as a celebrity, but back here folks didn't take much notice.

In May I was proud to receive an Outstanding Service Award from the Disabled American Veterans. Also, even though there was no public recognition, there was personal satisfaction in my prisoner rehabilitation success.

In the beginning I was probably the last law enforcement officer to have any interest or confidence in prisoner rehabilitation. My concern was victims' rights, and I thought that trying to rehabilitate prisoners was successful in only a small percentage of cases, 10 percent or less. The liberals would say it was worth it to pile millions of dollars into housing and rehabilitation even for a small success rate.

The courts pretty much dictated that the sheriff get involved in rehab on the county level. If you didn't, you were well criticized. There was the philosophy that it was cheaper to support rehabilitation and turn the short termers around than having to support them later on when they got some real jail time. Once they entered into the state prison system they were so experienced and educated in breaking laws, it was nearly impossible to change their attitudes.

Reluctantly I set up some rehabilitation programs. Once I got started, to be truthful, I enjoyed it and did much better than the 10 percent rate I predicted. Several of my prisoners participated in the Scars to Stars program I set up with the community college. With security guards, the inmates attended classes at the community college, working towards achieving their high school education with the GED. When they finished, we invited the families in on a weekend and held a little cap and gown ceremony with pictures and diplomas. We graduated two state certified mechanics; neither reentered the prison system; both became successful businessmen. For another graduate, I co-signed a loan so he could buy a gas station and garage. He also did very well, paying the loan off in full.

One inmate got religion, and wanted to work with young people when he got out. He had an opportunity to go out west and do counseling in a boys school, but didn't have any money, or a way to get to his job. I bought him an old car. It had many miles on it, but the body was sound. The guy got pretty emotional when I gave it to him, gave me a big hug before he left. I didn't know if I'd ever see him again, but didn't care. The man was a good inmate, studied hard, did his best to turn his life around.

It was two years later when that man was back at my door. Laying an envelope on my desk, he told me he had a job in Arizona and then thanked me for my help and said, "God Bless you." The envelope contained full payment for the old car I bought for him.

Another inmate, whom I will always remember, is still in the prison system. He was convicted of assault and rape when he was 17 years old. Having served 20 years, he was an adult when I met him. His attitude really impressed me. Committed to straightening out his life, he was very intelligent and extremely trustworthy. Being a high category trustee, he had access to the whole department and prison grounds. Responsible for changing the oil, cleaning and washing the patrol cars, he handled all the car keys, even those of our unmarked car. The man could have walked or driven away at will. Only at night was he locked back up.

When his parole request was reviewed, there was so much negative feedback from the community where he had committed the crime as a teenager, that the parole board turned him down.

After this happened he came to my office one day with a request. "I want to be locked up here or returned to prison."

"Why?" I asked.

"If you don't send me back, I am going to walk away."

Sometimes the prison system is not real fair. I'd seen murderers who had entered and been released in much less time than this man had served. I called a state prison representative. The answer was, "No, he's a low risk prisoner and that status will not change no matter what he says. He can't return to the prison and you can't lock him up all the time." This response I relayed to the prisoner.

One of this prisoner's duties was to push a candy cart between the cells each day. The inmates could use up to two dollars a day of their money to purchase some snacks to eat, or buy cigarettes. The cash box could easily contain three to five hundred dollars.

There was an envelope and the cash box at my door one morning. In my office I opened the envelope, reading the note inside.

"Thank you for treating me as a man. As much as I could, I enjoyed being in your jail. As the parole board could not see fit to release me, I told you I would walk away, and I did last night. I could have taken the unmarked car and been in Florida by the time you got to work, but I didn't. All the keys are in the cash box, along with five hundred dollars cash and a current inventory from the candy cart. I could not and would not take any money from that cart, knowing you would be personally held accountable for it. I've only taken my property. Good luck." Below he had signed his name.

I called the prison system and told them of the walkaway. There was no breakout, just a walkaway. Deep down in my heart I hoped he would not get caught, that there would be a new, happy life. However, after two years or so he was caught and returned to Jackson prison where he still resides.

While I was sheriff there were no guard injuries, or riots. Our jail had a reputation of treating inmates humanely. I'm proud of that and whatever rehabilitation we accomplished. I didn't judge these men and women by their crimes, but rather on how they conducted themselves in my jail. I had their respect; many had my respect. There never was worry for my safety, if I ate Christmas dinner with them or played a game of cards in a cell, or even my wife's safety, who worked in the jail.

My safety outside of the jail was another story, at least since the last board of commissioners' meeting. J.T Nelson knew that I was hot on his trail, bent on putting an end to his illegal operations. The state police charges against him merely touched on his activities. I was doing my best to put enough together to file charges against him myself, or at least nudge the state into filing charges against him. He, in turn, was probably doing everything possible to keep that from happening. Since the CETA investigations, I also had contact with the press, trying to bring public pressure down on him.

Finally, at a board of commissioners' meeting, the press showed up along with a TV channel's camera; it was an unexpected, shotgun type interview with J.T.

"Sheriff Blevins has said ... that you have been doing ... He's threatening to bring charges. What is your response to his allegations?"

"The sheriff ought to be careful. Some one of these days someone may shove a 357 up his ass and pull the trigger."

J.T.'s remark was pretty blatant and bold for a live television interview, but he meant it. He was real angry with me, to the point that, anytime we were in the same room, we almost came to blows.

After that, whenever I went to a board of commissioners' meeting, I always wore my second chance bulletproof vest. There seemed the possibility that one day J.T. would pull a 357 out of his briefcase and start blazing away.

Late one evening I was at home when the dispatch call came in requesting that I immediately get to the scene of a shooting, on M-37, involving a deputy. One person was dead. I lived on the east side of the county, at least eighteen miles from the shooting. I jumped in the patrol car and traveled the distance as fast as I possibly could.

Randy Gerke was a very good deputy. I had sent him to the academy; he was very knowledgeable and I had confidence in his ability. Randy pulled over the car when he saw that it was weaving on the road. The driver was James M. Nelson (no relation to J.T. Nelson), a large black man that was a resident of Webber Township, and a Lake County Road Commission employee. Mr. Nelson went to Grand

Rapids to pick up some airline tickets, was northbound on his way home when Randy pulled him over. Randy spotted an open bottle of vodka and a container of orange juice on the seat beside the driver. He could also see that Mr. Nelson's eyes were dilated, his speech was slurred. Randy asked him to get out of the car and perform two dexterity tests, picking up coins and walking a straight line. Mr. Nelson could not perform either.

Randy said, "I believe you're intoxicated. You're under arrest, I'm taking you to jail."

Mr. Nelson started yelling and screaming at Randy, saying that he wasn't going to be arrested. The noise brought witnesses from two separate homes out to see what was going on.

The man accused Randy of picking on him. A fight broke out with both men on the ground wrestling. Randy had his work cut out for him because the guy was probably a hundred pounds heavier. According to witnesses, the man pulled Randy's gun from the holster and pointed it at him. Randy and the two witnesses ducked around behind the patrol car, trying to circle to keep the car between them and Nelson. Nelson fired off three rounds in the men's direction. Randy finally pulled out a hideaway, snub-nose 38 and ordered Nelson to drop the gun. When Nelson took aim again, Randy fired some warning shots and then aimed to stop him with two shots. Nelson was hit in the head and chest, dying instantly.

I called in another medical examiner, as I only wanted to function as the sheriff in this case. Detectives were requested from both the Reed City and Newaygo State Police Posts, and an autopsy was later performed in Grand Rapids. The witnesses were sequestered in separate rooms at the sheriff's department. Randy was taken off duty with pay until the incident was completely investigated. This could be turned into a racial incident; I didn't want that to happen.

The autopsy revealed a hefty .24 alcohol content in Nelson's blood. That was more than double the state .10 limit of intoxication, plus he had a history of drinking problems. The witnesses more than substantiated Randy's account of what happened and the state police report found no wrongdoing on Randy's part.

Robert Williams could now see the opportunity to promote some unrest, and didn't delay in seizing the moment to level accusations against me and the Michigan State Police. It was police brutality against blacks again. Mr. Nelson's body was probably injected with alcohol to make it look like he was drunk. The witnesses, the police, everyone involved was white. It was just another conspiracy against blacks.

Though it was physically impossible to raise a person's blood alcohol through injections, as always, the press was more than willing to carry Mr. Williams' comments. And once something is in print, it makes it even more believable.

It was a few days later, the 5th of July. I worked until around 10 p.m., then headed home on US-10, driving my black and white patrol vehicle, the Dodge Ram. I was in full uniform. As I approached the village of Chase, a call came over the radio from a deputy that was involved in high speed pursuit of a motorcycle. This was taking place behind me on US-10, but they were coming towards me.

As I swung my vehicle around I radioed out, "I'm turning around and heading in your direction!"

The deputy was soon back on the radio, "He's abandoned the motorcycle, heading into the woods on foot."

"Don't pursue him by yourself. Wait until I get there."

On my way to assist the deputy, in my rear view mirror I noticed a vehicle gaining speed behind me. Its headlights were being flashed off and on. Obviously someone was trying to get my attention. I had been going at a pretty good clip to aid the deputy; now I slowed down to see what these folks wanted. From the mirror, I determined that it was a van or pickup truck, as it had the orange, teardrop clearance lights on the roof.

The vehicle was almost abreast of me. I quickly turned my head to the left, to see what they were trying to get my attention for, and at that instant there was a loud bang, an explosion. I thought they had driven too close and clipped my car. The next thing I knew, I was going off the road and had this sensation in my body and arm, like being touched by a hot furnace poker.

I quickly realized that I had been shot. My vehicle was going off the road, I couldn't control it. Down into the ditch, then back out again, through some trees and finally, the patrol car came to a dead stop against a good-sized tree. The vehicle on the road stopped, then the brake lights went out, and the backup lights came on. They were coming back to finish the job.

My windshield and the side window glass both were blown out. I should have called on the radio, but the only thing I could think of was getting out of the car and into the woods, an environment that could offer me some concealment. I grabbed the shotgun from the rack, left the vehicle and headed into the woods. The shotgun held enough rounds to hold them off for a little while. If I ran out of ammunition, I would be at their mercy. It was dumb not to call for help on the radio. No one knew I was out here.

The vehicle backed up and sat there for 30 seconds or so. Finally it sped off. Perhaps they dropped someone off. I listened intently for some noise, brush cracking, anything. It was dead quiet. After awhile, I convinced myself that they had left. Either they figured I was dead in the vehicle, or if they saw me get out, they weren't willing to try and find me.

Returning to the vehicle, I called the dispatcher, told her I was on US-10 near Idlewild, had been shot and to send backup. I was real lucky. The fact that my head was turned just before the shot was fired probably saved my life. My shoulder caught the buckshot along with the dashboard and windows, but not my head.

They took me to the Reed City hospital for treatment. It was a grazing wound. The flesh had been peeled back but there was no buckshot to remove.

The next day I called a news conference. I explained that I didn't want any manhunts started, no retaliation white versus black, no revenge. There would simply be an investigation to look for suspects. It probably was just a random shooting of a police officer, and unknown if the shooters were black or white.

That's what I told the press; I thought otherwise. Everyone in the county knew that I was the only one that drove the Dodge Ram patrol car. There was no doubt in my mind that this was an organized hit. The shooting of the black man, Mr. Nelson, gave someone the perfect opportunity to get me and make it look like a retaliation shooting.

There were some tough folks from the Luther area that were more than willing to head over to Idlewild and look for some black suspects. One of those characters came into my office, offering his services, just looking for the go-ahead signal from me.

I tried my best not to let this be turned into a racial thing. The press, on the other hand, went out of their way to make sure that the public knew that racial retaliation for Mr. Nelson's shooting could have been the motivation behind the shooting of Sheriff Blevins.

For several weeks I had been pursuing charges against J.T. Nelson. The day I was shot my secretary had received nine death threats against me. I remembered J.T.'s statement to the press about sticking the 357 up my ass and pulling the trigger. There was little question in my mind; somehow, J.T. Nelson was involved in the attempt to kill me.

I was concerned about my family's safety, but figured they would be out to get me again, rather than my family. I had Dee drive to work by herself, and on a different road than what I took. I changed my route to work every day. Someone with a rifle and scope could easily get me otherwise.

We lived in a rather remote area on a lake. I started taking security precautions around our home. State-of-the-art listening devices were installed on the perimeter of our property. From inside the house we could hear any unusual noise that occurred 360 degrees around the house. I was a light sleeper anyway, so from our bedroom I knew when a car was approaching or could hear someone walking up our gravel driveway.

I brought a watchdog that didn't like anybody, and eventually had a small flock of geese. Anyone who's been around geese knows that when they are disturbed, they can be a better alarm mechanism than anything.

For protection, I kept a Thompson semi-auto submachine gun setting in the corner of the bedroom, with two 30 round magazines. It was locked, cocked and ready if I needed to hold off trouble while reinforcements arrived. These were not temporary measures. This was the way we lived for over a year.

Meanwhile, Robert Williams was doing as much as he could to keep the shooting of the black man at the forefront of the news. The witnesses were white and couldn't be trusted. The Michigan State Police reports were flawed. Probably the black man hadn't intentionally fired the officer's revolver at all. It was merely involuntary finger contractions that were caused as the man was dying. The whole thing was a conspiracy against a black man.

On August 25th, Robert Williams and 50 black Lake County residents made a pilgrimage to the state capitol in Lansing. There they held a candlelight vigil on the state capitol grounds, protesting alleged brutality and racial discrimination by the Lake County Sheriff's Department and the Michigan State Police.

Governor Blanchard decided that there should be some action on his part to respond to the charges of the Williams' protest. He announced that there would be a panel appointed to probe these citizen concerns. An article in the Detroit Free Press on August 28th mentioned that Conrad Mallett Jr., legal advisor to the Governor, was heading this state probe.

The Michigan Sheriff's Association (MSA) was concerned about where this probe was headed. They arranged for Genesee County Sheriff John O'Brien, a member of the MSA's ethics committee, to participate and monitor the hearings and report back to the board immediately upon completion. Also, a special ongoing MSA monitoring committee was established for follow-up review. Dale E. Davis, MSA's deputy director, would also attend the probe.

There was no question in my mind about Robert Williams' intent; it was strictly aimed at stirring up racial unrest. His track record had proven that. Governor Blanchard, on the other hand, probably had more than concern for the citizens on his mind too. Preserving his political power base would be quite important.

Detroit's long-time Mayor Coleman Young was instrumental in Blanchard's election to Governor. Conrad Mallett Jr. was Coleman Young's executive assistant and political advisor, so obligingly after the election Blanchard named Mallett his legal advisor.

J.T. Nelson cultivated many political contacts during his tenure with the State and National Association of Counties. On more than one occasion he bragged about his close friendship with Mayor Coleman Young. If that was the case, I would guess that J.T. was doing his best to get Coleman to help relieve some of the pressure on him. Looking at things in that light, Conrad Mallett Jr. was an obvious choice to head the Lake County probe.

In April the State Police charged J.T. with tax fraud, but I wanted to nail him with a lot more than that. J.T. along with other township officials named in State Police charges were also facing a recall. A committee for better government in Webber based J.T.'s recall petition on four charges: 1) uncommendable language by a public official in a television interview 2) unethical conduct by a public official at the polls 3) being a paid advisor to the township while serving as its elected commissioner and 4) mismanagement of cemetery funds. J.T. threatened to sue those collecting signatures on the petitions, but that didn't work. In a letter to the *Star* editor, the committee publicly criticized J.T. for threatening people with physical harm if they should cross him. "Mr. Nelson has dealt in intimidation of voters and citizens in the district, most especially in Webber Township, in the past as evidenced at the recent hearings on the recall petition. He seemingly intends to try this tactic in response to this recall campaign. It is not surprising as this means of maintaining his position in the community has served him well in the past."

The committee letter went on to say, "Mr. Nelson has cultivated many contacts over his years in politics that serve him well. How many of us in this community have talked with someone who will not openly speak against Mr. Nelson or his close associates because they fear for their jobs or their position in the community? These people are both black and white."

On the 7th and 8th of September Governor Blanchard's task force held two days of public and private meetings looking into civil rights violations in Lake County. Mike O'Connor, the editor for the *Star*, provided observations and quotes for a front-page story on the proceedings.

"According to Gov. Blanchard's legal advisor, Conrad Mallett Jr., the purpose of the task force was to look into allegations of civil rights violations by the sheriff's department as well as failure in delivery of services in the areas of social services, civil rights and education. With few minor exceptions however, those opposed to Sheriff

Bob Blevins were far and away the most vocal during the hearings and neither the sheriff nor those opposed to him felt much was accomplished."

"One of the most outspoken members of the task force panel, Dale Davis, vice president of the Michigan Sheriff's Association, said he was bothered by the one-sided, anti-sheriff testimony presented at the hearing. Davis indicated that he felt those who might have been supportive of the sheriff's department may not have testified because they felt intimidated."

"There is a tremendous fear factor that certain people are advocating," Davis said. "I'm not sure that if I lived there I'd come forward to testify. I'm concerned about the apparent lack of questioning of some people who offered comments to us ... I was asked not to be so aggressive in my questioning."

Mary Trucks, director of Five Cap, was a spokesman for the People's Association of Human Rights (PAHR), Robert Williams' pet organization. She summed up the association's feelings about the sheriff by saying, "He's a symbol of the white community, that section of the white community who wants to see black people kept in their place."

I wondered if Mary Trucks knew that I had received complaints about Five Cap, the organization she headed. She might have perceived me as much of a threat as J.T. did. Although I couldn't verify it, I also heard rumors that she and Conrad Mallett Jr. were friends previous to the Governor's task force meeting.

Randy Gerke resigned on the Saturday following the hearings. His name had been at the forefront of many complaints aired during the task force's meetings, and he was named in four of the complaints the PAHR took to the state capitol. Randy said his resignation was not an admission of guilt, rather a realization to continue to try to work and serve as a law enforcement officer was impossible after the successful assassination of his character. Randy also said that he was leaving with a great deal of respect for the Lake County Sheriff's Department and its members. Randy also was quoted as saying, "I feel a great deal of frustration as all professional police officers must feel when their careers are ended by a tragic situation. It is sad to note that a small group of citizens could band together to distort and lie about nearly every arrest I've made in the past year, to systematically and deliberately change those facts to make me out to be a racist and brutal police officer."

Ernest Goodman who had represented Ruby Nelson in the shooting of Lonnie Deur now represented the PAHR in investigating the Randy Gerke shooting of Mr. Nelson. He said that the Association's belief was that the full story had not been told yet.

The October Michigan Sheriff's Association (MSA) bulletin pretty well summed up the results of the State probe: "The official investigative team held public and private meetings in Baldwin, the seat of Lake County, on September 7 and 8. It was a heated proceeding in which a vocal opposition only served to fan the fires of racial strife that have been lit in Baldwin. In light of those proceedings, Sheriff Blevins issued a statement recognizing that Lake County faces many problems, including social, educational and economic problems. He called for greater attention to be given to all aspects of the problems in Lake County, but recognized that the law enforcement issue was immediate."

Dale Davis didn't feel much of anything was going to come from the task force hearings. In an interview with the *Star*, Davis was quoted as saying, "If nothing else is to take place as a direct result of this task force, I must urge with all immediacy that a state and/or federal grand jury be seated." Davis noted that a grand jury would have the power to subpoena witnesses and that people would be required to testify under oath. That would allow the "silent majority" to be heard and avoid having the "violent vocal minority lead us into the wrong conclusions."

Davis continued, "I have reason to believe from some reliable sources that perhaps state and/or federal money has been misused for many years by either current or past people in authority in the county or township governments up there."

In a letter to Gov. James Blanchard, MSA Deputy Director Davis said a grand jury investigation is needed because alleged injustices "transcend all services by government to its citizenry."

When asked what I thought about Davis's letter to the Governor asking for a grand jury, I said, "Let the grand jury come. You won't be able to wipe the smile off my face with a two-by-four."

Conrad Mallett Jr., who headed the task force, was not so interested in calling for a grand jury. He wanted everybody to refrain from further action until his task force arrived at some sort of decision. The *Star* quoted him as saying, "You've got police communication problems and citizen communication problems. I don't know if the delivery of services is bad because it's racially motivated. The task force must weigh all the suggestions before any decisions are rendered."

Conrad Mallett Jr. also appeared before an MSA meeting, trying to smooth out relations with that organization. His speech tap-danced around the problems with my department and concentrated on the community communication problems.

In a letter to the *Star* editor from Robert Williams' People's Association the solution was simple. "So long as we have law-abiding citizens who live in terror and mistrust of the police; so long as we have a sheriff's department that concerns itself more with covering up misdeeds by twisting facts and truth; as long as we have a sheriff who condones and justifies the use of excessive force and intimidation, we will continue to have strife and disunity in our community. For this reason we feel that Mr. Blevins should resign or be removed from office."

There were a lot of other letters to the *Star* editor that really made me feel good. For instance from one reader, "To the citizens of Lake County: How many of you want to go back to the days of our 'do nothing' sheriffs that we had prior to Sheriff Blevins? It's about time the task force left town and let our sheriff do his job. I think it's about time to give Sheriff Blevins and his professional staff credit where due, instead of running him down. I go downtown to Baldwin almost every day. I see no racial problem everyone is talking about. It is a sad situation when a handful of citizens go to Lansing and bring a 'task force' to us citizens ... Let's support the man we elected sheriff."

Because of the Nelson shooting there was a call for special training for my department; training in areas such as sensitivity to minorities and non-lethal use of force.

I personally had a problem with requiring my department to have sensitivity training because of this incident. According to the *Sheriff's Star*, "A subsequent investigation by the Michigan State Police, as requested by the sheriff, disclosed no wrongdoing, and their findings were upheld by the Prosecutor and the Attorney General's office on July 22nd."

I believed in affirmative action and had gone out of my way to establish a racially balanced sheriff's department. I also think that my programs within the community had cemented solid relations with black residents, but now they were being torn apart by Robert Williams, and J.T. Nelson. No one knew what part Conrad Mallett Jr. would play. I balked at forcing my department to take sensitivity training because it created the impression that we had done something wrong.

The MSA told me to just go ahead and schedule the training. If I would do this, the politicians would be satisfied, and this would all go away.

For the most part, the sensitivity training was ridiculous. It might have looked good on paper, but we just sat there shaking our heads, suffering through it. I still remember one bit of instruction we were given. "If you arrest a white man it is all right to impound his car, but you shouldn't do that with a black man. He is a minority and it will be a financial strain just to pay his ticket, let alone pay for the towing and impounding of his vehicle. You should, instead, call his relative to pick it up."

As if I didn't have enough things going on with the governor's task force, now I was back in a battle with the board of commissioners. In late summer I received yet another call from Russ Webb. My application for the secondary road patrol grants had never been received by his office. He had sent a notification to me that the application was missing, but I was never notified. A late report was one thing; the yearly application was something else. Russ said it was out of his hands, and that there would be no funds for the first fiscal quarter (Oct-Nov-Dec). I would have to apply separately for the other quarters. If there wasn't money left over I might go all of '84 with no grant funding.

I was devastated. Sherri Holliman and Herman Smith's wages were paid entirely from this grant money. Neither deputy had worked long enough to qualify for unemployment, both had families to support and with the holidays coming up this was the absolutely worst time to lay anybody off.

I went to the board of commissioners with my problem. Nobody knew a thing about why my grant application hadn't been sent in or acted as if they cared. One member asked, "Well, are you going to lay these two deputies off?"

"No, I'm not going to lay them off. They have families to support, and they can't draw unemployment."

It was pointed out to me that there was no money in the budget to pay their wages. I said I still wasn't going to lay them off, and left the meeting.

There still was a possibility the board could somehow force me to lay them off, so I petitioned and received a temporary injunction from the Manistee Circuit Court to keep them from doing that.

For October, November and December I was paying the wages for the two deputies from my own savings account. I still have the canceled checks to prove that it cost me about $1600 a month to keep these two deputies on duty. Both were certainly worth it to me and the community. Sherri was the first female road deputy and Herman Smith the first black road deputy in the county's history. I thought the court might force the board to reimburse me but instead Judge Wickens lifted the injunction and sided with the board and against me. Fortunately there was enough public pressure on the board to approve reapplication for the road funds in '84 and keep both deputies on.

The task force probe seemed to be over, but it really wasn't. There was a chain of events that would soon begin which would continue well beyond the next election. Although I was becoming more accustomed to the unexpected, I hadn't seen anything yet.

It was on a Sunday night when Joe called, "Boss, do you mind if I use a patrol car to take a couple of the recruits back to the academy? Their wives picked them up for the weekend; this would save them the long trip back to drop the boys off."

"Joe, you know it's my policy not to drive the patrol car in civilian clothes."

"Don't worry, I'll wear a uniform."

On the following Monday morning Joe didn't show up for work.

Finally I called his home. His wife Lois answered. She didn't know where Joe was, hadn't seen him since Sunday.

I was beginning to worry about him. I called over to Delta College, where the academy was, and left a message for Larry McDuff to get a hold of me. Larry was one of the recruits in school there.

When Larry returned my call, I asked, "Have you seen Joe?"

There was a long silence, then Larry says, "I don't want to get him into trouble."

"Tell me anyway," I insisted.

"Well, when Joe picked us up he had been doing some drinking. In fact there was a bottle in the patrol car."

"He was drinking in uniform?"

"Yes."

"When Joe arrived at the academy he came up to the dormitory room for awhile, sat down with a bunch of recruits and told Vietnam war stories while polishing off what beer we had in the refrigerator. When that was gone he said he would go out for some more beer or liquor. His intention was to go in uniform in the patrol car. We talked him out of that and someone else went out for some beer. By early morning Joe was passed out. Rather than have him discovered by the instructors, we checked him into a motel. That's the last I saw of him."

"Great!"

Apparently Joe slipped off the wagon. That wasn't uncommon among alcoholics. By Tuesday Joe would be dragging in here with his tail between his legs. As long as he had worked for me, Joe had been sober. He obviously would need help again. I really didn't want to lose him.

Tuesday, Wednesday, finally it was Thursday and still no Joe. I called the motel where Larry took him. He wasn't there. I was ready to put his name on the LEIN machine as a missing person, when the sheriff from Sanilac County called, "Are you looking for your undersheriff and a patrol car?"

Hesitantly I answered, "Yes."

"I don't know how to tell you this, but I must write it up in a report; there's no way I can cover it up."

"What are you talking about?"

The Sanilac Sheriff starts to explain. "I'm setting here with my only on duty road deputy right beside me, when this lady from the liquor store calls."

"I want you to tell me why your deputy is drunk and trying to buy booze in my store."

"Lady, my only deputy on duty is doing paper work right here in front of me."

"Well, he just walked out the door, he's wearing a brown uniform and he got into a black and white patrol car."

"We'll be right over."

"The deputy and I jump into the patrol car, head towards the liquor store. We spot a black and white police car and followed it. Eventually the vehicle pulls into the drive of a private residence. The driver in uniform goes in the house."

"After awhile I go up and knock on the door. The guy in uniform opens up. He reeks of booze, smells sour, terrible from not bathing in awhile. He starts crying, blubbering about how he's finished and we probably want his badge and gun."

I tell the guy, "I'm not your sheriff, but I'm sure your sheriff will want to know why you're over in my county, intoxicated, driving a patrol car." With that, the guy flies into a rage. Now he goes out in the yard, strips down to his underwear. He leaves his gun, uniform, shoes, the whole works in a pile, then goes back in and slams the door. We put his stuff in a bag and call a wrecker to impound the car."

"I'll be over as soon as I can to pick up the car. Thanks for calling, I'm sorry for the trouble."

Mike Seelhoff, my sergeant, went with me over to Sanilac County to retrieve our vehicle and Joe's belongings. The sheriff said that he needed to make out the report to prove to his commissioners that it wasn't one of his deputies driving around drunk in a patrol car. He didn't want to hang for the actions of my undersheriff. I said that I fully understood his position and thanked him for the help.

At the time, I didn't know the real reason for Joe's drinking binge. I was extremely angry with him, but at the same time, I knew enough about alcoholism to know that even the strongest individual can be tempted and fall off the wagon. In a way I felt sorry for him. Probably his drinking triggered Vietnam flashbacks. I decided, unless something changed my mind, I'd give him the benefit of the doubt. Try to get him into a recovery program.

The next week, Joe sheepishly showed up in my office. "Guess I blew it. Am I fired?"

"No Joe. I'm not going to fire you, you need help. What caused this, Vietnam?"

"No, just tough times with the wife. Guess she's going through menopause or something. She's 48 now."

I bought Joe's story; hook, line and sinker.

Joe says, "Well where do we go from here?"

"I want you to get counseling for your drinking and counseling for your marriage."

"Will you set that up or should I?"

"Joe, you're a big boy. You're not one of my young deputies. Let me know what you need. I'll get you the time off."

In a few days Joe returned to my office. "I've got good news and bad news. For the alcohol problem I've got counseling set up for Wednesday and Friday of each week. My wife and I have the marriage counseling on Thursday. All the counseling is set up in the Grand Haven area where I'm familiar with the counselors."

"Joe, in other words, you're telling me that you can only work Monday and Tuesday with the rest of the week off for counseling?"

"Yep, that's the bad news."

"That's ok. I want you to get the counseling and be back as my undersheriff."

After a few weeks I get this anonymous phone call. "You better check on your undersheriff. He's not doing what you think he's doing."

The next Tuesday I checked Joe's house. He was home. On Wednesday I parked down from his house again. At daybreak Joe was up and packing bags in his car. I followed them to a cottage in Grand Haven. There, I watched from the weeds as they unloaded their things from the car.

In about two hours Joe came out with fishing gear, threw it into a boat and went out on the lake fishing. I went back to the office, but returned the next morning again. Their car was in the drive at 10 a.m. All sorts of things were going through my mind. Maybe he had an afternoon appointment. Joe spent his day on the dock with a fishing pole in one hand and a drink in the other. I stayed until 4 p.m. I didn't need to go back on Friday.

When Joe came to work on Monday I called him into my office. "How's the counseling going Joe?"

"It's gonna be awhile, there's some hard roads to cover."

I forget what else Joe said. It was his little dog and pony show to string me along.

"Joe, I've been watching you sit on the dock and fish for two days."

Joe's face got red. "Guess I'm finished?"

"I can fire you or you can resign. You've strung me along. I've got to have an undersheriff I can count on, not one that's going to lie to me."

Joe resigned and left, my third undersheriff. I would see and hear a lot more from Joe when the '84 election rolled around. At a future sheriffs' conference, I would learn from Bud Gryson what he hadn't told me before. Joe had been involved in the theft of cash from a booked inmate. He couldn't prove it for sure, but knew it was Joe. Bud said that he could see I was probably going to hire Joe no matter what he said.

In the days to follow a flood of memories came back. My grandfather's saying, "You can't take the spots off a leopard. And, thinking with your heart not your head." I should have listened to Bud Gryson. Bud had nothing to gain by telling me about Joe. Later when he was dying of cancer, I said to Bud, "My biggest regret is not listening to what you were trying to tell me about Joe Kuhlman."

Bud said, "I understood your comradery for a fellow Marine."

On November 1st there was a special election and the voters recalled J.T. Nelson from office. That certainly was good news. I was eager to take up where the voters left off, but shortly after his recall there was a disappointing and rather puzzling situation that took place at the jail involving Gwen Warren. I had promoted her to Lieutenant and put her in charge of running the jail schedule. She was a nice lady who seemed to be doing a good job keeping jail activities on a smooth course. Unfortunately, I witnessed a surprising incident in the jail involving her and conducted some counseling in my office. I asked if there were problems in her marriage or what had happened in her life to cause this. She didn't elaborate, but did admit to stress on a personal level. The next day when she came to work, I saw her walk by my office with her hand in a glass of water. Maybe she'd hurt her hand. From the hallway I could see her dabbing water on her face. Then, without much warning, she slid out of her chair to the floor, where she appeared to be unconscious. I called her husband. She would wake up, then pass out, then wake up again. Her husband took her home for two weeks' sick leave.

The next thing I knew she had an attorney filing a lawsuit against the county for work-related stress. It was equally hard for me to comprehend the out-of-court settlement of $25,000 that was agreed upon between her attorney and the board of commissioners. Not long after that she announced her candidacy for J.T.'s seat on the board of commissioners.

When the man that nominated Gwen to be the democratic candidate found out about the lawsuit settlement, he wrote a letter to the *Star* editor apologizing to the democratic voters for nominating her. He publicly withdrew his support for Gwen and backed her opponent. Gwen was a good employee, but at the time I had a little trouble following all the circumstances that led to her resignation.

Bob Smith, the bank president, called; he really sounded serious, "Bob, there were two State Police detectives in here. They wanted a copy of all the checks from your account. All of them since the account's been open."

"Did they have a warrant?"

"No, just showed me their badges, said they wanted the checks so I gave them copies. Thought you should know."

It wasn't long before I started getting calls from merchants around town. They all told the same story, two detectives wanted a list of things that I had purchased at their store.

Finally, two detectives and a State auditor showed up at my office. They introduced themselves; I invited them in. They were clean cut, wore cheap suits, said they were special investigators from the attorney general's office and were conducting an investigation and an audit of my affairs. They hoped I would cooperate with the auditor and open my books to him. Dee slipped in through the door to listen. "Why are you auditing me? In my opinion this is a big witch hunt."

One of the men asked, "Do you think Young or Blanchard will take a fall?" Then he answered his own question, "No, you can bet your ass a podunk sheriff will take the fall. Futhermore ... you will never get a job or work again."

Dee's eyes opened a little wider when she heard that remark.

I got up from my desk and stood by the door. "I haven't done anything wrong; your auditor is welcome to look at my books." They left.

When the state auditor finished, he came over to my office. "Sheriff can we talk off the record?"

"Sure."

"I'm retiring soon and I will deny saying this if I'm asked."

"Go ahead."

"I completed my audit, and couldn't find anything wrong. When I called the bosses and told them you were clean, they said they weren't satisfied, that there would need to be another audit. What you said earlier was the gospel. This is a witch hunt. They must want you out of office real bad."

I thanked him for being honest with me.

Less than a month later the same auditor came back. "I've been ordered to audit the department again. This time I know where to look; the contingency fund." He wasn't there very long and I had a feeling he found what he was supposed to have. He left without talking to me.

After Joe resigned I happened to look through his desk drawers to see if they had been cleaned out. In the back of the top drawer I found the tape. The recorded conversations when Joe had talked about being approached by two men who offered him $15,000 to give them some information on me. I wondered if Joe had accepted their offer, if he had something to do with this audit. All sorts of things were racing around in my mind, especially what the two state police detectives had said, and the auditor had confirmed.

I called my attorney and we set up a meeting with the Lake County Prosecutor, Mark Raven, for November 17th. At my request we met in Raven's private law office, not in the courthouse.

I played Raven the tape Joe Kuhlman had recorded during his meeting with Chaney and Finch, pointing out how Joe admitted on the tape to being approached by two men to get information on me for a year's salary. I explained that since the task force had been in town, the state police had visited the bank, acquired copies of all my personal checks, and a state auditor had found some gray area in our contingency fund. Something was obviously going on here and I asked Raven if he knew or had any ideas to shed some light on the situation. He said that he wasn't aware of any of this and didn't seem that interested in finding out either. I would find out later that Joe had visited Raven over a month earlier with "something on his chest," and Raven referred him to the Michigan State Police in Reed City. Of course Raven didn't mention this to us.

Whatever they were up to, I knew that I had done nothing wrong. Come hell or high water I intended to get J.T. Nelson convicted and investigate further into the other allegations I had received concerning local corruption. I also intended to get reelected in '84. This sheriff wasn't about to throw in the cards. Little did I realize that in this game the stakes were going to get higher.

Mike Seelhoff Leaves

There was an opening in the Reed City Police Department. I was in the frame of mind that I would jump for a job anywhere, just to get out of Lake County. In my resignation letter to Bob, I tried to explain that the overwhelming political environment was significantly interfering with my ability to be an effective police officer. For a long time Bob had tried to keep me insulated from the politics. I ran our 24 hour road patrol from the basement of the jail. Also I was responsible for our grant funding requests and the associated bookkeeping. Eventually I started getting dragged into the political mud so I asked Bob if he would relieve me from the sergeant's position and just let me be a plain old deputy again. He said he was unable to do that; there were no deputy positions, just the supervisor's slot.

How well Bob Blevins would be able to deal with the politics this time was the question in my mind. I watched when the local kingpins destroyed Lonnie Deur. And now, in Bob Blevin's situation, there were higher political powers plotting his demise in addition to the locals. Bob was good working the media, but sometimes when the media doesn't have a good headline story, they lose interest and instead of doing some good investigative journalism they quote the powers that be. In the end that's what the people are exposed to.

The political pressure on the sheriff's department had been mounting for some time. When Randy Gerke shot Mr. Nelson things got really bad.

We lived just down the road from where Randy pulled Nelson over. I had just let the dog out to piddle when I heard the shots. Next Randy's voice came over my scanner. He was screaming a 1035 code, meaning officer needs assistance. Bob Smith was my neighbor and I almost hit him backing out the driveway. I was the first officer on the scene. The radar on Randy's patrol car was still flashing 69 mph, the speed of Nelson's vehicle. Nelson lay on the ground, obviously dead. There was the smell of alcohol, and a bottle on the seat of Nelson's vehicle. Two or three people were there who witnessed what had happened. Randy tried to arrest Nelson for speeding and drunk driving. Nelson resisted, and there was a struggle. Nelson grabbed Randy's gun from the holster and started firing at him. Randy pulled a backup gun and killed Nelson.

Sheriff Blevins arrived and immediately called in the state police detectives. Since the shooting involved a black man, every effort was made to keep the investigation out of our department. Bob laid Randy off with pay while the investigation was underway. The state police report completely cleared Randy of any wrongdoing, but that didn't seem to matter. Robert Williams and his group were at the forefront of the protest. They made a pilgrimage to the state capitol to demand action. Williams said that this was just another example of a white prejudiced police officer attacking a black.

Shortly after Nelson was shot, Sheriff Blevins was ambushed and slightly wounded. We always received good intelligence from the two Smith brothers who ran the bank. They passed word to us that there would be retaliation against the Lake County Sheriff's Department. People would be out there to snipe some cops. There also would be the possibility of riots. Forces would be shipped up from Detroit and other places during the Trout Festival for that purpose.

We were all very scared. We managed to get volunteers so that two people would be in every patrol car. The tension and stress were unbelievable. The sheriff organized a reserve backup police response from both Muskegon and Kent County in case riots broke out. Calls for patrol response greatly increased. Many were hoax calls from Idlewild. After working in the Idlewild post office for some time, I knew a lot of the black residents and felt comfortable doing police work in that area alone. Not so now. All of a sudden we were being portrayed as the enemy. Even before Nelson's shooting, any arrest of a black resident by a white cop was fodder for Robert Williams to use for stirring up trouble. If you caught some black guys inside a locked store stealing at night, you still had to do a great deal of work to gain a conviction. According to Robert Williams it was still a white cop versus an innocent black situation.

The year before I had been involved in the shooting of a white man in circumstances less justifiable than Randy Gerke's. Nobody really cared because the guy was white.

Eventually Randy would be cleared of any wrongdoing by the State Police, local prosecutor, Attorney General's office, FBI and even the Justice Department, who had been dragged in. Despite all that, Randy left the county and removed himself entirely from police work. He didn't feel his family was safe, or that with this baggage he could be effective in police work again. The last I knew, he worked in construction in Grand Rapids.

During my few years in Lake County I received more police experience than I have in all the rest of my 20 plus years in police work. I remember the day that we all stood around the office waiting to be interviewed by the new sheriff Bob Blevins and his undersheriff Dale Gibbs. When it was my turn he asked me if I felt that I could work for him. When I said I could, he said, "Good." He slid over some keys, saying, "You start tonight." For one reason or another the rest of the deputies were fired. For quite a period of time it seemed that I was on duty all the time, but eventually Bob started to rebuild the department and bring some new people in. We were catching lots of criminals. Probably not just because we were so good, but because there were so many criminals to catch that it was hard to miss.

Bob was a good public relations sheriff. He enlisted the support of local civic groups and went out of his way to get the cooperation of the county residents to enforce the laws. Bob got our department involved in some good community service projects. Toys for Tots was one of them. I remember one year we had a truck pick up a whole load of new toys in Grand Rapids and Bob's friend, the rock musician Ted Nugent, wrote a check for the whole load. We would get a list of poor families from Social Services and on our off duty time distributed toys. Probably these were the only presents some of those kids ever got.

The Michigan State Police Post in Reed City got a new Commander by the name of McGhee. McGhee lived in Baldwin and through his friendship with Bob there became a close working relationship and a noticeable presence of state police cruisers in our area. Previous to this, the state police pretty much stayed out of Baldwin, unless they were called in.

We didn't have the funding for a detective position anymore, but occasionally I had the opportunity to do some investigative work. Whenever there was time, even on my days off I would watch the state police detectives at work. They were professionals.

David Hill, the prosecutor, came to Lake County to get some experience. With all our crime, I think he ended up getting a lot more experience than he wanted. David knew I was struggling with the detective work, and he did his best to point out exactly what kind of evidence he would need in each case to get a conviction. Whenever we got in over our heads we would give the state police a call and they would give us a hand. One of their detectives, George Pratt, did a lot of good work in our area.

Bob brought in a new black deputy from Grand Rapids by the name of Count Fisher. Along with being a county deputy he was also sworn in as Chief of Police for Idlewild. The Count was not tall but he was built like a bull. He bragged of having a 7th degree black belt in the martial arts, but he really didn't need it, because he was fearless and so strong. I'll never forget when he first got here. He brought this big bad black dude in from Idlewild. He sounded all out of breath and excited on the radio and when he got to the jail he drove into the security garage so fast I thought he'd crash out the other side. The Count did a good job of cleaning up Idlewild and was respected by the people there even though they might not have approved of his methods.

The Count and I became close friends and worked on several cases together. I was Bob's road sergeant then and the Count always called me Sergeant. He had not had a lot of formal police training, but he knew a lot of people and through his contacts bad guys were apprehended that otherwise would have probably remained at large.

J.T. Nelson pressured Bob into hiring this one deputy and the deputy was on duty the night a local murder suspect accidentally escaped from our jail. The Count nosed around and sometime later we had ourselves an informant who was willing to give the escapee's where-abouts. To provide the information, the informant wanted a pack of Kool cigarettes and $200. David Hill, the prosecutor, had a fund for this type of thing so he gave me the $200. The informant was sup-posed to deliver the Kools to the fugitive that was holed up in a house near the Cloud Nine Motel. The Count figured that because the guy was a murderer, we needed to kick the door in. When the Count hit the door it not only came open, it came off of the hinges and flew over on top of the bed where this guy had been sleeping. The guy was suddenly not a bad guy anymore. He was shaking like a wet puppy as he looked into the barrels of two 357 revolvers. I called Bob on the phone and said the Count and I had an early Christmas present for him. When we told him who we had he said, "That's the best present I could get." He had been quite embarrassed over the escape from his jail.

The Count's only shortcoming was in administrative procedures. So although he brought in a lot of offenders, many didn't get con-victed because of a lack of paper work. After awhile when he got a hold of someone he would call for "Sgt. Seelhoff" and we'd fill out all the necessary reports that were needed to make the charges stick.

As I said the Count was a real asset for the county sheriff's depart-ment and for Idlewild, but unfortunately the job didn't pay much. Before the trouble of 1983 started happening the Count had returned to Grand Rapids.

The Lake County deputies actually made more money each year than the sheriff. Our basic wage was the same as his; however we got paid for overtime and the sheriff didn't. He had a $3,000 housing allowance, but the board threatened to take that away from him. For a long time the yearly base wage was $14,000. This qualified me for low income housing. Eventually, through the persistence of our union, we forced the county to raise the wages to around $18,000 plus overtime. We still made more than the sheriff. This certainly wasn't right, because he was shouldering a terrible burden, at least in this county.

When Bob's honeymoon period was over with J.T. and the board of commissioners, they did their best to use the budget weapon against us. This was not a surprise for they had done the same when Lonnie

Deur was sheriff. Bob temporarily got an injunction keeping them from removing our secondary road patrol funding, but that didn't last long. Cindy Keefer was the county clerk when Bob started and she was extremely helpful in making sure all our grant paper work got in on time. Her husband, Art Speers, was the sheriff in Osceola County and he was over here all the time helping Bob get started. Art lost his election in November of 1979. Shortly after that the relationship with Cindy Keefer went sour and we were receiving no help from her. In fact Cindy became one of our most vocal opponents at the board of commissioners' meetings.

The financial reputation of the sheriff's department really began to suffer because the courthouse went out of their way not to pay our bills on time. It became almost impossible to order any equipment because the vendors weren't getting paid.

After about a year Dale Gibbs was replaced with Bruce Finch as undersheriff. Bruce was a good undersheriff and investigator, plus he liked hunting and fishing so I thought he'd be in that position for some time. It didn't work out that way though. Through some embarrassing marital problems he was kinda shamed out of the area.

Bob asked if I would consider being his undersheriff. I thought that would put me closer to the politics than I wanted to be, so I recommended Joe Kuhlman. Joe hadn't been there too long, but he seemed like a capable person, being a retired Marine Master Sergeant.

My first impression was that Joe was going to be a pretty good undersheriff. The only trouble was he liked to drink. More than once he would come to work in the morning hung over. I don't know whether Bob noticed this or not, but if he did he never mentioned it and I didn't either. My wife and I spent several evenings socializing with Joe and his wife Lois. Mary and I both liked them, but I got this underlying feeling that Joe was trying to enlist my support against Bob.

In the last part of 1983 Joe took one of the new recruits back to the academy at Bay City. Joe was in uniform, drinking in the patrol car and at the academy. The boys there finally got him into a motel. Joe didn't show up for work and in a day or so Bob got a call from the Sanilac sheriff to pick up our patrol car. They had impounded it after

Joe was turned in for trying to purchase liquor in uniform while intoxicated. I rode over with Bob to drive Joe's patrol car back. We drank a lot of coffee that morning and I really needed to use the bathroom, but I was so embarrassed by the whole situation that rather than use the Sanilac jail bathroom I just got the keys and drove to a local restaurant to relieve myself.

The voters finally got smart and recalled J.T. Nelson. Shortly after that, Gwen Warren quit her job with the sheriff's department. She had been there since Guy Lee was sheriff. Bob had promoted her to Lieutenant and put her in charge of jail administration. He seemed to have a lot of faith in her, but personally I thought she was in over her head. Anyway she came up with a stress attack, sued the county for a nice settlement then ran for J.T.'s old position on the board of commissioners. Not many people knew why she left the sheriff's department. I remember reading a letter to the editor of the *Star* from the man that nominated her. He found out that she had sued the county for stress and he didn't think that was right. He withdrew his support for her and backed her opponent. She was still elected.

When Gwen's desk drawers were cleaned out we found literally hundreds of forms that were to have been filled out and returned to the state police so that criminal records could be established in the computer network. Consequently a lot of offenders that should have had records didn't have them. Maybe she just didn't know what to do with them.

Joe Kuhlman finally resigned. That fall a governor's task force came in to examine the problems in Lake County. There were some investigators who said that they wanted to go through all the accounting for the secondary road patrol grant. They seemed quite satisfied with everything, but they also were quite intent on learning something about the sheriff. They weren't too specific, but indicated that they didn't want to drag the department or me through anything, mostly they were concerned with the sheriff.

The thing that really bothered me was that, after having such close relations with the Reed City State Police Post for so long, now they didn't seem to want to associate with us. I would have thought that they would have been real interested in helping to investigate after

Bob was shot, but that was not the case.

It was in this time frame that I decided to leave the Lake County Sheriff's Department. Mary and I were having problems and it was all more than I wanted to continue with. In talking to her after we were separated, she told me that there were some people really out to get Bob.

Shit Hits the Fan

The governor's task force findings and recommendations were released the middle of November. There was something in their report to please everyone. It recommended that the governor direct the Michigan Department of Treasury's Bureau of Local Government Services to immediately audit governmental units in the county that are alleged to have misspent public monies. The task force further recommended criminal prosecution and/or impaneling a grand jury if allegations are found to be true. The task force didn't specify the government agencies to be investigated, but according to Mike O'Connor from the *Star*, "Complaints at the September 9th public hearing were leveled against the county government, township governments, the Department of Social Services and the Baldwin Community Schools." It appeared that the investigation would be far-reaching and thorough. There were 9 more recommendations that centered around the sheriff's department. Although I wasn't happy over the requirement for department sensitivity training, I didn't really mind having the department put under the microscope because I didn't feel we had anything to hide or be ashamed of. The Michigan Office of Criminal Justice was charged with reviewing and evaluating our policies and procedures and along with the Michigan Sheriff's Association evaluate our day-to-day operation with special emphasis on citizen contact. There was even a recommendation that the County Board of Commissioners review a National Criminal Justice Association evaluation which involved the department's low pay, and to take appropriate budgetary action.

It certainly looked like the governor's task force recommendations would address the problems of Lake County, but in reality it was only the machinery assembled to get rid of the sheriff. As far as I know my department was the only county agency ever audited and inspected and the request for our audit went out two weeks prior to the task force recommendations even being made public.

Shortly after Mark Raven was appointed prosecutor, all felony charges against J.T. Nelson were dropped. In Judge Farabaugh's 78th District Court J.T. was allowed to plead guilty to using false property tax credits. He paid a $500 fine and a $1000 settlement. This to me was unbelievable. A State Police detective by the name of Pratt had prepared and presented the charges to the former prosecutor David Hill, and Hill was actively pursuing the charges. Pratt was not a rookie. Indeed, he was an expert, a knowledgeable professional of the highest caliber.

J.T. Nelson was temporarily off the hook. I didn't intend for him to remain that way long. I conscientiously prepared a very comprehensive package for the new prosecutor. This time I was requesting that he prepare and present a request for a grand jury to examine a multitude of charges aimed at J.T. Nelson. This package was thoroughly prepared, containing depositions and the results of long investigations.

I thought if I could get a grand jury seated there would be a lot of folks coming out of the woodwork to testify against him. After all the bad publicity I had received in the last year it would certainly help my reelection too.

Mark Raven needed some time to adjust to his position as county prosecutor. I was anxious to proceed with the case against Nelson, but I tried to be patient. Raven didn't contact me, so occasionally I would mention my grand jury request when I ran into him. He just was always too busy to talk to me. After several months I was convinced that the prosecutor did not intend to follow through on my request. He never gave me an answer, one way or the other, just sat on the package. Why he did this, I'm not sure. Maybe he wasn't familiar with the procedure for drawing up such a request, and didn't want to admit it. Possibly he was intimidated with threats as David Hill had been, or quite possibly there was other incentive to encourage him not to proceed. Whichever was the case, between the charges the Michigan State Police filed and the package I gave him, there was plenty of reason for any prosecutor to jump in and get the wheels of justice in motion.

The auditor's report became available to the state police in January and I found out that there were 3 checks deposited in the contingency fund that should have instead been deposited in the general fund. I didn't know how they got there but on my attorney's advice I went to Mark Raven and offered to reimburse the county's general fund for that amount from my personal account. I didn't authorize the money to be put in the wrong account, but I did have overall responsibility for the handling of those funds even though the contingency fund had been delegated to the undersheriff. Kuhlman had to be involved here some way, but as yet I wasn't aware of the details. Raven refused to accept my offer. He didn't say why or give me any idea who was doing what with this misdeposited money information, just that he couldn't allow me to pay back the money.

Next, I was invited down to the Rockford State Police Post to meet with the two special investigators, Beck and Holt. My attorney Charles Probert accompanied me to the meeting. These two had originally picked up copics of my personal checks from the bank and informed me of the audit. Now they were flipping through my personal checks asking what I had purchased with various checks. It was a game to intimidate me. I wasn't intimidated and I told them that they had obtained those canceled checks illegally and that there was no authorization from anyone to audit my personal account. They said they really wouldn't have to use them anyway, the charges alone would be enough to sink me politically. That certainly seemed like an open threat, but I had more important things to do than play games with these guys.

Instead of waiting any longer for Raven to take action on my request, I decided to bypass him and take my case directly to Judge Cooper. To do this I would need a properly drafted grand jury request. I solicited the services of an old friend, David Sawyer.

David Sawyer was the prosecutor in Kent County. When I was a deputy in Kent County and David was an assistant prosecutor, he used to spend many hours riding around in the patrol car with me. I explained that I was having trouble prosecuting some individuals in my county, that I was in dire need of a properly drafted grand jury request. David said that in the past he had drawn up seven such requests and would be glad to help me out.

Meticulously, I prepared my grand jury request package and on January 31st delivered it to Cooper. Hopefully he would agree to this request and a grand jury could be formed to consider all the illegal activities going on in this county.

In my grand jury request to Cooper I asked that a special prosecutor be brought in to replace Raven. I was hoping he might consider appointing Jim Talaske, the prosecutor from Reed City. Jim had acted as a special prosecutor in Lake County before, and had a reputation of being a no-nonsense prosecutor. Besides that, he was a friend, someone to be trusted.

Circuit Court Judge Rick Cooper was from the local area. As a young attorney, he was very athletic, and somewhat of a hippie, with quite long hair. During his campaign he used to jog the county roads wearing a T-shirt that said "RUNNING FOR CIRCUIT JUDGE" on it.

One day my wife Dee came into the office and caught one of Judge Cooper's assistants taking improper liberties with my female dispatcher. I met with Judge Cooper in a local restaurant to relate what his assistant had been up to. Cooper erupted like a volcano, called me a liar and created a mini scene in the restaurant. I decided that I would have to tread lightly with this guy. I certainly did hope that now he would forget the past, and objectively consider my grand jury request.

In February I received the 1983 Uniform Crime Report compiled by the Michigan State Police. Once again for the third consecutive year the crime rate in Lake County had gone down. Ours was the only county in state history to have done this three years in a row. It seemed like the perfect time to announce my reelection bid for sheriff.

In the February 7th edition of the *Star* I announced my candidacy and quoted the latest crime report figures for the county; murder down 100%, robbery down 60%, burglary down 58%, larceny down 1%, motor vehicle theft down 50%, arson down 66%. Only aggravated assault went up and rape stayed the same. Why had we lowered crime again? In my view because 1) My deputies were in fact professionals, and they have a strong never-give-up attitude, 2) The strong assistance and support given by Lt. Gary McGhee and his State Police Post in Reed City, 3) The many citizen support groups such as Neighborhood Watch that have been organized all over Lake County. And, the reserve and posse units that are unpaid volunteers that assist whenever needed.

I stated that much had happened in my opinion to force this department as of late to put law enforcement on the back burner. "Many people would agree that it's difficult to concentrate on draining the swamp when you're up to your pits in alligators."

It seemed like the right time to strike out at my adversaries, "the alligators in the pond," and I let them have it.

— "I'm disillusioned by the governor's office for allowing himself to be hoodwinked by so few, to believing there was and is a breakdown in communication between the sheriff's department and the community. If this were true, then five deputies would be responsible for lowering crime three years in a row in a county over 500 square miles … no sir? We didn't do it without help from the community."

— "And now since the governor has admitted (at least privately) he should have never stuck his nose in Lake County, he seems to have a problem. Someone must be offered up for sacrifice. Someone must take the fall so the governor doesn't look stupid. His choice is simple: Either his own man Conrad Mallett Jr., or a small-time sheriff in a little dirt water town. Seem like a difficult choice?"

— "The governor has gone to the head of his private police (Michigan State Police) and ordered him to investigate, audit and turn this department inside-out and upside-down. Former prisoners, every former employee has been interviewed. They were asked, 'Did you ever see the sheriff take a quart of oil or put gas in his private car? Can you give us anything on the sheriff????' Every business in the community that has provided a service or a product has been interrogated and their records checked. Not since Watergate has such a thorough investigation been done by so many. These detectives even know where I buy my Fruit-of-the-Loom."

— "And yet, even with all this controversy and the controversy to come, I intend to continue to fight! It doesn't matter that I am now paid the same $18,000 per year as my deputies are, and it doesn't matter that the Board of Commissioners have voted to decrease my pay next year by $3,000 by removing my housing allowance. My work as sheriff is not finished! Therefore, I announce today that I intend to seek a third term as your sheriff."

— "Further, I charge voters of Lake County, Republican or Democrat, to remove any elected official that is not responsive to the needs and desires of the people that place them in elected office. It is not enough to reelect me your sheriff and then force me to be crippled by the same county clerk, and members of the Board of Commissioners. Some of them (not all) must go."

Joe Kuhlman was next to announce his candidacy for sheriff. He gave a little dissertation on why he resigned, but it wasn't close to the real reason. I had never brought up Joe's drinking problem and when he left I did my best to smooth things over for him. I said that we had parted as friends and that Joe was just tired and couldn't see that the department was going any place.

There were several men that we had under investigation for local area theft. All of them had prior records and all were black; I needed a black undercover officer to see if we could get some hard evidence on them by wearing a hidden listening device. I hired the officer from another department so he was totally unknown in the Lake County area.

The undercover officer was in his forties and seemed to have a good head on his shoulders. I secured legal authorization for the wire and furnished my officer with mug shots of the men we were looking for. The Sportsman's Bar was where he started. We had an additional man in the parking lot who was listening and recording the conversations that our undercover man's listening device picked up. We were about to get a real earful.

My new deputy proved to be very smooth and polished in his undercover role. Flashing a little money around it didn't take him long to attract some attention. We were interested in where the stolen property was warehoused and who was connected to it. Unexpectedly, our man gets waved to a table.

J.T. Nelson and some of his friends were enjoying a few beers and J.T. spots this new guy in the bar and invites him over to his table. J.T.

had been drinking a lot and after a short period of time it was evident he was trying to impress our deputy with his power in the community.

J.T. reminisced about his friendship with Coleman Young. They were pals. If he ever needed anything, he could go right to Coleman. Coleman could press a few buttons and always get things done. J.T. went on to brag that one of the buttons Coleman could push was Governor Blanchard. Lake County could get grants or monies, anything they needed.

In the course of the conversation, and his bragging to the undercover deputy, my name came up. What a S.O.B. I was. Wouldn't it be nice if I was in his pocket like the Detroit Chief of Police was in with Coleman. Someone at the table asked, "What do ya mean J.T.?"

"Well, Coleman's got this secret fund for whatever he wants or needs. The Chief of Police does exactly what Coleman wants because the Chief is one of the recipients. His daughter lives in a fancy condo in California, with the condo's lease payments coming from Coleman's fund."

J.T. went on to say that it was too bad he couldn't work the same deal in Lake County. Lake County was a poor county. Because of that, there were all kinds of monies available from programs and welfare, if only the sheriff would be a little more cooperative. With a little laugh, J.T. said, "I should have asked for more than 500 dollars to take care of the sheriff, maybe we could have gotten a better shot, could have hired someone who could shoot better." Everybody at the table has a good laugh over that one.

When I heard the tape I didn't laugh. It only confirmed in my mind that there was a lot of monkey business going on at the State level. Coleman Young was a definite force in the State government, and it didn't seem just a coincidence that his old political advisor Conrad Mallett Jr. happened to become the chief legal advisor to Blanchard. It also might be the reason we were having a difficult time pursuing charges against J.T. Nelson.

Coleman Young's secret fund became news a couple years after we first heard about it. It was discovered that the Chief of Police's daughter's lease was paid from the fund. The chief was fired, but Coleman Young got off clean.

Even though J.T. was under the influence of alcohol during our tape recording, and who's to know what was bullshit and what was truth, I was pretty well convinced that J.T. Nelson was responsible for setting up the hit on me. Possibly, the 500 dollars J.T. supposedly asked for the hit, came from Coleman Young's secret fund. Although I couldn't prove it, I thought I knew who the shooter was.

Willie Lathimer was a strong arm for Robert Williams' group. When they boarded their vehicles to head down to the State capitol to protest, Willie was in the lead of the convoy, driving his big Harley chopper. Willie was a big man, 250 pounds plus. He had been involved in several scrapes, but did not have a felony record.

Not long after my shooting Willie came to my office to fill out an application for permission to purchase a 357 revolver. I checked; he had no record, so legally I couldn't refuse him. However, I did ask why he wanted the permit. A half grin formed on Willie's face, "I'm a much better shot with a pistol."

That statement raised the hair on the back of my neck. I felt Willie was taunting me. As close as he could, he was telling me he was the shooter. I had no way of proving that Willie Lathimer tried to kill me, but personally I was convinced that while someone else drove the truck, Willie had fired the shotgun buckshot that was meant to kill me. I never would find out for sure because later his motorcycle collided head on with another car and he was killed.

In March the Board of Commissioners was forced by an arbitrator to pay my deputies for the overtime they had accrued in the previous year. They also discussed The National Criminal Justice Association review which pointed out that the sheriff's department pay scale was a major deterrent to the operation of a stable law enforcement agency. At one of their meetings in January I had stated that my budget was almost $42,000 short, and that also was mentioned. Perhaps now, with the elections coming up, they were becoming more aware of citizen concern for law enforcement support.

I was still receiving recognition for my efforts to reduce crime. In March of '84 the Baldwin Community Schools presented me with a commendation for my efforts in improving the quality of life in the community. There was also the Certificate of Appreciation from the United States Jaycees.

In April the County Board of Commissioners received word from the State that there was no reason for the State to audit any other county agencies. Board Chairman Stark expressed profound disappointment in the State's decision as the audits would have cleared the air and eliminated any mistrust in county government. The very next month the Board of Commissioners organized a testimonial dinner to honor their former Chairman, J.T. Nelson. Perhaps it was a celebration not only for J.T., but the fact that county government was not going to be in for any audits and that only the sheriff would be the focus of attention. To me it was ironic that people could honor a man who had left office in disgrace. It would, however, be the last of the happy times for J.T.

She seemed rather young to be J.T.'s wife. He must have been over sixty then. I'm not sure what he did to piss her off, but when she came into my office she was hell bent on helping me put J.T. behind bars. In fact she said she could give me enough evidence from over the years to put J.T. in prison for the rest of his life. Her only request was to be put in the witness relocation program. I explained to her that the County had no funding for a relocation program that could provide her new identification and an income.

Maybe some of her information could lead to federal charges. I called Dan Bidwell, my FBI friend. Dan arranged for her to be transported to a safe house in Newaygo. It wasn't far away, but at least she was protected, away from J.T.

After J.T.'s wife came up missing he filed for divorce. Four days after that on May 26th his wife got tired of being confined in the safe house. She slipped out to a local bar for a little night life. One of the patrons recognized her, gave J.T. a call and said, "Hey J.T., the wife you've been looking for is over here having a good time."

J.T. stuffed a snub-nosed 38 in his waistband, filled his pockets with shells, then checked the lever action 30/30. There already was a shell in the chamber.

His wife was still in the bar when he burst through the front door. J.T. shouted her name, took aim and pulled the trigger. The hammer fell on a spent cartridge. Before he could rack in another shell several patrons subdued him. The Newaygo sheriff was called.

J.T. was finally in jail, but when he developed or faked chest pains he was rushed to the Reed City hospital. He never returned to jail again, for that's where he died a couple of months later. No one's sure exactly what caused J.T.'s death, but it was rumored to be from a drug overdose. I heard that he broke into a drug locker and ingested a whole bottle full of nitro tablets. That's something I can't confirm. What I do know is that once again James T. Nelson had escaped the wheels of justice. He had died with his secrets. On the front page of the *Star* there was a nice picture of J.T., and a glowing obituary. They failed to mention how he'd attempted to kill his wife, his recall or even alluded to his dishonesty.

The primary election would soon be coming up. I was busy campaigning, making speeches at town halls and at every community meeting that I could. Lake County was being portrayed as a racially torn community. That was just window dressing to mask the real problem, which was organized crime. During the campaign I was trying to get the focus back on the organized crime issue.

My opponent in the primary was Joe Kuhlman. Joe was telling the folks that he felt compelled to leave his undersheriff position because he didn't want to be a part of the mismanagement that was taking place under my leadership.

The Democratic opposition in the general election would be Peter Loucks, a retired State Police Motor Vehicle Department Lieutenant. He would be a write-in candidate.

Early in the campaign, friends in the black community told me a State agency printing office in Ludington was being used as a campaign headquarters for the Democratic candidate, Peter Loucks. The telephones there were used to encourage Lake residents to write in Loucks' name. More than that, the state-purchased copy machines were pumping out his campaign literature. People who had actually been in the office and observed this furnished the information. It was not a rumor.

I informed the election commission that a state office's resources were being used to launch a private campaign. They said that they would need documented proof that this was happening before they could go to the Attorney General. They suggested that maybe the state police might be able to help me. Right, fat chance. They were too busy investigating me.

There was plenty of rumor and speculation, but I still didn't know what would happen with the results of the audit and investigation by the state police. The *Star* editor Mike O'Connor was certainly not a big supporter of mine, though he too wondered what was going on. In an editorial on June 26th he said, "I really think it's time for the state to quit … ing around and time for it to … or get off the pot in regards to the sheriff's department."

O'Connor went on to say, "You see, depending on which 'source' you speak to in the state police in Lansing or the attorney general's office, you get one of two stories. The first story, which has been published in at least one larger city newspaper, is that Sheriff Blevins has done nothing wrong, there are no problems with any contingency fund, no monies are unaccountable for, etc., etc. But other sources in the same offices will tell you—off the record of course—that they've got enough evidence to issue warrants against the sheriff but they're being held up in another office. That 'other' office said every two weeks for the last six months, 'in two weeks we'll have something for you'."

O'Connor said that he didn't think it was fair to the people of Lake County and certainly not fair to me, and that the Michigan State Police and the Attorney General's office and whoever might be involved needed to put the rumors to rest. If I was guilty of misdemeanors or even felonies, the electorate certainly has the right to know before the election. If I had done no wrong then O'Connor said that I deserved to have that fact made public so that I didn't have to campaign under a dark cloud of doubt.

He ended his editorial by saying, "I'm beginning to agree with his (the sheriff's) original assessment of the task force and its various spin-offs as a 'dog and pony show', and that's too bad, because the people of Lake County deserve better."

A report issued jointly by the Michigan Sheriff's Association and the Michigan Office of Criminal Justice said that with a few minor exceptions the Lake County Sheriff's Department is doing a commendable job despite being understaffed. The report came out at the end of June and was issued in response to a task force recommendation to review and evaluate the day-to-day operation. The bottom line of the report was that the county board of commissioners should seriously consider providing funds for additional deputies.

When I was asked for my reaction to the report I said, "I'm glad the truth's finally out." Commission Chairman Robert Stark called the report "a damned cover-up."

The report was based on interviews with many of my employees, plus the reporter spent about 25 hours riding with four deputies on road patrol. The report added that despite critical staffing levels and long hours, citizen contact was very acceptable. Additionally, based on an interview with one former deputy the high turnover rate was due to excessive pressure and low pay.

Judge Cooper never responded to my request for a grand jury. I again got in touch with Vince Persante, head of the State Organized Crime Task Force. I explained to Vince that we were getting no place with the Lake County judicial system; he said that it was time to take the package to the Attorney General, Frank Kelley. Vince reviewed all my material. New to the package was information gathered on the tape recorded conversation of J.T. Nelson, Mayor Young's secret fund, his relationship with the Detroit Chief of Police and Governor Blanchard. Because J.T. wasn't specifically mentioned in our wiretap authorization I didn't think his conversation could be used for legal proceedings, but it certainly should get Kelley's interest.

On previous occasions I tried to get information from the Attorney General, but could never make it through his buffer zone of state attorneys. Vince Persante knew Kelley on a more personal level, and arranged the meeting with him.

Two other attorneys reviewed the package that I was to present Frank Kelley. One was David Neff, a private attorney familiar with criminal law; the other was Charles Probert. Charles Probert was the former Police Chief of Allegan. Charles had worked his way through law school and became an assistant prosecutor in Kent County. He was a brilliant prosecutor who compiled a record of many hundred successful prosecutions and only a couple of losses. Later he became a very colorful Wyoming municipal judge. He was a policeman's kind of judge. If the apprehending officer suggested leniency the Judge would be lenient. On the other hand, if the policeman suggested throwing the book at the bad guy, the Judge would give the maximum sentence. The tenure commission objected to Probert's sentencing record among other things and removed him from office. Charles also had some severe personal problems to deal with, but eventually he came back to private practice. Like many others, Charles Probert spent a lot of hours in my patrol car. We were good friends.

Vince Persante, Charles Probert, David Neff and myself entered the State Attorney General's office. Personally I was overwhelmed with being in the presence of all the big dogs in law enforcement and I was more than happy to let Charles Probert, my attorney, take the lead in the proceedings. A former prosecutor and judge, he felt comfortable doing that. Charles waded through our material and played a portion of J.T.'s conversation from our wiretap. I tried to focus on the gray-haired attorney general and what questions he might ask. We told Kelley that we had been getting no help from either the prosecutor or circuit court judge in our requests for a grand jury and that we needed his help to clean up Lake County. Kelley listened intently and I thought it would be difficult for him not to take some kind of action.

Kelley now spoke, "Gentlemen, do you realize everything you're asking me to do is of a political nature?"

Probert questioned Kelley's question, "What do you mean, of a political nature?"

"Everything you're presenting is against Democrats. Allegations against a Democratic governor, mayor, former board of commissioners chairman, all Democrats. Looks like a Republican sheriff trying to eliminate his Democratic opponents. There's nothing concrete here. Nothing I can go forward with."

I couldn't believe I was hearing this from the Michigan Attorney General. He was insinuating that this whole thing was political fodder.

Kelley got up from his desk, walked over and opened the door, "Gentlemen, have a nice day. You were never here."

When we walked out of Kelley's office I was numb. There must have been something wrong with our presentation. There was the feeling inside me that somehow we had insulted the Attorney General. Kelley seemed very angry with me personally. Somehow I had forced him into a corner. I hoped that when he reconsidered our presentation, there would be some action taken. It was a false hope.

Vince Persante later turned in his badge to Frank Kelley. That was the end of the State Organized Crime Task Force. He said that he felt the failure of Frank Kelley to write warrants on Democrats for political reasons made his job a waste of time. If the people that broke the law had to be categorized as far as their politics, and only the Republican bad guys could be prosecuted, he didn't want any part of it. There was more than one case that brought him to his resignation. I remember reading about his resignation in one small paragraph inside some paper the Michigan Sheriff's Association sent out. There was no mention of it in the state news media.

The meeting with Kelley took place less than a week before I received the call. When I first heard the familiar voice there was a momentary expectation that perhaps the Attorney General was finally taking some action. That thought quickly went by the wayside, for there was not a good news sound in his voice.

"Do you know that the Attorney General is conducting an investigation on you?"

"Well, there have been two State Police detectives snooping around here, wanting to know where I buy my underwear and when I brush my teeth. It's a big witch hunt. I'm not sure who sent them, but I'm not too concerned."

"If I were you, I would be concerned. I'm calling to offer you a deal."

I was starting to get angry. "A deal. What kind of a deal are you offering me?"

He continued, "I've been told by the Attorney General that if you resign from office, the investigation will cease. You can cite any reason you want for not running in the next election, tired of politics, stress, whatever."

"But I have nothing to hide. I have not done anything wrong!"

"Bob I've been around politics enough to know that nobody's that squeaky clean. If they look, they'll find something. I'm told that when they finish with you, not only will you not be reelected, you will also be charged with something."

Now I was enraged. "Tell Mr. Kelley that I decline his deal."

"If I were you, I would genuinely be concerned, Bob."

I tried to find out more from him about what Kelley might try to charge me with. He was evasive, and ended the conversation by saying, "Well, I've just delivered a message and you have given me your answer. Good Luck."

It was obvious to me now that Frank Kelley was not just reluctant to prosecute Democrats, he was actively trying to get me out of the sheriff's office, and probably that's why the State Police detectives started investigating me in the first place. Later, I learned that there were two other sheriffs who were offered deals in return for leaving office and dropping investigations.

There were at least a couple of good things that happened for me. U.S. Representative Guy VanderJagt, in recognition of my record in lowering crime, had my name read into the Congressional Record. He presented me with the flag that flew over our nation's capital on July 10th, 1984.

In August, I received a letter of appreciation from Tom Johnson, head of the Michigan Office of Criminal Justice.

"Sheriff Blevins has maximized his resources by involving members of the community in a crime prevention program ... i.e. Neighborhood Watch/Child Watch/Operation Identification. This effort seems to have paid off. Before the sheriff took office, the crime rate was 64% higher than the city of Detroit. In the past five years, during a period when the state had an increase of 20%, Sheriff Blevins has lowered crime 51%.

The ironic thing about the statistics was that they were all compiled by the Michigan State Police and furnished to the Office of Criminal Justice.

In a phone conversation with Tom Johnson, he said, "Isn't it strange that a few years ago they were patting you on the back for lowering crime by 14 percent, and now, when you have lowered it 51 percent over five years, they want to nail your butt to the cross."

To finish up my primary campaign, I had the letter of appreciation from the Office of Criminal Justice printed in the newspaper. The primary election was held, and I won by 93 percent of the vote. At least the voters were smart enough to see what was going on.

The primary vote sent a clear message. If I was to be prevented from being elected sheriff they would have to think of something else to stop me.

It had been two or three years since Dee and I had taken any kind of vacation. We needed a little break. The election campaign, along with my frustration over Kelley's deal and lack of support, was starting to wear me down. We decided to take some days off after the primary and visit Mackinac Island. I notified the board of commissioners that I would be on Mackinac Island for a few days. My new undersheriff was Tom Byrne. I felt confident he could handle anything that came up.

Dee and I spent the night in Mackinaw City. We purchased tickets on the Arnold Line for a morning trip to Mackinac Island. It would be a nice day to walk around the island, just relax, see the sights and forget about our troubles back home. There was some time to kill before our ferry left, so we stopped into a donut shop that a high school buddy of mine owned. He was working that morning, and after saying hello we sat enjoying some coffee and donuts, casually watching the TV that was mounted high behind the counter.

Halfway through our coffee, the TV channel transmitted a series of beeps, the kind that proceed a severe weather announcement or special news bulletin. The regular program is interrupted and this news guy is on. "This news release is from the Michigan Attorney General's office. According to various sources we have learned that Lake County Sheriff Robert Blevins and his wife have fled the United States with an undetermined amount of county funds. A warrant is being sought for his arrest." A 1-800 number was flashed on the screen for anyone with information as to my whereabouts.

My buddy hollers from across the restaurant, "Hey Bob, they're talking about ya!"

Dee and I are stunned. We leave our coffee, return to the motel and pack our bags. All kinds of things are racing around in my mind. We must get back to Lake County as quickly as possible, but not break any speed limits in the process. The last thing I need is to get pulled over by some energetic state trooper and have a gun in my face. Maybe they've already got people out looking for us now. It certainly wasn't any secret where we were going, so it wouldn't be too difficult for them to find us.

I dropped Dee off at home, then immediately went to my office and called Charles Probert and David Neff, my attorney friends that were with me in the Kelley meeting. They advised me that we should press for an immediate arraignment. My question, of course, was what happens if the State Police show up at my doorstep and try to arrest me. My attorney said that, due to a quirk in Michigan law, only the county medical examiner can arrest the sheriff. The sheriff is the chief law enforcement officer of the county; the State Police don't have the authority to arrest him within the county. I don't think I was necessarily convinced that this was the case. In my mind I knew that if a couple of state troopers waltzed into my office, they would have their work cut out for them if they tried to take my gun and put handcuffs on me. Hell, I didn't have any idea of what they would be trying to arrest me for anyway. I didn't know of a law that I'd broken, but then my friend's words came back to me. "Nobody's that squeaky clean. They'll find something."

On August 23rd, a week before my arraignment, I called a press conference at the Grand Rapids Press Club. It was time for me to take the initiative and get some information out to the public to counter the impressions that the attorney general's office had created with their state-wide bulletin for my arrest.

Several television stations and newspapers were represented when I read my prepared statement. I pointed out how underhanded the Attorney General's office had been to announce the state-wide arrest warrant on all the television and radio stations immediately after they knew I had left for a short vacation with my wife to Mackinac Island. It also was clearly evident that the intent of the Governor and Attorney General's offices was to hold these charges until election time to destroy my chances for reelection.

I reviewed the narrow investigations that resulted from the Governor's misguided task force the previous fall and my offer to repay the funds that were inadvertently deposited in the wrong account. Also, I charged that Judge Cooper had sat on my request for a grand jury to investigate widespread improper allocation of funds and county sources of income.

Someone got the word to Judge Cooper about my press conference, for he was present in the press room also. When I finished reading my statement, Cooper got up and handed out a prepared
statement of his own. He also offered to answer questions about my
grand jury request.

Cooper stated that immediately after he had received my request he
had contacted Lt. Gary McGhee of the Michigan State Police, Raven
(prosecutor) and the Local Audit Division of the Department of
Treasury. They all indicated that they knew of no problems in the
investigation that would warrant a grand jury. Cooper stated that "It
should be recognized that grand jury proceedings are subject to the
spreading of unfounded rumors." Cooper also went into a dissertation
that was aimed at shredding any credibility I might possibly have in
drawing up a grand jury request.

When Cooper mentioned that the grand jury request was not
properly drafted or submitted I stood up and said, "Judge Cooper, I'd
like you to meet the man that was responsible for making sure this
request was drafted correctly." I asked David Sawyer to stand up.
"This is the prosecutor from Kent County and he has drawn up seven
grand jury requests."

David said, "I have known Bob Blevins for many years. Bob has
had difficulty prosecuting cases in Lake County and as a favor I personally drafted the grand jury request that he gave to the Lake County
Circuit Judge."

Unbelievable anger flashed over the face of Judge Cooper, but he
stuck to his position that there was no reason for a grand jury.

The State Police didn't come to my office. However, when we
walked into the Manistee District Court building for the arraignment,
the same detectives who had been investigating me in Lake County
walked towards me. One pulled out his handcuffs, the other asked for
my gun and made the formal, "You're under arrest," statement.

"I'm not giving you my gun, nor are you arresting me and putting
handcuffs on me. I'm in this courthouse to appear for arraignment; it
is not necessary to create a dog and pony show and treat me like a
common criminal."

If those detectives would have taken a couple more steps towards me, there certainly would have been a confrontation. Before they could have taken my gun or attempted to place handcuffs on me, one or both of them would be on their butts.

My attorney stepped between us. "Bob go into the courtroom and set down." That sounded like good advice to me. Other court proceedings were going on as I sat down in one of the back benches. My attorney soon came through the door and sat down beside me. He was pretty sure the detectives wouldn't try anything in here.

Finally it was my turn. "State of Michigan vs Robert Blevins." It was a clear and crisp announcement that sent a mini tremor through my body. We moved down in front of the bench; the charges were read: three felony counts of misusing public funds, three misdemeanor counts of co-mingling public money and three counts of willful neglect of duty. The arrest warrant had been personally signed by Frank Kelley.

"How do you plead?" My attorney answered, "Not guilty."

"Until trial, you are still the sheriff and chief law enforcement officer of Lake County. I'm releasing you on a personal recognizance bond."

When the judge finished, my attorney spoke. "Your honor, there are two state policemen waiting in the hallway to handcuff the sheriff and take him to a state police post for fingerprinting and booking. My client has his own booking and receiving department in his own jail. Would it be possible for his own employees to do the booking and send the paper work and mug shot to the Post for their records?"

The judge agreed and so ordered. Now the two detectives were really pissed. They had planned to be leading me outside in handcuffs, for all to see. Well, they could at least salvage the situation and walk out beside me.

In a fairly loud voice, so the detectives could hear, my attorney said, "Bob we need to have a meeting in one of those conference rooms, right now." He motioned to a door towards the front of the courtroom.

As soon as we entered the room and closed the door behind us, Charles took the lead. "Follow me." There was a back door that led straight to a parking lot behind the jail. The three of us climbed into his car and pulled out of the parking lot. No one was there to see us get in the car or pull out of the parking lot.

We drove close enough to the front of the courthouse to see the reporters and TV cameras lined up on both sides of the steps. The state police car was positioned in front of the courthouse so the cameras could easily catch the whole scene, the guilty sheriff being led in handcuffs down to the patrol car. There would be a nice close-up of the sheriff as a trooper put his hand on the prisoner's head, guiding him into the back seat.

The scene had been carefully orchestrated for maximum effect, but thanks to my quick-thinking attorney, we had thwarted their best laid plans. It was pretty obvious now, somebody was directing the Governor's police force in an effort to publicly humiliate me. The auditor had finally found something in his second trip to Lake County. That was almost a year ago. For maximum political effect, it was all saved until election time to discredit me. When I didn't accept the deal from the Attorney General's office, the dogs were turned loose.

What really hurt me was that all my years of friendship and cooperation with the Michigan State Police were now being flushed down the toilet.

Less than a month after my arraignment the Board of Commissioners voted unanimously to request that I take a paid leave of absence and turn my county vehicle back to them. The board chairman, Robert Stark, was the initiator of this action and I told him no on both his requests. In the *Star* I also took great relish in rendering my evaluation of Robert Stark. "Quite frankly Robert Stark in my opinion is about as useful as dive brakes on a buzzard as far as the Lake County Board of Commissioners is concerned, because of course Robert Stark lost the primary election to every candidate that ran against him and he is now a lame duck representative."

My attorneys requested a pretrial hearing. In a pretrial hearing a judge would listen to the State's charges and witnesses would be called. My attorneys could cross-examine the witnesses, and from these proceedings the judge would determine whether enough evidence existed to support the charges and whether or not it seemed possible for me to have committed these crimes. If so, he would schedule a trial.

The pretrial took place October 10th, less than a month before the general election. The pretrial or preliminary exam took place in Manistee before Judge Stapleton. Judge Farabaugh from Lake County's 78th District Court had already disqualified himself from the proceedings and Judge Stapleton was appointed the visiting judge. I was represented by David A. Neff and Charles Probert. The State was represented by the Assistant Attorney General, Thomas Johnson.

The preliminary exam was an all-day event with 12 witnesses being called to testify. The proceedings certainly gave me a much clearer picture of Joe Kuhlman's involvement in all this. In fact, it was evident from his testimony and that of other witnesses that my ex-undersheriff, my fellow Marine had supplied the state what they needed to bring charges against me and now under oath he was lying through his teeth.

Apparently it all started back in the summer of 1982 when I went on vacation. Joe contacted the Genesee County Sheriff's Department and told them that I wanted the checks for housing their prisoners to be made out to our sheriff's department rather than the County Treasurer. He was in charge while I was on vacation and he told Mary Seelhoff that from then on he would personally handle any checks coming in from Genesee County. Even after I came back, according to Mary Seelhoff, Joe was always looking for those checks. Well, without my knowledge Joe deposited those checks in the contingency fund instead of the general fund. I reviewed the contingency fund ledger, but the checks totaling $3223 from Genesee County were never recorded. Meanwhile over a period of time Joe purchased shotguns and extra equipment which included fatigue uniforms from the contingency fund. The ledger amount recorded for the checks used for the purchase was listed for some small item while the actual check was written for a much larger amount. This way he could draw from these misdeposited funds without me knowing it. I noticed that a few of the shotguns in the patrol cars looked new, but I didn't think much about it. I had certainly wanted the fatigue uniforms, for the cleaning bills for regular uniforms were extremely high. The fatigue uniforms could be washed at home saving us some money from the budget. It should have dawned on me that something was going on, but it didn't.

On October 3rd of 1983, about three weeks after the task force hearings, Joe went to Prosecutor Mark Raven with information about the Genesee County checks being deposited in the contingency fund instead of the county general fund. Raven sent Joe to the Michigan State Police at Reed City with his story and almost immediately state police headquarters sent special investigators to interview Joe.

Joe testified that under my specific direction he had told the Genesee County jail administrator to make out the checks to our department. Then he said that when the checks arrived and Mary Seelhoff had brought them to him unopened, he had shown them to me and at my direction deposited them into our contingency fund so that we could purchase shotguns and fatigue uniforms. At the first part of his testimony he stated that he had no knowledge of any fatigue uniforms being purchased, but later under cross-examination he contradicted himself by saying he was aware of the purchase. Also under cross-examination Joe's credibility and character were questioned and brought into focus by my attorney. He entered into the court proceedings Joe's disciplinary record in Ottawa County before I hired him, the drinking incident at the academy, plus the Sanilac County Sheriff's report where Joe's uniform, gun and patrol car were impounded. Also noted was the fact that counter to department written policy Joe had been regularly borrowing money from the deputies that worked under his command. That also had occurred in Ottawa County when he worked there. It was now well-known that I had Joe's taped conversation with Finch and Chaney. This too was acknowledged under oath by Joe. So, the fact that he said he was offered money to find information against me was validated by admitting to the tape.

Joe's last minutes on the stand went like this. Probert: "I'm curious, Mr. Kuhlman, when did you first start cooperating with the State Police investigation of Sheriff Blevins?" Kuhlman: "October 20th 1983." Probert: "How do you recall that date?" Kuhlman: "Because it was probably the second hardest thing I've ever done in my life."

With this revelation my attorney seemed very sincere and moved close to Joe. Probert: "And what was the hardest?" Kuhlman: "Leaving three marines on a hill to die."

Probert seemed to be in deep thought for a moment, then he turned away from Joe. Abruptly he turned back around and looked directly into Joe's eyes. "Well, I would submit you're probably doing the same thing here today."

Besides the Genesee checks there were two other checks uncovered by the auditor that were deposited in the contingency fund and should have been deposited in the general fund. One was a check back from overpayment for repairs to one of our patrol cars. It was for $311.33. Another was for around $25 from one of our sheriff's auctions. These were deposited back when Chaney and Finch had control of the fund. Both men had been interviewed extensively by the State Police and testified in court that I had told them to deposit the money in the contingency fund. Finch even admitted that he knew it was against the law to do this, but did it under my direction.

I had no knowledge of either of the checks that Finch and Chaney had deposited, nor did I know for sure why they were involved. Maybe the auditor had found the checks and they were just covering their butts, or maybe they remembered the checks and worked it out with the State Police. They both had been interviewed by the State Police and the Assistant Attorney General. Each acknowledged their taped conversation with Joe. There wasn't much else they could do about that.

Probert pretty well discredited Chaney when he had his boss from Mason County on the stand. Sheriff Carrier stated that he had to fire Chaney because he couldn't follow orders. Eventually because he was a Veteran he got his job back, but Carrier wouldn't give him any meaningful work. Just made him set at his desk all day doing nothing.

The judge determined that although none of the money questioned was personally used for my gain, according to statutes, it was a crime to put the money in the contingency fund and spend it for departmental items. I was the person responsible for that fund and how the money out of it was spent. The case would go to trial.

Of course the press and TV channels were keeping right up on the course of events. The coverage was not of a positive nature. Attorney General Frank Kelley had charged Blevins with three felony counts, three misdemeanor counts and three counts of willful neglect of duty. Blevins would go on trial and it would be difficult for him to ever be reelected.

I wasn't about to throw in the towel and my campaigning only intensified. To any news representative that would listen, I told them about my 51% reduction in crime over five years and that there was a lot more in Lake County that I would go after if I was reelected. If I couldn't get the cooperation of the State Attorney General I would expose the whole mess on 60 Minutes or 20/20.

Another nice letter and certificate came in for me from Fredrick Pierson, the Director of the National Police Academy. That organization also recognized my 51% reduction in the Lake County crime rate. With this and the primary win, I was feeling a whole lot better even though somewhere down the road there would be a trial. Individual citizens were voicing their support and my attorneys felt that I would have a solid case in court. In two terms, only three checks had not been accounted for properly. Adding to that was the fact that the monies had only benefited the sheriff's department, not me personally.

From the Forgotten Man Mission, I received a letter of appreciation for what I accomplished in the county jail. Specifically mentioned were the jail library, chaplain's office, a staff of deputized chaplains, G.E.D. completion program for inmates, a yearly memorial service for fallen police officers, jail chaplaincy training and the establishment of an exercise and weight room for officers and inmates.

I appointed Tom Byrne my undersheriff. Tom ran against me in the primary election but he had a good solid law enforcement background and I liked the man. In a letter to the *Star* editor, Tom asked Lake County folks to give me their support in the general election. He also made what some folks might have thought was a strange statement. "To the best of my knowledge John Wayne loved and respected his parents, and I am sure John Wayne's parents would have supported him for sheriff and not have an eight foot Blevins sign in their yard."

What Tom was getting at was that the parents of Pete Loucks, my opponent, had erected a four by eight foot VOTE FOR BOB BLEVINS sign in their front yard. That sure generated a few chuckles around town.

As general election time rolled around I felt reasonably confident in my chances for reelection. I was still very security conscious. There were several people that didn't want to see me in office for political reasons, but there were others who didn't want me as sheriff again because their livelihood of crime would be threatened. Those were the people I needed to watch out for. Other than the future trial on the Attorney General's charges, there hadn't been any new developments politically.

There was one disappointment though. In the last edition of the *Star* before the election, the editor, Mike O'Connor, made some very damaging comments. He questioned why so many people could be against me, the judges, prosecutor and board of commissioners. He asked why the voters should believe as I had charged that the governor, attorney general and state police should dig up charges to ruin my reelection campaign. It was difficult for him to think that there was an undercurrent of corruption in Lake County, which was the reason for so much of my opposition. He didn't come right out and tell people not to vote for me, but he might just as well have when he said, "What I'm trying to say, I guess, is that I just can't buy this political conspiracy bit—it's way too much for me to handle ... and if I can't buy both doors, I won't buy the car."

The polls were closed. At last count I was well ahead of Loucks. Some of my staff, a reporter or two, and some other community citizens were in a room at the Courthouse, waiting for the official results to be announced. I was in a good mood. My supporters were jubilant.

Finally, Cindy Keefer enters and makes the announcement that the computer is broken. The prosecutor's secretary, Vivian, and she must work through the night counting the rest of the ballots by hand. Those ballots from Webber and Yates Townships must be tabulated. The results would be announced at 9 or 10 the next morning.

Several TV stations and the Grand Rapids Press declared me the probable winner. There was a small party to celebrate, then I went to bed. I was satisfied, but still a little apprehensive, even as good as things looked.

In the morning, a lady by the name of Carol called. She said that the chairman of the Board of Commissioners wanted her to call and inform me that I had lost the election by 274 votes. I said there would have to be a recount of the ballots. With the votes I received in the primary and my lead when the polls closed, it just didn't add up. She said that the only thing that could be done was to run the ballots through again.

"Run the ballots through the computer again? I thought it was broken."

"No, the computer wasn't broken. All the folks in those two townships had merely folded their ballots wrong. They couldn't be fed through the computer the way they were, so we repunched new ballots so they could be fed through. We just didn't want to embarrass those folks by letting it get out that they'd folded their ballots wrong, so we said the computer was broken."

"Unbelievable! Then I want a count of the misfolded ballots."

"We can't do that. They were destroyed."

"Destroyed!"

I put through a call to the Board of Canvassers in Lansing. Their spokesman said, "Yes, altering or destroying ballots before they can be verified by the Board of Canvassers is a state felony. There is however no provision in the constitution to set aside the election results or require a new election."

"What are you telling me?"

"We can only run the ballots that are available through the computer again, to verify the results. If the ballots were altered or destroyed, it is up to the prosecutor to bring charges."

Mark Raven won his election by 10 votes. It was rather inconceivable to think he would prosecute his own secretary and Cindy, especially for my benefit. He was a wishy-washy prosecutor. I never missed an attempt to expose him in the newspaper for dismissing or plea-bargaining cases which would have been simple to prosecute. I'm sure he was happy to see me out of office. Mark's public statement on the matter was that it had been an honest mistake, an oversight. Cindy Keefer's honesty was beyond reproach. She had accurately repunched the ballots and I had overwhelmingly lost the election. Life goes on. I was not the only candidate to challenge the election results and there were other irregularities that made the election results questionable to many of us.

A recount verified that I lost the election. This was expected but I demanded it anyway. Mark Raven refused to prosecute Cindy and his secretary for destroying ballots.

Convinced that I really hadn't lost the election, I was angry and felt that I'd been cheated out of a victory. People rallied around me through the controversies; they saw through Kelley's barrage. I had not let them down in my determination to hang in there. The people had not lost faith in me.

The election was lost and there wasn't a damn thing I could do about it. Now I knew how women felt after being raped. There was no one to turn to, no one to make things better. My choice was to come back and run again in four years.

As if the lost election and future trial weren't enough to have on my mind, my mother's health was becoming an increasing worry. After my stepfather passed away she moved to an area across the expressway from Butterworth Hospital. There was a small group of her Chippewa Indian friends living there. Now she was in the hospital and the doctors didn't expect her ever to leave. First it was the stroke, then the two small heart attacks. With diabetes, high blood pressure and hardening of the arteries, Mother didn't have a lot going for her. She was depressed, full of anxiety and overly concerned for me. Mother watched the evening news and it looked to her as if I'd soon be in jail or shot. I tried to reassure her that nothing bad was going to happen to me, but she wasn't convinced, "Tell me the truth."

All during the summer, right up to the election, I had been driving back and forth to visit her. Now after the lost election Mother kept asking if there wasn't something I could do to make this trouble with Kelley go away, cut a deal or something.

There was about $20,000 in attorney fees that would be coming out of my pocket. Who knows how high that was going to go when we went to trial. Even if we went to trial, there wouldn't be any change in the election results. That was already water under the bridge.

Someone from the Attorney General's office contacted my attorney, promising to drop all other charges if I plead no contest to willful neglect of duty, which was a misdemeanor charge. It was a take it or leave it offer, otherwise they were ready to go to trial. My attorney relayed their offer to me, but advised against accepting it. He said we still had a very strong case. I was beginning to think otherwise.

When I told Mother about the deal I was being offered, she asked, "With a misdemeanor on your record, can you still be a police officer?"

"Yes."

Mother took my hand, "You have nothing to gain. All you are going to do is spend more money. A misdemeanor is not going to hurt you. You can still be a police officer. Make it stop. Make it end." I promised I would take the plea-bargain deal.

The news media used to refer to me as "The Bulldog." I didn't know what giving up or letting up meant when I had a goal in mind. That's the attitude I nurtured on the Sparta football team, as a Marine, a Kent County deputy and paramedic, the same as the Lake County sheriff. The fight and determination of "The Bulldog" was nearly gone. I was at rock bottom. Out of money, out of a job at the end of the year and responsible for the worry that was slowly taking my mother's life. Hell, I didn't even feel comfortable down at the local coffee shop. People seemed to glance at me with the "He's got himself in a heap of trouble" look in their eyes. For the first time, I felt like a failure. Lost was my self-respect, drive, desire, even credibility. If people see it on the news enough, they begin to believe what they hear and see. I had fallen into that trap with my initial opinion of Lonnie Deur, the sheriff that was recalled. The news media, both state and national, covered my story for many months. I don't think their coverage left a very good impression of me in the people's mind.

Before, I could never understand how someone would consider taking their own life. Now I understood. Those thoughts did travel through my mind. This time in my life, there was no grandfather to turn to, no one to give me advice and support except Dee. And Dee was exactly what I needed. I had to confide in her and to lean on her. She was as solid as a rock. Her presence and attitude gave me the strength and will to survive.

When I informed my attorney that I would accept Kelley's deal, he said he didn't think that was the thing to do, but I told him to do it anyway.

Circuit Court Judge Cooper disqualified himself from my proposed trial, so now the sentencing in my plea-bargain was bound over to the 76th District Court in Mount Pleasant. Judge Peter J. O'Connell was the sentencing judge. He stated that the pre-sentencing investigations indicated that the contingency fund was set up by myself, not for personal benefit, but in all good intent to benefit the sheriff's department and the court could find no evidence whatever that I gained from this matter one cent, "one iota." He then proceeded to tell me that if he had found that I had personally benefited from it, that he would have given me a tough sentence. He then gave me a lecture on the fact that even oak trees have to give some to the wind, and that I had made things difficult for myself by not bending some. I guess I kinda took that comment as a compliment, though I know he didn't mean it that way.

My fine was $500 and court costs $500, plus I borrowed some money and wrote the county a $3,736.83 check for the money that had been spent on police equipment that was supposed to go into the general fund. It did seem like the Lake County folks had gotten their money's worth out of me. I had paid two deputies' wages out of my pocket for the last quarter of 1983, about $4,000, spent over a year as ambulance director without pay, saving $11,000. Acted as assistant medical examiner free and for my two terms in office I worked for the same wages as my deputies. In all actuality they made more than I did because they received overtime and I didn't. In addition to all this my dedication to being sheriff put a terrible burden of stress on my family.

Why did I do it? Because I loved the job. First I wanted to become a Marine and then when I had achieved that I left the Marines, because I wanted to be a law officer. It seemed that was what I was meant to be and it was my life. If I had more than two days off I was anxious to get back to work. I never took a sick day even when I was sick. In the Kent County days I would fear that something would happen to my partner if I didn't show up. In Lake County I was sheriff all the time, not just when I felt like showing up. Even when I took some vacation, after a week I was restless and ready to get back to my job.

Dee is convinced that losing the election probably saved my life. If I would have entered my third term, she feels that with my determination to go after the big fish of organized crime, I would have been dead within six months.

In the end, cutting the deal with Kelley didn't help my mother at all. She died before my plea was entered.

My road deputies all came in together and turned in their resignations, to be effective when my term ended. I told them not to throw away their own careers, they should reconsider. Most of them did decide to leave however, and for the next month I did my best to get them job interviews, and gave them my highest recommendation in hopes that they might land a job at another police department.

My term ended with a going away party given by my employees. They presented me with a beautiful plaque inscribed to:

"THE BULLDOG"
FOR DISTINGUISHED AND DEVOTED SERVICE.

At the bottom of the plaque they had mounted one of my sheriff badges. When I tried to thank them, it was very difficult getting the words past the huge lump in my throat.

Of all my awards, letters and certificates of appreciation, that plaque was my greatest honor.

A great deal of satisfaction came a year later when this letter appeared in the *Star*. It was a letter to the editor from a black resident that I had known. While I was sheriff he had been a very vocal citizen, not always agreeing with what I was doing.

SOLUTION
VOTED OUT

To the Editor:

For a month or more now, every time I pick up the *Star* I read about the crime in Lake County. Break-ins, people are crying because their house was broken into (cleaned out) no easy solution to the problem.

Seems to me the people got rid of the solution a year ago. He was nicknamed Bulldog (maybe he was). We did not have people on the courthouse steps waiting to discuss the crime problem with the board of commissioners.

My thoughts are, he was gotten rid of 'cause he stepped on the wrong toes.

Jasper Smith
Baldwin

Picking Up the Pieces

The first few months of 1985 found me spending a good share of my time moping around, licking my wounds, just basically feeling sorry for myself. Dee and the kids were very supportive for awhile, but eventually they grew tired of my bad moods and pouting. I could feel them shrinking away.

There was so much anger and pain inside, I really didn't know what to do or where to turn for help. If I had been a drinker, I surely would have climbed into the bottle for an escape. I have always held strong religious beliefs, but I really didn't ask the Lord for help. He didn't have anything to do with the problem, so I didn't involve him in the solution. He gave me a brain and the power to use it. Finding the answer to the problem was my job. I just needed to set down and figure out what to do.

One thing was certain, with no income the bills were not getting paid and we were behind in the house payments too. Our savings was used for attorney fees. The only thing we had left was the equity in our house, plus our personal possessions.

The home was our pride and joy. It sat on 40 acres with a half-mile of lake frontage. There was a two-acre trout pond, fully stocked with five pound Rainbows. A 40-foot waterfall cascaded past our bedroom window to the lake. We had four bedrooms, three baths, a four-stall garage, large recreation room and two horses. On top of that, the property provided me with some outstanding White Tail deer hunting.

As much as we loved this place, if we couldn't make the payments we couldn't stay. With all of my experience in law enforcement, there must be some sheriff's department out there that could use my services. It was time to stop moping around and get busy finding another job.

Dee and I organized all my certificates, awards, letters of appreciation and commendations, along with pictures, to form a nice three-
ring notebook type resume. We went to a printing office and had 300
copies printed. This cost us close to $10 apiece, but the resume was
quite impressive. On the cover of each booklet was a stick-on photo-
copy of the sheriff's oath of office. When I arrived in Lake County,
there was no oath of office for the sheriff. I composed this one, and it
was officially used to swear me into office.

Next, Dee and I got out a map of the United States and circled
places we might like to settle. We concentrated our efforts on the West
and Southwest. Some of the places were familiar to us because we had
previously lived in or traveled through them. Others were picked
because they just looked like they might provide an interesting area to
live in. With the National Sheriff's Directory in hand, we addressed
and mailed out the resumes to the circled cities. It was a nice resume, I
had fine credentials; we expected a good response.

It wasn't long before we were starting to get phone calls and letter
responses requesting interviews. I bought two new suits and ties for
the interviews. To save on motel bills we purchased a small travel
trailer. We were now off to visit and interview with all the sheriff's
departments that had responded to my resume. All we had to do now
was to decide which job would offer the best pay and most opportunity
for advancement.

The kids were out of the household now. Our daughter lived in
Nebraska, our son was married and going into the Army. It was just
Dee and me traveling around the Western United States looking for
the best job and most attractive place to settle. From cities in Texas, to
Golden Colorado, Angel Fire Arizona, Redding California, Wash-
ington, Nebraska and many more.

Along with the interviews was usually a written exam, sometimes a
physical fitness test. I did well on all of them, even in California,
which is supposed to have one of the most difficult exams. We spent a
good share of 1985 traveling around for these interviews. In the process, we spent a great deal of the little money we had left, confident
that a job was just around the corner.

The first few letters that came in were form letters turning me down. I disregarded them, threw them away. More kept coming, so I filled a shoe box, then two shoe boxes. They all said the same thing, "Thank you for the interest in our department. Sorry to inform you that we are no longer considering your application for employment." We didn't have to worry about spending much time sorting out the best job offers, because there were no job offers.

If I couldn't get a job in law enforcement, I would go out to Wyoming and become a hunting guide and outfitter. I considered myself an expert in hunting and fishing, with a natural ability to survive in the wild. There was some good money to be made as a guide.

Finally in the spring of 1986 reality hit. We were broke and in danger of losing everything we had. We listed our home with a realtor for $125,000. By that summer we had virtually no lookers. No one in the local area had that kind of money to spend on a house. We were too far North of Grand Rapids to draw people who could afford it. If I was to get my resume around and obtain a guide license out West, we would need to be leaving Michigan soon. Finally we sold the house to my wife's sister for what we still owed the bank, $80,000.

An auctioneer was contracted to sell the rest of our belongings, along with most of my lifelong collection of guns. Except for a few keepsakes, our dishes and clothes, everything else was dragged out in an open field and auctioned off.

We drove away from Michigan with everything we owned packed in a horse trailer, behind our pickup truck. After 22 years of marriage that was the sum total of our worldly possessions. Fortunately, we had each other.

A friend of mine had a guide business near Cody, Wyoming. We left his address for our mail to be forwarded, so when we arrived in Wyoming so did the notice for the registered letter.

When it became apparent that there would be no law enforcement employment generated by the 300 resumes, out of desperation I wrote a letter to the Commandant of the Marine Corps. I explained my situation and asked if I might be considered for active duty again with the Marines. For quite some time I had participated as a Marine reservist, and before that, of course, was my six-year active duty tour. I really didn't expect a positive reply; after all, I was well into my 40's now.

I contacted someone back home to pick up the registered letter, put it in an envelope and mail it out to me. The letter was from the Marine Commandant. If I was interested, there was a two-year active duty tour at the Alemeda Naval Air Station as the Training Programs Chief for 600 Marines.

Happiness prevailed. This was the first time in almost two years that I had been offered a job. Somebody actually wanted me. My friend in Wyoming offered to take me on as a guide with his outfitting business, but I told him that I must take advantage of the Marine offer and maybe I would be back.

Once again, with the horse trailer packed with our belongings, we headed to California. Outside of the Naval Air Station, we rented a small two bedroom home for $1,125 per month. That was pretty steep rent, but there was no on base housing available.

It seemed strange being on active duty again. I did enjoy my work and respected my boss Col. Ondrake. The Colonel was kind of a crotchety old fellow that no one seemed to like. He was rather old-fashioned in his way of doing things. I understood that, and didn't have to be told twice what to do. That impressed him, the fact that things got done without him having to repeat himself. He seemed to like me and I thought he was a great boss. I also thought it was great having a paycheck. Even though we were paying through the nose for rent, we were able to start putting some money away in savings.

The job I held was not a career position. It was part of a support program offered to reservists. The program enabled reservists to reenter the Marine Corps for a one or two year active duty tour. Although my tour originally was set up for two years, I was informed that the program would not be available past the first year. I'm sure that when money was tight for the Marines, these were the first programs to go.

We liked the Cody Wyoming area and planned to return there pursuing the hunting guide opportunities. Unfortunately, the hunting season only lasted for three months in the fall. I sent resume letters to different police departments around Cody. Maybe I would luck out and find something. People seemed friendly in Wyoming. It was a large state, with the entire population about the same as the city of

Grand Rapids, in Michigan.

When my one-year tour with the Marines ended, we returned to Cody and settled in. We managed to scrape up enough cash to put a down payment on a home out on Lower South Fork Road. Until the hunting season returned, we both had to work at rather low-paying jobs to make ends meet. Dee worked in the shoe department of K-Mart for $4.50 an hour. I worked in a local hospital's physical therapy section for about $5.50 per hour. Not a lot, but we could survive on that.

Greybull was a neighboring town. I learned that they were looking for a patrolman. With my resume and scrapbook in hand, I sat down with Greybull's Chief of Police, Sam Bull. Sam had been an under-sheriff in California and understood how politics worked. I explained everything that happened to me back in Michigan, even the misde-meanor on my records. He would occasionally nod his head and smile, like he had heard it all before. Well, Sam hired me on the spot. Didn't do a background check or anything.

I was on cloud nine, on top of the world. I was going to be a cop again. With a newly purchased cowboy hat, and my star, I was back in a patrol car again.

Even though I was an outsider, the community seemed to welcome me, and I felt as though I would fit right in. The local merchants were friendly, and I took a liking to them.

Sam explained that they didn't have much crime in Greybull, just the normal kid problems. I asked him how many murders they had a year. That question put him deep in thought. "Guess it's been about 10 years or more since we had a murder in the County."

"How do you manage to keep the crime down?" I remembered how many murders occurred back in Lake County. Baldwin and Greybull were similar in size too.

The Chief walked over to the window. "Come over here and take a look down Main Street. You'll see the answer to your question."

The only thing I could see was a bunch of pickup trucks going back and forth. After a few moments, he pointed something else out. "What do you see in the back window of those pickups?"

There didn't seem to be anything special, just a rifle rack with a gun in it. Bingo, the gun in the rifle rack is what the Chief's talking about.

"Anybody try to rob our bank, we don't have to call for backup. It's right out there all the time."

The Chief thought they had some unruly kids in Greybull, but I got along with them just fine. I could communicate and reason with them, they seemed to listen to me. The kids really weren't unruly or violent, they were just kids.

After four months on the job, I noticed that the Chief was missing some work. At first it was just a day or two, then a couple of weeks. He finally called both the Sergeant and myself into his office and closed the door.

"I want to talk to both of you. One of you may have the opportunity to be Greybull's next Chief of Police. Due to ill health I'm going to retire soon. You should both put your applications in for the job. I think you're both qualified and good candidates. I'll give you each an equally high recommendation to the Council. I'll also advertise for the position. There will be fingerprinting and background checks too. From there it will be up to the Council to choose the next Police Chief."

I asked the Chief if he thought my no contest plea to the misdemeanor charge would hurt my chances. He said it wouldn't, plus he had already explained to the Council that I had just got trapped in bad politics.

For the first time in a long time, our future started to look bright. It seemed I had been accepted by the community of Greybull, and I would now be considered for the Police Chief position. And, even if the council decided to appoint the Sergeant as the new Chief, I would surely end up with the Sergeant's job. I had been commuting 150 miles round-trip from our home in Cody until I felt confident this job was solid. We had a beautiful home in Cody, and didn't want to suffer yet another sacrifice. At this point in our lives each move meant a loss. Each loss dictated that we could afford less and less as we started over. However, being considered for this promotion sent a signal that we should sell and relocate in Greybull and begin to build our future. The house was sold; we walked away, losing only a few thousand dollars this time. Soon we were hauling everything we owned in a moving van. We found a nice home that was for sale. We couldn't come up with a down payment, so the owners decided to allow us to rent until we could get on our feet.

The moving van was backed into the driveway and Dee and I were taking boxes into the house. I looked out the window and saw a patrol car pull into the driveway. The Sergeant was driving and the town Mayor was with him. I thought to myself, either they came to help or tell me I had been chosen as the new police chief.

Seconds after they knocked, I opened the door, "Come on in."

"We can't stay." There was seriousness in their voices.

My mind was racing at top speed now. Obviously I hadn't gotten the job, that's why they were here. They knew I'd be disappointed. Nice of them to come and tell me anyway. The Sergeant probably got the job and I'd be working for him. Hell, maybe I'd be the sergeant now.

There was an uncomfortable moment for all of us. Finally, I asked, "What's the problem?"

"According to your application, you lied. Your application said you pleaded no contest to a misdemeanor; you lied."

"What do you mean I lied?"

"We're not going to discuss it with you. If you want, you can talk about it with the city attorney. We are here to get your badge, gun and ID card."

I could see spots flying around the room. I was dizzy; I knew I was going to faint or throw up. This was the first time I had ever been fired. Walking past my wife to the bedroom she knew something was wrong, "What's the matter?" I couldn't answer her. Just continued on into the bedroom to get my gun and take the badge off my shirt.

When they walked away from the house, I just broke down and turned to jello. For days I tried to get an appointment with the city attorney, but he wouldn't see me. Finally he called on the phone. He said, "There is no reason for a face to face meeting. There is nothing you can do or say to change the facts. Goodbye."

Dee and I spent a lot of time trying to decide what to do. We could go back to our five dollar an hour jobs and pursue the hunting guide idea; however our home in Cody was now sold. Hunting season was over for the year. Christmas was coming up, and in our period of grief over losing this job, being with the family seemed even more important. Our daughter had married, our son was out of the Army. They both were in Michigan with kids, our grandkids. We wanted to be near them.

I was determined not to sink into a dark stage of depression, as I had when the election was lost in Lake County. This was only a job I had held for a few months, nothing more. How they figured I lied, I didn't know, and they wouldn't give me an explanation. Still, it was just a job that didn't work out. At least, I tried to convince myself of that. We decided to leave Wyoming.

All the new belongings we collected in Wyoming and the year in the Marines were hauled out on the front lawn for a yard sale. With that money we rented a U-Haul truck, packed what few possessions we had, and returned to Michigan.

Tim Petty was the owner of Classic Chevrolet in Grand Rapids. We had been friends since I worked in the Kent County Sheriff's Department. He started out as a meat salesman for Frost Pack. Then he became a very successful car salesman on 28th Street, eventually owning his own dealership.

Tim thought I'd be good at selling cars. I had reservations about that, never having been a salesman before. He said that I was good at talking to people, and that's all it took to be a good salesperson. I needed a job, so if Tim thought I could sell cars, I'd sure give it a try.

We started to look for a place to rent or buy. A lady we knew recently lost her husband and was moving to Arizona. She had a small two-bedroom house on 10 acres near Cedar Springs. The house looked comfortable, but it really wasn't the type of place I wanted to live in long term. She was willing to sell it on a land contract, requiring very little for the down payment. I managed to borrow enough for the down payment and we moved in.

I proved to be a crackerjack car salesman. The first couple of months I sold 25 to 30 cars each month. My first paycheck, which included my commission money, was $7000. That was more than I made for a whole year when I started as a police officer back in 1968. It also was more than I could earn each month at any law enforcement job in the state. Tim was happy with my work and put me in charge of new car sales.

It was my good fortune in life to always have worked at something I enjoyed. When Monday morning rolled around I looked forward to going back to the job. Seldom did I take more than a week's vacation. My work provided all the satisfaction in life that I needed. This job was different. Tim Petty treated me very well and with a year's experience, I was proving to be a successful salesman; the trouble was I didn't like selling cars.

There was an ad for a job opening in Eaton County. The sheriff was looking for a jail corrections officer. The sheriff was Art Kelsey, a former State Trooper who used to work in the Kent County area. We were friends, having worked the roads together back when I was a Kent County Deputy.

I drove over to see Art. "Art I'd like one of those jobs you got advertised."

"You mean, after being the sheriff, being on top, that you want to start at the bottom and work for me turning the keys as a jailer?"

"Yes, I'm willing to start over, repay my dues, whatever it takes."

"Ok, here's what you'll have to do. I have a union here, so you'll have to fill out an application, take the test, pass the interview and if your name comes across my desk, I'll hire you."

"Great."

I filled out the application. In the past criminal record section, I checked NO in the felony box and YES in the misdemeanor box. For remarks on the YES box I wrote, No contest plea to willful neglect of duty.

A week later Art Kelsey is on the phone. "Can you come over and see me?" I could tell by the sound of his voice that something was wrong. "Sure I'll be right over."

"Bob, I can't hire you. It's the result of the background check. Legally I can't discuss any findings of this nature. There is a huge fine, $20,000, if I ever get caught divulging this type of information."

Art picked up the manila folder, looked through it and held it so that I was able to see my name on the outside. He then closed it up and laid it in the center of his desk. "I'm going to slip out for a cup of coffee. I'll be back in five minutes or so."

Art couldn't discuss what was in my records, but he was giving me the opportunity to see for myself. I opened the folder and there on the top of the papers was the results of the LEIN criminal history inquiry.

According to the LEIN report, I was convicted of felony fraud and sentenced to five years in Jackson Prison.

Since I was fired from my job in Greybull, Wyoming, this was the first time that I again felt extreme pressure and frustration. My heart was beating real fast, my blood pressure sky high. There were the stars again, and the feeling that I might pass out.

For five years I had been carrying this false conviction in my records. No wonder I couldn't get a job in law enforcement. The boxes full of "no thank you letters" from the 300 resumes I sent out were the result of this. The only reason I got the job in Greybull was because the Chief was impressed enough with my resume and scrapbook to hire me without a background check. When they finally found this in my records at Greybull, the five year sentence wouldn't even have been up. Maybe they thought I had escaped from Jackson Prison. They probably were a little embarrassed they had even hired me, and more than happy when I left town. It would even have been difficult to get a job with a large company, for they do criminal background checks too.

In my mind, Frank Kelley wanted to make sure I never got a job in law enforcement again. At least not in the United States. All the money and time Dee and I had spent traveling around the States interviewing for jobs was a complete waste. The Attorney General wasn't satisfied in just ruining me politically, he had taken away my chances of ever being employed as a police officer again. In my mind, this had been done willfully, with intent, and illegally. I was angry and enraged when I thought about the Attorney General's "let's make a deal" proposition. It was all a smoke screen. No matter what I pled guilty to, they were going to put this felony fraud conviction on my records. I had never been convicted of a felony or fraud. They had just plucked this from the air to slam dunk me, and make sure that I never again got a job that required a background check.

The action that was taken from the Attorney General's office was more than just trying to remove me from the political picture. It was a get-even action. I can only guess whether Frank Kelley did this on his own, was directed to, or was just acting in concert with Governor Blanchard and Mayor Coleman Young. I was a threat to all of them, and they didn't want to see me surface again. Both Dee and I remember what the State Police detective said in my office, "If you think Blanchard or Coleman Young will take the fall, you're wrong. The podunk sheriff will take the fall, and you'll never get a job again." That statement held much more meaning now.

I immediately contacted an attorney to see what I needed to do to get this false conviction information removed from my record. He said I should start with a letter to the Lake County prosecutor.

Mark Raven was not the prosecutor anymore. The prosecutor was a man by the name of Mike Riley. I wrote a letter to Riley explaining the situation. Soon I received a letter back apologizing for the "error." Riley said that neither he nor the former prosecutor had anything to do with what was entered into my record, but he had contacted the Attorney General's office and steps were being taken to change this. Later I received word from him that the error in my records had been corrected and everything was taken care of. But it hadn't. My records still indicated that I originally had been charged with fraud, which was not true. This time my attorney exchanged several letters with the Attorney General's office to remove the fraud charges. They explained that fraud was used as a generic term to cover the co-mingling of monies and depositing in the wrong account charges. The word fraud would remain in my records, along with the negligent duty conviction.

Once again, a sizeable chunk of money was spent on attorney fees to get this false felony conviction removed from my record. The Attorney General's Office was quick to point out that the error on my records was an honest mistake. To prove otherwise, I would need to bring a lawsuit against the State and prove deliberate malicious intent on behalf of the Attorney General. I knew that I didn't have a snowball's chance in hell of ever accomplishing that. With a wealth of evidence, I couldn't get the small fish of organized crime prosecuted. How could I ever prove that Frank Kelley intentionally planted a false conviction on my record. I would be a damn fool to think my odds were any better now.

Although my record was now void of any felony conviction, I had neither the time nor money to conduct another massive job search. Also, I was crowding the age of 50, which was old to start as a cop again.

My adversaries had won. The false felony conviction had hung around my neck for over five years and effectively prevented me from entering law enforcement. All the attorney fees strapped me financially, and I knew that to continue a legal fight would be beyond my means and too costly on my marriage too. Finally, it was time to close this chapter in my life and move on. As Grandad used to say, "You gotta play the cards you're dealt."

The Call-Up

My record was clear of the felony conviction, but I was still selling cars. There was just too much political baggage for me to try law enforcement again. Several of my friends were still sheriffs, but I didn't feel comfortable asking them for jobs. It was better that I shelve the idea for good.

I was still active in the Grand Rapids Marine reserve unit, and served as the Alpha Company First Sergeant. Now in my middle forties, I was considered the old man in the Company. Captain Punter was the Company Commander. He was only three or four years older than my son, but we had a fine working relationship. I respected his leadership abilities and he seemed to appreciate my experience and wisdom.

Alpha Company was a rifle company composed of 200 Marine reservists. I called them my kids. They were young, tough and I was proud of them. As in any military outfit, the First Sergeant plays a key role in making things run right. He must wear several hats to do this. First and foremost he is the commander's advisor. The First Sergeant is a manager, the commander's chief representative with his Marines, and in turn he is the senior enlisted man, the non commissioned officer that must be concerned with the welfare and morale of those Marines. I felt very comfortable with these dual responsibilities. It was my goal to be the best First Sergeant that I could possibly be. Not only in the Commander's eyes, but also to achieve that same respect from those young reservists.

It was late 1990, and things were heating up in the Middle East. The problem was of a large enough scale that reserve military units would be called up and deployed to the area of conflict. I certainly did not wish for war, but I did welcome the opportunity to be activated and serve with these fine Marine reservists. It just felt good to be needed, a welcome retreat from the past and a more fulfilling endeavor than selling automobiles.

There was a full-time active duty First Sergeant assigned to the Inspectors/Instructor(I&I) staff. He did much of the day to day administrative paper work. This allowed the reserve First Sergeant to spend more time with the Commander or the troops, rather that being buried in the office. It was easy to avoid the paper work and let the I&I First Sergeant do it all, but I knew if we were activated I would be expected to be competent in all aspects of the job, including the paper work.

On a regular basis, I started coming into the Reserve Center for a few hours each day. First Sergeant Mike Bachus was the active duty First Sergeant assigned to the I&I staff. Mike was a good friend. He led me through all the regulations and administrative procedures I would need to be familiar with, plus he shared with me his experience in how to make sure things ran smooth administratively.

It was the first part of December when Alpha Company received its activation orders. Each man left his home and civilian employment to become a full-time Marine. Six days were spent at the Grand Rapids Reserve Center, going through the mobilization processing. Medical and administrative records were checked, new active duty I.D. cards were issued and all war gear inventoried. It seemed like the reserve auditorium was filled with Marines standing in a variety of lines for one thing or another. There was even a line to complete the paper work for dependent I.D. cards.

Of course this same process was taking place in other Michigan cities as the other Companies of our Battalion were being prepared for their war time mission. Bravo Company was located in Saginaw, Charlie in Lansing, with the Headquarters and Support Companies coming from Detroit. A Weapons Company from Toledo, Ohio, was also undergoing their mobilization and were part of the Battalion.

A feeling of excitement prevailed in the reserve center. Years were spent training for a war that might never come. Now, unexpectedly, something big was happening and we were needed. It felt good to be needed, to have a purpose in life. I am not a warmonger, but I was happy to be here with these 200 young Marines. For the first time in quite awhile, I had some control over my life, and at the same time, as First Sergeant, I welcomed the responsibility of looking after these Marines, my kids. They were all unique as individuals, but probably the Marine that stood out the most was the Company Clerk, Lance Corporal Han Lin.

Corporal Lin finished basic at about the same time that I took over as First Sergeant. Han did very well in training. He completed recruit training as the platoon honor man, which meant that he topped his class in academics and on the rifle range. For this, he was promoted to Private First Class (PFC) and given a dress blue uniform. In basic infantry training, Han did just as well and earned another promotion.

Becoming a Marine was Han's idea. His father was paying for his education, but the Marine program allowed Han to help out. It also was a way of achieving some independence, for his father had thus far been responsible for structuring his course in life.

Han Lin was extremely intelligent. His college education had been extensive. When he finished that, his father sent him back to the Orient for martial arts training. Han studied under several masters in achieving high degrees and black belts in most of the Martial Arts.

I compared Han Lin with the character Radar on the TV program MASH. Not only was he extremely intelligent, he could do the administrative work of four other people. He knew what to do and when to do it. He also was very perceptive, an expert in reading my moods. He knew exactly the right time to approach me with an idea or question and, on the other hand, the time to give me space if something else was occupying my mind. In short, Lance Corporal Lin was indispensable to Alpha Company.

For their day-to-day business, Marine Companies were hooked into the central Marine computer system via what we commonly referred to as the "Green Machine." The Battalion Commander, LtCol. Davis, assured us that upon reaching our deployment base we would have our own Green Machine capability. I had been on active duty and in the Marine Reserves long enough to always question guarantees. I happened to mention to Lance Corporal Lin that we probably should have a backup plan if the Green Machine capability wasn't available. In a day or so Han came to me with a brand new lap top computer and printer. His father had given it to him for our deployment.

Han's father was an Executive Director at Amway. He had been instrumental in setting up Amway's business ties with mainland China. In fact, his father might well be in China when we deploy overseas.

Han said that he'd like to put the Green Machine data into his lap top so that we wouldn't have to lug a lot of backup paper work on the deployment. He sure had my permission to do that; however our computer technicians said that the two computers weren't compatible and it couldn't be done. Well Han was light years ahead of these guys. In short order he had stored all the Company administrative data we needed, in addition to that of the Battalion.

On the 12th of December we lined up to board the buses which would take us to Detroit. There was more than just a feeling of excitement now. Reality was setting in. The Company was leaving familiar surroundings and there were plenty of unknowns ahead. The news from the Middle East did not sound good. Sadam Hussein of Iraq was positioning his armies on the Kuwait border. President Bush was lining up support among many other countries to form an opposing force. It looked like the whole thing was going to be much more than mere saber rattling, and there were fears that any war in the Middle East might quickly grow and spread to involve the whole area.

The television news was saturated with commentary and speculation about what would happen if war broke out. Would our military men and women actually be fighting the Iraqi troops? Would chemical or nuclear weapons be used if war broke out? The families of our Marines were all well aware of the danger that might lie ahead. Mothers and fathers, wives, children and friends were all there to say goodbye as we boarded the buses. It was a tearful departure. More than one parent or wife came up and asked that I look after their loved one and bring him home safe. I said that I would do my best.

Our flight from Detroit delivered us to Camp Pendleton in California. We were the first Marine reserve unit to arrive, but there would be many more to follow. There still was additional processing to complete, shots, physicals and record checks. After four days of this we found ourselves camped out in tents on the desert floor. Not only were we getting acclimated to the environment we might have to fight in, we were getting used to eating MREs (Meals Ready To Eat). Those dried and condensed rations were compact and easy to carry but very short on quality. I think a dog might turn his nose up at some of them.

At Pendleton we had an opportunity to spend a few days at the firing range, then we fought our way through a simulated combat town. The town seemed pretty realistic. We set up an infiltration course to attack, then went door to door, routing out the adversary. Active duty Marine training instructors evaluated our progress and other Marines filled the role of the enemy that was holed up in the town.

As the Company First Sergeant, I had access to a Hummer and a driver. I also had the privilege of eating at the main chow hall if I desired. When the training day was complete the active duty instructors would head back for a nice hot meal at the base. I was amazed at the variety and quality of food that was served there. It certainly wouldn't impress my men if I headed in for a hot meal with the instructors, while they set out in the field eating those damn MREs. For that reason I had no intention of eating anything different than my men. But after a week of eating MREs, we were all ready for something else. There must be some way to get my kids into the chow hall or get some real food out to them. I had an idea that was worth trying.

Jim Flynn was the Company Gunnery Sergeant. He and the driver accompanied me to the main chow hall on my little food mission. I explained to them, "I've got an idea on how to get some food out to the boys. You guys keep a straight face and play along with whatever I do or say."

To one of the young men working there I asked, "Who's in charge of the mess hall?"

"Master Sergeant Wilson."

"What's his first name?"

"Gee, I don't know. I just call him Top or Master Sergeant Wilson."

"Ok. Thanks Private."

There was a lady civilian employee working there. I asked her, "What's Top Wilson's first name?"

"It's Ben."

Some one pointed out where I could find MSgt Ben Wilson, and I moved directly towards him. "Ben Wilson? Ben Wilson, is that you? You old S.O.B., I haven't seen you since Camp Lejune. Hell, that must have been 15 years ago, and you would have had to be a Lance Corporal then. How the hell are you? I thought you must have fallen off the face of the earth."

Ben looked a little puzzled as his mind worked overtime to remember my face or the name Blevins, which was plain to see on my fatigues. I didn't let up, "Don't you remember me, Ben? It's Bob Blevins."

There were others around listening, and I guess Ben decided to himself that he didn't want to be embarrassed in not remembering me, so he finally responded. "Bob, sure I remember you. How are ya?"

"Well I'm just fine."

From that point on I filled Ben in on my military background and offered him a blend of fact and fiction. Before he had an opportunity to ask any questions, I asked him what he had been up to for the last 15 years.

He did a nice job on filling me in on his background, so it was now easier to blend my story with his. By the time I was through old Ben seemed to be convinced that I was a long-lost friend. Now I could get to discussing what I really wanted.

"Ben what are they going to do with all these scum bag reservists? I know it looks like we're going to war, but why do we need these guys?"

"Well, they didn't use them in Vietnam or much in Korea. Guess they plan to use them this time."

I sank the hook a little deeper. "I guess that's the case Ben. Some of my men are deployed out in the field, training these guys, trying to bring them up to speed. I know one thing. My men are getting tired of eating those damn MREs. I wish I could get them some of this great chow you're putting out here."

"I can't feed those Reserves, but I sure can fix your active duty boys up. What do ya need?"

"Gee Ben, that would be super. See you got some homemade bread over there. A few loaves of that, and some stuff to make sandwiches with would be great."

Good old Ben threw in four six inch by two foot long rolls of salami, blocks of cheese and some chili along with 30 loaves of bread! While my driver and Gunny Flynn were loading things up, Ben and I reminisced some more. Bars, people, Paris Island; we had a lot in common as most Marines do. Ben said that I could come back and get food any day I wanted to. He'd see that it was all put up for me.

At the end of the day, after the active duty instructors left, we broke out all the food that Ben had provided. That sure improved morale. Ben kept true to his word and everyday thereafter when we arrived at the chow hall he had insulated pans filled with whatever food they were serving that day. We had beef stew, goulash, chili, lots of bread, fresh fruit, whatever I wanted, Ben was willing to provide for my active Marines in the field. Ben's good work sure made life more tolerable for the boys in Alpha Company, and it didn't hurt my prestige in the Company either. Even though I bent the truth in a few places, my conscience was clear. Part of my job was to keep morale high and make sure the boys were adequately fed.

The day before Christmas our whole reserve battalion celebrated Christmas in the chow hall. Someone came up with a big old pine tree and we decorated it with pop cans, cigarette packages, whatever we could find for decorations. Christmas carols were sung and we thought about our families back home. Our celebration feast was cookies and juice. There were a few speeches and then the festivities ended up with a good laugh.

From the back, a young Private asked to be recognized by the Commander.

"Sir, I just figured out what U.S.M.C. stands for."

"What's that?"

"Sir, it means You Suckers Missed Christmas."

Laughter from a thousand voices filled the hall. It was a warm unanimous response. The cohesiveness and comradery were in place. We were all brothers in an organization we loved and with the common purpose of serving our country. We felt ready for war.

Christmas Day found us on the flight line at Norton Air Force Base. On December 27th, we were on another Air Force ramp, that at Kadena Air Force Base on the island of Okinawa.

By bus we were transported from Kadena to Camp Hanson. Our

barracks were not very impressive. They had been targeted for demolition because asbestos was used for insulation. All the beds and lockers were in a junk heap outside, and little of the plumbing and electrical fixtures worked. It was deemed sufficient for us, at least until something else became available.

It was amazing how everyone pitched in to make these barracks liveable. The beds and lockers were pulled out of the junk pile and carried in. The place was cleaned up and we did our best to repair the electrical and plumbing breaks.

We were now assigned to the Striking 9th Marine Regiment. The Commander was Colonel Strickland. Col. Strickland felt that the reserve First Sergeants should be replaced with their active duty counterparts. The active duty First Sergeants worked with all the administrative matters on a daily basis. They were more knowledge-able on regulations and procedures, which in the long run would allow for a smoother operation and a better mating with the active duty regiment. Before he replaced the First Sergeants, he would allow each to be interviewed individually by his advisor, Sergeant Major Gorczewski.

My interview went extremely well. I had done a good job prepar-ing myself before we deployed. The more questions Gorczewski asked me, the more correct answers I came up with. Points of military law, regulation references, administrative procedures; the more the questions, the harder they became. He asked me a lot of situational questions. If this happens, what are you going to do? Which form will you use? How would you discipline the man? Finally, after what seemed like a never ending session, he said, "You know more than any active duty First Sergeant in my Battalion."

When Colonel Strickland came in, the six reserve First Sergeants snapped to attention. "Men, I respect your positions, and we will have a job and things for you to do, but we are relieving you as First Sergeants and replacing you with active duty First Sergeants. These men have been doing this job on a daily basis, and we must be able to offer your Company Commanders correct and accurate advice and administrative support."

The faces of the First Sergeants were sad and depressed looking.

Colonel Strickland continued, "SgtMaj Gorczewski has recommended that I retain as First Sergeant, First Sergeant Blevins. He has accrued more active duty time and enough expertise to handle this job."

I tried not to show it too much, but I was bursting with joy inside. Things got even better soon, for SgtMaj Gorczewski wrote a letter to the promotion board giving me his strongest recommendation for promotion to E-8. That was the pay grade associated with the First Sergeant position. Although I had been filling the First Sergeant position for some time, there hadn't been a position or pay vacancy for seven years, so none of the reserve First Sergeants were receiving that pay or wearing the stripes.

SgtMaj Gorczewski rated me in the top two percent of all the First Sergeants he had ever known. The promotion board went along with his recommendation as expected, and I was the first promotion to the pay grade and rank of first sergeant in eight years. Between the three stripes facing down and the three facing up, I now wore the diamond of the First Sergeant.

As I might have guessed, there was no Green Machine capability at the Company level, or in fact, at the Battalion level on Okinawa. Regiment had a computer which we could use, when it was free. That meant standing in line when time might be critical.

I took Lance Corporal Lin with me to see the Battalion Commander, LtCol. Davis. I assured Davis that we had the Green Machine capability in the Company and could grant any administrative wish he had at Battalion level. He was doubtful, but when I had Lin run off a Battalion form, the Colonel was convinced. Lin was promoted to full Corporal, and for his foresight in providing computer capability to the Battalion, he was awarded the Navy Achievement Medal.

That was only the beginning of Corporal Lin's notoriety in the Marines. SgtMaj Gorczewski called me one day and said they were having a competition for a Regimental Non Commission Officer (NCO) of the quarter. The competition would start at the company level, then the battalion and finally among the battalions at the regimental level. He didn't know how the reservists would stack up among the active duty boys, but we were welcome to compete. I also was invited to set as a board member at the regimental level, to determine the winner.

Corporal Lin was entered in the competition. It didn't take him long to sweep up our Company and Battalion top spot. Up at Regiment there were some sharp NCO's entered in the final competition. During the oral examination, board members fired questions on weapons component nomenclature, tactics, weapons identification, ballistics, camouflage and concealment, strategy, everything a rifle company NCO would give his right arm to know. Many of the questions a Company commander would have had problems answering. Lin not only had intelligence and a photographic memory, he was respectful with not a cocky bone in his body. He carried no air of superiority and was easy to talk to. Needless to say, the board was duly impressed and Corporal Lin was Regimental NCO of the quarter.

Colonel Strickland came to my office one day, "I hear you have a martial arts expert in your company."

"Yes sir. Corporal Lin has had extensive martial arts training."

"Could you give me a little demonstration?"

I called Corporal Lin into the office. Lin was Chinese, 5'8" and about 130 pounds. He certainly didn't look overly tough. "Corporal Lin, I'd like you to face me and rest your shoulder on that bulkhead (walls are called bulkheads in the Marines). Now, without removing your shoulder from the bulkhead, I'd like you to kill that spider above your head with one of your feet."

In a split second there was a blur and the sound of a smack. A boot print and squashed spider appeared on the bulkhead. Outside I had Lin lay down in front of a standing Marine. On command he was up, flipped over the Marine and with a roundhouse foot swing, he took the Marines's cover (cover is a cap) off without touching the Marine's head.

That settled it. The Colonel wanted Corporal Lin to represent the Regiment in a competition on Taiwan. That idea wasn't very acceptable to me as I needed him here, especially if we were to be deployed to Saudi Arabia. The Marine we sent was half as good as Lin, but still won the competition, so the Colonel was happy.

From the 6th to the 12th of January, we had the luxury of a week on the firing range. Before, as reservists, we were lucky to get one weekend on the firing range and that was only every few months. All the practice paid off and our qualification scores were very good. Thirty six percent fired expert. The average for a Marine company was twenty percent. What made the scores even more impressive was that we qualified during a rainstorm. It was raining so much we had to continually blow the water from our peep sites. Targets were blowing out of the holders and the range flags were blowing in different directions from the changeable, gusty winds.

An active duty Marine unit vacated their quarters and were deployed to the Saudi desert. We moved into their barracks, which were much nicer than ours.

At Camp Foster we tested our swimming skills. Out of the company, 59 were classified as first class swimmers and 29 received a WSQ rating (water survival qualification). The WSQ rating was a tough one to achieve, and I proved that the old man was still fit, for I was one of the 29 that qualified.

We were prepared and ready to board ship for combat, but Desert Storm was over in a matter of hours. The decision was made to keep our Battalion on Okinawa and conduct training in special operations. Representative Marines were sent to Scout Swimmers School and others to an assault climbing course.

The remaining members of Alpha Company had the distinct honor of participating in Forest Light, a cold weather training exercise in northern Japan. This was the first time a Marine reserve company had ever participated. Most of the active duty units were deployed and not available. We felt fortunate to be picked because it was both a training and diplomatic mission. We would be working side by side with the Japanese Ground Defense Force.

Being from Michigan, we were accustomed to living in a cold weather climate. Previously, the company had also participated in a NATO exercise in Northern Norway. Even with this experience, we would find out that there was a lot to learn from the Japanese. It did seem ironic though, since the unit was activated, all our training was in preparation for desert warfare. Soon we would be training in knee-deep snow.

As we approached the road leading into Camp Iate, Japanese

troops from the 21st Infantry Regiment lined both sides of the road. They were clapping and waving as our buses rolled by. Some weren't smiling though. It was probably mandatory that they be there. We had the feeling they were unsure and mistrustful of us, and probably we felt a little of that towards them.

The ski equipment we were issued was really quite old and inadequate, compared to what the Japanese had. Nevertheless, we did our best and if nothing else our clumsiness on this equipment broke the ice with our Japanese instructors.

The Japanese were always so polite and proper compared to the Marines who were outgoing and energetic. Our instructors would laugh and relate through interpreters that they thought it was humorous how the Marines would fall, then get right back up and want to go faster and faster.

Their equipment was sure superior to ours and they looked much more graceful skiing. We managed to talk them into trading equipment for awhile and then they were the ones that looked clumsy. At least they could see what we had to put up with.

Captain Punter planned to conduct physical training during our spare time, but the Japanese Commander just laughed and said that there would be neither time nor energy for this. We were on their schedule now, and the training would prove demanding enough without extra exercise.

Our Regimental Commander, Colonel Strickland, arrived and there was a joint fire demonstration. It was held right in the middle of a light snowstorm. As the Colonel moved down the firing line he would relate to the Marines that while our visibility was poor due to the snow, the Marines in the Saudi desert were having similar visibility problems with the sandstorms.

The training was certainly not all work, for there were frequent parties. At the end of each training day small groups of Marines would visit the tents of their Japanese counterparts. They were given gifts plus some instruction on the proper way to consume saki, the rice wine. There was a little ritual to it. The uncorked bottle was set in a tin of hot water to warm. The guests were always served first and there would be some sort of toast. Saki was not meant to be gulped, it was for sipping only. The Marines would return from these get togethers with gifts from the Japanese along with a significant buzz from the saki.

The Marines proved they could be just as generous when the Japanese men visited our tents. They were given T-shirts, insignias, caps and other trinkets, but what our friends liked best was our whiskey.

The 1st Platoon introduced a little game that involved trying to bounce a quarter off the table into a glass of whiskey. Each person would take a turn at it, and when one was successful he could designate someone at the table who would have to gulp a glassful down. Sometimes the group would concentrate on one poor individual, just to see him get silly and finally drunk.

This was dangerous territory for me, because I rarely drank anymore. There was enough pressure from my boys that occasionally I felt obligated to participate anyway. The Japanese were quickly catching on to our game and becoming quite proficient at it. On this one evening, I lucked out and on my first try the quarter went right into the glass. I pointed at my counterpart, a Japanese Sergeant Major, to down his glass of whiskey. Being the loser, he gets the next opportunity to bounce the quarter. Well this guy bounces the quarter in the glass about fifteen times in a row and each time he points to me.

I don't remember much about that night. Guess I had a good time, at least that's what my kids said. They brought me back and zipped me up in my sleeping bag, checking occasionally to make sure I was still alive. The next morning I wasn't sure I wanted to be alive. The hangover was terrible. It happened to others too. More than once I ran into a Marine trying to navigate his way back to his tent without crashing into something.

Before we arrived here, we were made well aware that one of the most important aspects of our visit was a diplomatic one. Not only were we to cultivate good relations with our Japanese allies, but also the civilian community. In small groups we were invited for home visits with Japanese families. The people went out of their way to be gracious to us and provide a feast of their native delicacies. Of course the raw fish, seaweed and other exotic dishes were quite a variation from our meat and potato diets. I was proud of the way my boys conducted themselves and there was nothing but compliments on their behavior.

The joint winter training ended with a three day war game exercise. The Japanese were our allies and a small group of Marines served as aggressors. The weather really turned bad, with lots of wind, snow and freezing rain. We started out with a forced march on skis, pulling our gear in Akio sleds behind us. We wore white overcoats and they were frozen stiff.

It was difficult to keep warm or dry. Somehow we mustered the energy to reach our LOD (line of departure). The falling snow was becoming a blizzard. All day it had been whipped against us by the strong winds. Our faces felt raw and the snow was starting to hurt our eyes. The platoons set up their tents wherever they could find any kind of shelter, the south side of a hill, the trees, whatever. The tents were put up in small groups of four or five, nothing organized, just an attempt to gain refuge from the blinding snow. The weather seemed to be getting worse, if that was possible. I tried to map out where all the tents were set up, so they could be found if the snow covered them. I heard rumors that some of the Japanese troops were heading off the mountain. I did see a few of them myself, but it was no mass retreat.

This weather situation and the safety of our Marines worried me. There was no telling how much snow would fall. I knew our boys were wet and cold, probably burning candles, sterno, anything they could find to keep a little warm. If snow completely covered the tents, cutting off ventilation, there was the danger of asphyxiation. If they were outside, frostbite and hyperthermia were possible. Hell, the wind chill must be at least 50 below. The war was over. It would be dumb to have casualties from a training exercise. I knew if I advised Captain Punter to cancel out and pull the troops off the mountain, he'd do that. I also knew if he did that, his career might be in jeopardy, because the Battalion Commander was hell bent to prove how tough these Reserve Marines were. After all we were the first reserve unit ever to be chosen to train with the Japanese. How would it look if we were first to throw in the towel. Probably the Japanese exercise Commander, likewise, wanted to prove the mettle of his troops. I made the decision to take my chances with the exercise Commander. If there were any repercussions I could absorb them. There wasn't that much time left in my career anyway.

I had access to a SnoCat for my duties as First Sergeant. It had skis on the front and two wide hard rubber tracks on the back. I headed the SnoCat down the mountain, directly to the exercise command post. Inside, the big old Japanese Colonel sat in his easy chair while his staff officers and aides scurried about. He spoke good English. I asked him if I could speak to him in private, so he dismissed his aides.

"Sir I would like to appeal to your wisdom. I believe since we are in your country only you have the power to stop this training exercise. You can use your wisdom and ability to stop it without losing face, for the safety and welfare of the troops both Japanese and American. Call it a success and call it off because of the hazardous weather conditions. Let us get the troops in clean dry clothes, let them warm up, then have a little celebration. Everyone will be happy and everyone a hero."

The Colonel reached over and put his hand on my shoulder, "You are the strongest Marine." I wasn't sure what he meant, but he stood up, smiled and nodded. Then he walked into the radio room where he shouted an order in Japanese to call off the exercise. The SnoCats started assembling outside the exercise command post. Soon they started up the mountain to start removing the troops. I was on my way back to Alpha Company. The SnoCat I had would hold about 15 troops. That could be used for the evacuation too.

Everyone was accounted for except the 2nd Platoon. Captain Punter said he had no idea where they were. I said that I had placed them and knew the location. When I reached the spot, there were no tents in sight. It was still snowing hard, but I finally spotted part of a tent sticking out of the snow. Captain Punter and I started digging the tent out with our hands, and finally got to the opening and helped the Platoon Sergeant out. He acted a little oozy, like maybe the sterno fumes were affecting him. We got busy and continued to dig out the rest of the Platoon. The snow was so blinding that we tethered the men together. On snowshoes, we headed down the mountain, myself in the lead, with Captain Punter bringing up the rear. The Japanese brought five ton trucks up as far as they could. We climbed into the back and under the canvas. The 2nd Platoon huddled together to keep warm.

There was some minor frostbite and hyperthermia, but no casualties. Soon, everyone was in warm quarters and later joined together for a farewell party.

When we were ready to leave, there was some last minute exchanging of gifts. The Japanese Major, who commanded the troops we were allied with, got on our bus. He moved to where I sat and handed me an insulated canteen for hot drinks. Then he removed some wings from his uniform and presented them to me saying, "You are the strongest Marine." I took my Ft. Benning jump wings off my uniform and gave them to him.

This time when we headed down the camp road, all the Japanese troops were clapping and smiling.

We were back on Okinawa now, at Camp Hanson. The Marines we sent to the Assault Climbers and Scout Swimmers Schools had returned, so we were ready to start our Special Operations (SOC) training. Our instructors were active duty Marines assisted by our assault school graduates.

Most of our amphibious training took place off of Kin Blue Beach and we spent many hours getting used to Zodiak rafts. They were black rubber rafts that held four Marines on each side, plus the coxswain in the rear. If there wasn't a motor the four on each side paddled. The idea was for SOC troops to be dropped off out of land's sight and then navigate to their objective. We practiced navigation at night and during the day. The training also involved long distance swimming in combat gear, for the rafts might not all be available to get back to the ship after an attack. In the water we practiced installing pontoons to the rafts, and righting them if they were upside down. Even though we wore wet suits, after several hours in the water we would be cold. Sometimes we held Zodiak paddling races to warm up.

In the assault water training phase there was only one mishap. Sgt. Cris Nepper dropped a pair of night vision goggles (NVGs) into about 75 feet of water. These goggles were valued at $10,000 and the Battalion sent word down that money would be deducted from Nepper's paycheck until they were paid for. Cris knew exactly where the goggles went down, so we tried to get the Navy frogmen stationed there to conduct a little search. They declined our request. The first reason being that the area was used extensively by Okinawan fishermen. They would have to clear the area to search, and that would be unpopular with the fishermen. Secondly, there were dangerous currents and thirdly the area was inhabited by Great White Sharks. In fact, one of the largest Great Whites ever taken was caught in this area. We decided to make a little search on our own, so we donned our suits, tanks and slipped into the water undetected.

A water chemical light was thrown in the rocks on shore to mark where the NVGs sank. We found the chemical light on the rocks below. Hopefully the goggles would not be far away. There were volcanic blowholes on the bottom with warm water shooting out. There were some tricky currents too. We could see the fishing boat bottoms going overhead, but that wasn't a problem. We looked and looked; however we just couldn't find the goggles. Fortunately, we didn't see any Great Whites, just a few Blowfish.

Headquarters decided not to charge Nepper for the goggles after all. They just wrote it off as a training loss.

From March 9th through the 15th we switched from assaulting the beaches to climbing the cliffs. There were a few of us jump qualified, but the majority did not have experience with heights and for some that took some getting used to. We started repelling on 70 footers, then finally progressed to 100 and 150 foot cliffs. Up and down the cliffs we went, during both daylight and nighttime. There were a variety of different rope positions to be used, depending on the circumstances. We even practiced taking a wounded Marine down. Cliffs were not the only thing we went down. From the roofs and windows of buildings, to various methods of leaving a helicopter, we did it all.

Next came the northern training area with its triple-canopy jungle. Rarely did we see the sun or sky through the jungle foliage, and seldom did it stop raining long enough for us to dry out or even remember what it felt like to be dry. Red clay clung to our boots. This training area was at the north end of Okinawa, and used to practice advanced land navigation skills. It was also home of the Habu Snakes. They were like our Diamond Backs, but without rattles and more aggressive.

Finally, after all this training, we set out on a four-mile endurance trek to test all our skills. From a sheer 70-foot cliff, crossing a river on a rope, fast repelling down another cliff, walking through a muck pit, then finishing up with a 1000-meter litter carry. The course record was two hours. Our 1st Platoon did it in one hour 45 minutes.

Now, the final exam that would decide if the Battalion was to be officially designated a SOC Battalion. There was a scenario depicting an island government takeover. Missiles had fallen into the adversary's hands and could be a threat. We were to slip on the island, plant charges on the missiles and retreat by way of the sea. Captain Punter and the 2nd Platoon Sergeant, SSgt Dermeyer, were removed from the exercise in a simulated helicopter shoot down. The Exectutive Officer (XO), Captain Longworth, would command our company.

Right from the start the operation was plagued by high winds and 20-foot waves. The scout swimmers had problems reaching the beach, but finally made it in and secured it. After spending five hours in open, choppy water, the men were cold, even though they had wet suits on. Clouds obscured the moon as the men in the Zodiak rafts slid ashore. The first element secured the beach and the Zodiaks were pulled under the vegetation. The second element was the headquarters support element, which was followed by the assault element, whose job it was to proceed to the objective and plant charges on the missiles. The opposing enemy force had expected us to arrive on the other side of the island, so there was no resistance until the assault element reached the missile storage area. There, a fierce fire fight evolved. The assault element was successful in planting the explosive charges, and as they retreated the enemy defense force mounted a counterattack. Our perimeter defense element held them off while the assault and headquarters elements slipped out to sea. The assault swimmers offered token resistance while the defense element pushed out in their Zodiaks, then the swimmers left as they had arrived, under their own power in the sea. The exercise was a resounding success. Our Battalion was the first Reserve Battalion ever to be SOC qualified. We were extremely proud.

From Okinawa the Battalion was assigned to the Philippines MAGTAF (Military Air Ground Task Force). For a couple of weeks the company performed guard duty at Subic Naval Base and from there to the jungle for more intensive survival and operational training.

Our barracks in the Philippines were at Camp Tamez, 23 miles from Mt. Pinatubo. There had been some tremors and rumbling going on, but none of the locals thought anything of it. Mt. Pinatubo hadn't erupted for 600 years. The warnings were out though, and on June 15th I was just getting into my vehicle when she blew. Fortunately I had my camera along and got some good pictures of the mushroom cloud formed in the initial eruption. It was 10 a.m. By noon the clouds had obscured the sun and it was so dark that, if you held your hand out in front, you couldn't see your fingers move. It stayed that dark for two days. During those first few days there were about 60 earthquakes, and some typhoons generated by the eruption. It looked like the whole sky was on fire over the mountain.

We didn't know whether we were to live or die. In total darkness there was no electricity or communication. The closest town was completely destroyed. The Marines were busy trying to pull Filipinos from the wreckage of their homes. They were using their bare hands to dig folks out. Buildings were not constructed to withstand the weight of the ash. It was like falling wet concrete, weighing about 60 pounds per square foot. One village close to the mountain was reportedly buried under 700 feet of ash. Of course no one lived.

Our barracks were round topped quonset huts. They survived well. At night the earthquakes would cause the bunks to move across the floor and bump into each other and the wall. Some authority compared the Mt. St. Helens eruption with this one. On a scale of one to ten, ten being the worst, Mt. St. Helen reportedly would be rated a 6 and Mt. Pinatubo a 10.

People didn't have food or water so we headed into the village with a five ton truck loaded with MREs. The village Mayor was very political and tied to the anti-American sentiment that some people held. He told us to go back to the base. He didn't want our help. As soon as the mayor wasn't around, we drove down the side streets unloading cases of the MREs for the folks. We knew that the people needed the food and were planning to help out whether the mayor liked it or not. Several Marines stayed in town to continue to aid in digging out those trapped in their shacks.

The death and destruction was unbelievable. The three canopy jungle was flattened. All the foliage was buried under several feet of ash. There were dead animals everywhere. And, of course, with the destruction came the looters and pillaging. Many were armed and in the night there were shots fired at the guards we posted.

When things started to get organized, we were assigned to evacuate the servicemen and their families from Clark Air Force Base and Subic Bay, the Navy Base. These people were only allowed to take 40 pounds per person with them. All the rest of their possessions were left behind for the looters or whatever more destruction the volcano could dish out. Aircraft hangers at Clark were completely flattened. They had been constructed to withstand heavy rains and wind, but not the weight of the ash.

After a couple of weeks we finally got some generators in and a chow hall up and running. Our folks could have at least one hot meal a day.

The Sky Club was a large recreational facility run by civilians for the U.S. servicemen. The building collapsed under the weight of the ash, but there was a huge freezer inside that was full of food which would be spoiling soon. I asked the Colonel if I could have some of the men to remove the food and use it before it all spoiled. We retrieved hams, roasts, fish, fruits and vegetables, but ate none of it. To a man, the Marines agreed that the Filipinos needed the food worse than we did. They lost their homes and possessions, plus many of their loved ones had been killed. We had one meal in the chow hall and our MREs. A little smuggling operation was set up and the Marines distributed all the Sky Club food to the civilian population. We felt better, the people were better off, to hell with the mayor.

On June 26th it was our turn to leave the Philippines. We sailed out on the USS Midway, waving goodbye to the people and US facilities. Three days later we were back on Okinawa to start the demobilization process. In August of 1991, our company arrived back in Grand Rapids. Our Battalion was the last of all the reserve units to return home after the Desert Storm call-up. The yellow ribbons tied around trees and utility posts weren't as fresh looking as they had been before. Except for the families, most people assumed we had returned months before.

Top: Company Clerk Lin "Radar"

Bottom: Bob on left, his Company Commander Ken Punter on right

Top: My good friend the Japanese Defense Force Commander

Bottom: Company on skis at Akita, Japan

Top: Bob and Sgt Nepper after a dive for the lost night goggles

Bottom: Zodiak training on Kin Blue Beach Okinawa

Top: Jim Flynn, Company Gunny — Bob, First Sergeant

Bottom: Our quarters in the Philippines

June 14-16, 1991: First eruption of Mt. Pinatubo in 600 years. Twice the size of Mt. St. Helen

Top: Mt. Pinatubo destruction

Bottom: Our quarters after eruption

Volcano destruction

Standing on a pile of volcano ash waiting for U.S.S. Midway to take us home

The Beginning of the End

My active duty tour was less than a year, but it seemed like a lifetime. With winter coming, I needed to get as much done as possible on our new home, so that we could spend the winter there. We purchased the property after returning home from the Desert Storm call-up. The previous owner had started to build the house, but he hadn't gone much beyond some rough framing in.

Dee and I loved the place. Eighty acres of tall pines, a beaver pond, lots of deer and the setting for the house blended in perfectly with it all. We were well off the main road and literally in our own little world of peace and quiet, with nature's creatures as our neighbors.

By the time that cold weather arrived, I had the house liveable with water inside for the kitchen sink and toilet. There was electricity for lights and the cook stove, so we were pretty much ready for winter. There was still plenty of work to do inside, but I could take my time, working on that through the winter.

We moved so many times, and always had been so busy, that we never really took time out to enjoy each other and our home. In Lake County we had planned to settle for good, but the unfortunate circumstances there turned our dream home into a place where we didn't feel safe. Our new home here was meant to fulfill our need for peace, security and permanence. It was to be a haven for us and an environment that would give enriching experiences to our Grandchildren.

Although I was working long hours on the house, I still allotted time to enjoy both the bow and rifle deer hunting season. It was absolutely great being out in the woods again. Everything seemed to come into perspective and have meaning.

In the spring I knew that I would soon have to start looking for some type of employment. I wasn't sure what kind of work I'd do, but I figured something would come along. I was crowding fifty, and wondering how much demand there would be. Oh well, looking and filling out applications would be the thing to do.

Surprisingly enough, in June something did come along and it was quite unexpected. The Marines could use my services again. It was a full Colonel on the phone, a Colonel Creed from Kansas City. Said he was a good friend of Colonel Strickland. He needed a First Sergeant to manage a large number of Marines involved in drug interdiction on the Mexican border. Would I be interested in some active duty?

Wow, the Marines had come through again. It seemed like every time that I was in need, the Marines had been there for me. Certainly I wasn't looking forward to leaving my home or Dee, but the offer was one I would have trouble turning down. Not only did the work sound interesting, when a Colonel calls up and specifically asks for your help, based on the recommendation of another Colonel, it is difficult to refuse. The pay would be excellent and we would again have all the extra privileges and pay benefits of the active duty force.

I couldn't take Dee with me again this time, and wondered how content she'd be out here by herself. Our daughter and son were not that far away, both were married with children. That should provide her company and an option to stay with them if she wished. She did enjoy our new home as much as I did and didn't seem to mind its remoteness.

My post was out in the desert near Deming, New Mexico. A force of Marines were working on shifts to man observation posts along the border. It was a high density drug crossing area and the Marines were operating in conjunction with DEA (Drug Enforcement Agency) and the Border Patrol to interdict the heavy flow of drugs across the Mexican border. My work as First Sergeant was concentrated in the base camp area, with my boss Colonel Creed.

That changed one day when Colonel Creed called me into his office. He had relieved a platoon Commander for negligence in the field. He needed someone to take the Captain's platoon out until he could bring in another officer. Could I do the job? "Sure," was my reply.

That night I led the platoon out to man the listening posts spread along in the mountainous region next to the border. The command post for the sector was a mile back from the border. Before the Marines left for their individual positions, I received a call from Colonel Creed.

"There is a strong possibility of drug activity in your sector tonight. Put every available man out in the field. Be extremely watchful."

I was already minus a few men because of sickness. As far back as our command post was, probably just myself and the radio operator would be sufficient there. Everyone else must man the posts. Our job was to be the eyes and ears. If we detected activity, a call was placed to the DEA agents and they set up the interdiction and made the arrests.

Once everyone was in position, I told the radio operator that he and I would work in shifts. I'd get some rest in the early part of the evening, then relieve him around midnight. This was a rather snake infested area, so I just didn't feel comfortable rolling out my bedroll on the desert floor. Instead, I climbed in the back of one of our trucks. Just nicely settled in, but not asleep, the radio operator was soon beside me. "First Sergeant, First Sergeant," he said in a whisper. For some reason he was alarmed. We hadn't been there that long.

"What's up?"

"One of the post observation leaders is on the radio. Wants to talk to you."

I picked up the radio, "What's the problem? What do you need?"

"What's going on there?" he said.

"Nothing's going on here. There's no activity. I was just trying to get some sleep."

"Well, I see an occasional light out there, and it looks like it's coming toward your position."

I didn't see any lights, but just to make sure, I took the night vision goggles up on the back of a five ton truck to look in the direction the Marine was talking about. He was right. There was a column of 12 people coming right towards us. Less than a football field away, they were carrying packs and were armed with automatic weapons, Uzzis or Mac 10s.

First Sergeants weren't issued rifles. I had only a 9mm pistol with 2 clips. The radio operator had a rifle, but carried only 20 rounds. We wouldn't be a very good match against a dozen automatic weapons. Somehow, I needed to head them off before they overran us. Trying to simulate a sizeable force ahead, I moved from vehicle to vehicle, and then to several other positions, each time yelling, "Marines ahead. Halt." I changed my voice to sound different each time.

Back on the truck, I again used the night goggles to check the whereabouts of our guests. All my yelling hadn't deterred them at all. They weren't about to drop their packs of cocaine or marijuana, or for that matter it didn't appear that they were abandoning the objective of reaching their drop-off point. In fact, they had spread apart, forming a skirmish line, and were moving towards us, firing their weapons in our direction. I told the radio operator that it was time for us to get out of Dodge.

I moved to the back of the truck and slipped off the night goggles. Everything was suddenly dark out, so when I jumped off the truck I wasn't aware that I was going to land on some rocks. My right foot seemed to explode. There was intense pain. Pain or no pain, the foot was still attached and we needed to leave the area immediately. Strapping on the radio gear, we took out across the desert in an escape and evasion posture. When we had put some distance between us and the convoy I made contact and alerted the DEA. The other Marines were moving out of their positions toward the shooting, so finally, the drug smugglers apparently reevaluated their chances of reaching the drop-off point, and headed back across the border.

After all the running, my foot was really hurting now. There was something radically wrong with it. Back at the base camp they helped remove my boot and gave me something for the pain. The next day I was flown to the Army hospital in El Paso for X-rays. Bone fragments were revealed and there was tendon damage. The doctor shot cortisone in different areas of my foot and supplied me with more pain-killing pills for the trip to the Navy hospital in Chigago.

The doctors at the Naval hospital performed surgery, removing the bone fragments. Because there was so much damage done to the tendons, they removed that part of the achilles tendon that connects the heel to the toes. Probably I caused all the damage to the tendons by running with the broken bones in my foot. Hell, what was I supposed to do? Get shot or captured by some damn drug runners? I remember the story about the famous Naval hero, John Paul Jones. In later years he was riding a paddle wheel steamer during a storm on the Mississippi. The boat was taking on water and John Paul was heard making the statement, "After sailing the Seven Seas, am I destined to drown in a swamp?"

Similar thoughts went through my mind. For all the risk I had been exposed to, from diving, jumping out of airplanes, repelling down cliffs, to riding bulls, ending my Marine career with a jump off a truck seemed like an unglorified way to go. But I knew that this injury would, in fact, bring my Marine years to a halt. Fortunately I was kept on active duty status with pay until my foot healed and I could walk on it. I also was allowed a 20 percent combat-related disability payment.

During the recovery from my injury there was lots of pain, and there was also lots of time to think and reflect. It seemed that, through a collection of random circumstances, I had been associated with some pretty remarkable people in my lifetime. One of them, I didn't even want to meet.

Early one evening I was out on patrol with one of my reserve deputies, just cruising some of Lake County's back roads. One of my regular deputies was on vacation, so I was filling in. I sat on the passenger side, a little lost in thought as the deputy slowed the patrol car.

"The gate's open. He must be up this weekend."

"Who's up this weekend?" I asked.

"Ted Nugent."

"Who's Ted Nugent?"

"You know, the famous rock singer. He owns several hundred acres in Lake County. I know him. Want to go in and say hello?"

"Is that the screamer with the long hair that goes nuts on stage? No thanks. I'll pass."

"Ted's not what you think. He's a real family man who's not into drugs or booze and is a dedicated hunter. I think you'd like him."

"I'll bet. I don't like rock music now, and didn't when I was a kid."

The deputy had already made his mind up to go in and say hello, whether I wanted to or not. It wasn't long after I shook hands with Nugent that my preconceived opinion of him began to crumble. He was intelligent, personable and obviously dedicated to hunting. But what was most impressive was that, more than just being a hunter, Ted was remarkably knowledgeable on everything in the out-of-doors. Game animals, birds, trees, flowers and plants, all seemed to fascinate him. Watching Ted react with his two children convinced me that he was also very dedicated to his family.

I didn't have an occasion to see Ted Nugent again until deer hunting season that fall. His concert schedule never interfered with hunting; he just wouldn't let that happen. We did hunt together some, and made plans for a joint bear hunt the following spring. The more we were together, the stronger the friendship became. Love of hunting and family seemed to create the bond.

Ted had a real nice four wheel drive hunting vehicle. It was all painted a camouflaged color and was lined with speakers inside. I figured he probably listened to a lot of rock music. We were driving some place and out of curiosity I opened up his tape storage compartment to see if he might have some country western music. That's what I liked.

"What ya lookin' for?"

"Thought you might have a country tape."

"Nope. I make music, but I don't listen to it."

The compartment was full of hunting and out-of-door instructional tapes. Everything from animal behavior to calling and techniques of hunting them.

I did get to hear some of Ted's music. But, instead of rock, he sang ballads around the campfire. I asked why he didn't record and market some of those songs, because they were very good. He said he didn't want to change his image. Rock music was supporting him very well.

I remember one song that Ted wrote. It was right after his wife was killed in an auto accident. Their daughter was taking her mother's death very hard, so Ted wrote the song for her. Basically it recalled the positive things in her mother's life, and was meant to be uplifting. Listening to it sure brought tears to my eyes though.

Ted Nugent was responsible for my meeting with Fred Bear, another extraordinary individual. Ted invited me to go with him bow hunting to a 3000 acre hunting preserve by Rose City, called Grouse Haven. A General Motors Vice President by the name of Bill Boyer was the original owner and Arthur Godfrey held interest in it too.

Arthur Godfrey was quite a famous radio and TV personality. He became acquainted with Fred Bear when Fred came to the camp to promote his hunting bows. Godfrey was so impressed with Bear that he personally funded Bear's bow hunting adventures to Africa, where he killed elephants and other big game with the bows and arrows he manufactured. Fred traveled to Alaska for the big Kodiak Bears, and many other places across the globe, wherever big game animals were available to hunt. He established himself as a world renowned bow hunter, proving that anything that could be killed with a gun could also be dropped with a bow and arrow.

I had bow hunted some before, but as I became more acquainted with Fred Bear, my interest in the sport increased tremendously. There was some real good chemistry between Fred and myself. He took pride in the fact that, over the years of our friendship, my expertise with the bow equalled my ability with a gun. Fred made sure that I had the best of his bows to hunt with, and they were gifts. I guess it was compensation enough for Fred to see the pleasure his equipment gave me. He treated my son just as nicely as he treated me, and gave him a bow too.

The hunt at Grouse Haven was by invitation only. Over the years it became known as the proving grounds for Bear Archery. It was a reward to specifically selected dealers and distributors of Bear Archery to be invited to the preserve and hunt with Fred Bear. Even in his later years, when Fred's failing health prevented him from actually hunting, he still came to the camp for a month to be with the hunters.

Fred was an excellent storyteller too, and after each evening meal he would set at the head of the table and tell about his hunting trips all over the world. The stories were more than just tales of adventure, for Fred had such a fine sense of humor. It was a wonderful experience just to set in his presence.

Grouse Haven also entertained its share of celebrity hunters. I can remember sharing the supper table with Jim Lovell, the Commander of Apollo 13. Neil Armstrong was there one time too, along with many others. After my first visit, Fred made sure that I had my own personal invitation each year. We even talked on the phone through the year to trade our latest jokes.

During one of the last hunts at Grouse Haven, Fred told us that he would autograph some bows after the evening meal. I had a box of unopened cedar arrows that held the Bear Company broadheads with a razor insert. I brought six of these arrows with me for an autograph. When Fred spotted the arrows, he said, "What have you got those here for?"

"I thought maybe you would sign them."

"Nope. I don't sign arrows. The shaft is just too damn small and my eyesight isn't the best."

"That's OK Fred. They probably would be too hard to sign. I just wasn't thinking."

About two in the morning I needed to make a trip from my cabin to the outside toilet. On my way back I could see a light on in the main bunk house. I thought that was a little strange, so I walked over and peeked in the window. Fred was setting at the big kitchen table, and he was using steel wool to rub the finish off of a portion of my cedar arrow shafts. His jar of India ink sat nearby with his quill. I knew what he was up to, so I didn't interrupt him. Still, I sure felt good inside.

The next morning when I sat down for breakfast, there on my plate were the six arrows. Fred had signed each one and put a coat of shellac over his signature. When Fred came in, he had kind of a mischievous look on his face. "You know Bob, I have signed thousands of bows, but less than a dozen arrows. Six of them are lying on your plate."

I gave one of the signed arrows to Ted Nugent, one to my son and another to Doug Walker, editor of Western Bow Hunting Magazine. Doug used to be head of national sales for Bear Archery and was always a close friend of Fred's.

The last hunt at Grouse Haven was in October of 1989. As per a preestablished agreement, one year after Fred Bear's death the preserve would be closed and sold off in 300 acre parcels. At mealtime that year, the chair at the end of the table was left empty in honor of Fred Bear. In some ways, it was a rather sad time, for there certainly was an emptiness without Fred. We resolved ourselves into treating it as a celebration of all the good times we had been privileged to have there, and a final tribute to Fred Bear. He did more to promote the sport of bow hunting than any other single man ever has.

Of all the fine Marines I had ever known, Sergeant Major Peter Gorczewski stood the tallest in my mind. He was the one responsible for retaining me as the Company First Sergeant during active duty, and making sure that I was promoted to that rank. I can still remember verbatim the last paragraph of his letter to the promotion board.

"I have been the 9th Marines Sergeant Major in excess of three years and comparing GySgt Blevins to all of the First Sergeants I have known, I would rate him in the top two percent. A truly outstanding SNCO. It is satisfying to know that there are Marines like him to take my place. Therefore my bottom line statement would be, if you only promote one Marine to First Sergeant, promote the Gunny."

In the Marines, Peter Gorczewski was a legend. Probably he was one of the most decorated Vietnam Veterans. He served for five straight years in Vietnam and never came out for leave or R&R (rest and recuperation). He was wounded five times and immediately pulled out of combat. By regulation, a Marine cannot be assigned to combat after his fifth Purple Heart. Gorczewski appealed to the Commandant of the Marine Corps to please put him back into combat with his fellow Marines. The Commandant made an exception and allowed Gorczewski back into the war. Peter was wounded two more times and received a chest full of medals.

Peter Gorczewski's office wall was covered with souvenirs from his Marine years. The NCO sword, certificates, awards, pictures and an E-Tool. An E-Tool is a folding shovel used primarily for digging fox holes. When Peter's company was overrun, most everyone was killed. Peter was wounded and out of bullets, but when they found him after the battle, he was still alive and there were many enemy soldiers lying dead around his fox hole. He had successfully defended his position and killed them with his only weapon, the E-Tool. He was physically a small man, but one of the toughest that ever put on a Marine uniform.

Pete Hector was a highly respected writer that did free lance reporting for the Grand Rapids Press and the Big Rapids Pioneer. In his column he reported about the lawlessness in Lake County, and the fact that the crime and murder rate per capita was 60 percent higher than the city of Detroit. The murder rate alone was higher than any other county in any State in the United States. He also talked about how many sheriffs Lake County had gone through, something like 11 in 15 years. I remember reading some of his articles before I became sheriff. Maybe his writing had something to do with my decision to run for sheriff there. It certainly was evident that something needed to be done.

Pete Hector was a very opinionated reporter, but he was very thorough and accurate in researching his reports. He was a key figure in bringing about the investigation of the CETA funds collected by J.T. Nelson to refurbish black cemeteries. He was also instrumental in exposing the fact that Nelson was collecting welfare rent on vacant lots and that dead people were mysteriously voting.

About the time I was trying to secure a Grand Jury on all these charges, Pete Hector and his wife came up missing. I found out that he was renting a house not far from my home. When I approached the house, I could see that the door was partially open. I walked in and could see dishes left on the dining table with moldy food still in them. There were also clothes in the closet. It sure appeared that Pete and his wife had left in a big hurry, either on their own accord, or otherwise. I knew because of all his investigative work he probably had received death threats like the rest of us.

Missing person reports were filed, but Pete didn't show up and no one knew where he was. After a period of time, I was pretty well convinced that one day they would find Pete's body in a swamp with some bullet holes in it.

It was a couple years later when I walked into a restaurant and there was Pete setting at a table eating.

"Pete, I thought you were dead."

"Nope, it was time to get out of Dodge, so I vanished into thin air."

That was the only explanation he offered. There must have been a lot of pressure all of a sudden. Pete had been doing his investigative work up there for some time, and I'm sure he had received threats earlier too. The next time I saw Pete he suggested that maybe we should get together and write a book about all the things that had transpired in Lake County. In the next few months we discussed the book possibility a little more seriously, but before anything could be started, Pete passed away.

My foot was healed up now, and I was off of active duty status. It was time to be gainfully employed again. Dee had injured her back during my last year as sheriff. She suffered back pain on a regular basis, so it was not logical for her to go back to work and do even more damage to it. I put law enforcement out of my mind for good, and the Marines were not an option this time. There would have to be some other way to make a living.

If I could get a Federal or State job, my 12 years of active duty would count towards retirement. I checked at the local post office. The Post Mistress, Kay Smith, assured me that if they had an opening she would be sure to let me know.

After some time, there was a job opening at the Big Rapids Post Office. I applied and took the civil service test, doing quite well on it. About a week or so later the Postmaster called me up. "Mr. Blevins, I'm calling to tell you that you and one other applicant had equally impressive credentials and test scores. I felt the other applicant was just a little better qualified so he was hired. I'm sorry, but thanks for applying." I was disappointed, but that's life. The next time I was in the local post office I told Kay that I had missed out on the job in Big Rapids. She had kind of a funny look on her face. I'm pretty good at reading people, so I could tell that she already knew that I hadn't been hired and that there was more to the story that she didn't feel comfortable talking about. Now, I was a little bit curious about the individual that beat me out for the job. I called up to the Big Rapids Post Office and asked to speak to the guy. When he came to the phone, I told him that I was from the local newspaper and that I wanted to do a bit of a profile on him for the community. He seemed receptive to the idea and willingly answered my questions.

No, he was not a veteran, and never did serve in the military. Yes, he lived in the local area with his parents and was 23 years old. His last employment was working part-time trimming Christmas trees. I thanked him for his time. That was all the information I needed. So much for the job selection process at that post office.

I kept scanning the newspapers for other jobs that might pay a reasonable wage, and also that would utilize my experience as a manager. Such an ad did appear. They advertised for an experienced mid-level manager that could work with a large staff. A college degree was desired or at least extensive management experience. When I called about the job, they said there was a test involved for the applicants. The test results would be used only in the final screening process. The previous job experience and applicant credentials would narrow the field down to a dozen or so. I was welcome to apply and take the test.

Taking another test certainly did not hold much appeal for me. I was sick of it. Back in 1985, we toured the country for a year as I tried to get a law enforcement job. Almost all interviews required some type of test. I didn't seem to mind then, it was a challenge. I was enthusiastic about the prospect of getting back into law enforcement and living in a new part of the country. Taking tests now was pure drudgery. I certainly didn't have my heart into taking this one, but I felt that I must stick with it. The alternative choice was to flip hamburgers, push a broom or maybe be a greeter at Wal-Mart.

Dee picked up the mail that day, and I could tell she was happy about something. She was waving this one letter at me from the car. It was the test results from my last job interview. My score was well over 90 percent. That certainly made me feel good, as I wasn't sure how well I did. My mood changed considerably when I received the call from the lady involved in the hiring process. She said that they had reviewed all the applications and based on that information were eliminating applicants. They felt that since my only experience in the military was as a sergeant, I just wouldn't have the management experience they were looking for.

I knew damn well I wasn't going to get to first base on this job either, but I was really frosted now, because obviously this lady didn't have the foggiest idea what a First Sergeant in the military did. I tried to be cool, and even though it wasn't going to help me, I made up my mind to give her an earful on the duties and responsibilities of a Marine First Sergeant.

"Miss, I was a Sergeant in the Marines before I was twenty years old. The difference between being a Sergeant and a First Sergeant is like the difference between night and day. The First Sergeant is the Commander's chief advisor. The Commander relies on the First Sergeant's wisdom in making decisions and the First Sergeant is responsible for making sure the Commander's directives are understood and carried out. The First Sergeant also must be aware of and honestly represent the feelings and concerns of the enlisted Marines to the Commander. He is like a union chief without strike power. There is great respect generated for the First Sergeant in the Marine Corps. He is even protected from retribution from a Commander that doesn't like his advice. Only the Commandant of the Marine Corps can bust a First Sergeant. Yes, he can be moved to another job, but he can't be demoted to a lesser rank by the Commander. The First Sergeant must be free to speak his mind. He is the sounding board for the Commander's decisions and the interpreter of the Commander's decisions as well as his chief disciplinarian."

This lady was probably tired of hearing my spiel, if she was even listening. Oh well, I felt better getting it all off my chest and was ready for another rejection.

"Mr. Blevins, we'll leave you on the roster. It seems that you have had a great deal of management experience after all. You will be notified about the next phase in the screening. Thank you."

I was flabbergasted. She had actually listened to me and I was still in contention for the job.

Next, I was called in for an oral examination. There were three interviewers in the room. They took turns asking me questions about how I would solve specific situational problems involving employees. I did my best to answer each question, but when it was over, I really didn't have a good feel on how well I performed.

It wasn't long after the interview that a man called and informed me that I had been selected for the position. They were impressed with the results of my oral examination and felt that I was the most qualified person interviewed. One lady had a higher written test score, but didn't have the practical experience in her background. I won out over several applicants with college degrees, and I was told that one man with a Masters even failed the written test.

Dee and I were ecstatic. This job would have regular hours, health insurance and nice retirement benefits. We could pay the bills, keep our home and finally see some sunshine in a sky that had been cloudy for a long time. Once again my persistence had paid off. Inside the resolve was growing; I was going to be the best damn manager this place ever had.

I wished Granddad was still around. It would have been a comfort to have his advice during the days when I was really down and out. I wondered what he'd say if he knew about all the bad things that had happened to me. My next thought had a smile attached. Shucks, I knew what he'd say, "Son, if it doesn't kill ya, it will make you stronger."

Top: Bob and his two favorite hunting partners, Ted Nugent on his left and Fred Bear on the right

Bottom: Nice buck shot with a bow Fred Bear gave Bob

Top: Our home outside of Cody Wyoming

Bottom: Bob and his 26 year old horse Guy on their last elk hunt on Carter Mountain (12,000 ft.) in Wyoming

Top: Bob's mother Ruth on right and Helen Welch who adopted her from the Chippewa Reservation

Bottom: Bob with his son Buck and daughter Karen

Top: Dee and Bob at his retirement, Jan 1995

Bottom: Bob with sisters Judy (left), Nancy on right

LOOKING TO THE FUTURE

REMEMBERING THE PAST

Top: Bob and Grandkids — Nikki, Heather, Julie, Alisha, Cody, Amber not pictured

Bottom: Bob's Grandfather — Lawrence Friedrich

Last Thoughts

From the Author
Gordon Galloway

Frank Kelley continues as Michigan Attorney General, a position he has held for over 30 years. A large and talented staff of State lawyers closely examine each issue and provide Kelley with pro and con options to base his opinions. When an opinion will enhance his image with the public, it is not unusual to see both a State and National news release. On other more controversial cases, there is plenty of insulation within his office to protect him from negative press.

After serving two terms as Governor, Jim Blanchard was appointed as the U.S. Ambassador to Canada by President Clinton. During the campaign of '96 Blanchard quit his post to work on President Clinton's reelection campaign. Early in '97 he was considered the leading contender for the chairmanship of the Democratic National Party.

Conrad Mallett Jr., who served as legal advisor to Blanchard while he was Governor and political director/executive assistant to former Mayor Coleman Young, was appointed to the Michigan Supreme Court three days before Blanchard left office. The appointment was criticized as political because Mallett had no judicial experience before joining the bench. Democrats now hold a 4-3 majority on the panel and in 1996 elected Mallett Chief Justice.

One has to ask, "How many more people were intimidated? How many were investigated and threatened with false charges to protect the political reputations of others? Were there many who had false criminal convictions entered in their records as Bob Blevins did?" If only the walls could talk.

The Attorney David A. Neff
(excerpt from a 1991 letter)

He remains to this day something that he always was—a good cop! I can think of few people who could have endured the disappointment, rejection and financial deprivation of the past six years as well as Bob has. I think it is a testament to his courage, his sound emotional health and frankly shows his absolute innocence of any wrongdoing that he persists in returning to his proper calling—law enforcement.

The Deputy Count Fisher

He treated me as well as you could treat a person. That man is a man. He was granite. When the going got rough he wasn't setting up in the office, he was first in.

I'm 73 now, but if Bob called tomorrow and needed the Count, I'd be there to back him up. They did that man so bad. I never knew why. I pray to God somebody tells me why they didn't want him in office. Why would the bigwigs want him out of office? What was his crime? I saw and went with Bob Blevins. He was honest. Nobody had Bob except the people and they were so stupid they didn't realize it.

Bob Blevins

Being out of law enforcement has left kind of an empty hole in my life. I guess given a choice I would have traded a few more years as sheriff for the rest of my life.

A SPECIAL THANK YOU
TO THE AUTHOR GORDON GALLOWAY

When all the positive events of my life came crashing down with the climax of Lake County, I was devastated, hurt, and extremely close to a nervous breakdown. I withdrew, and other than my wife never spoke of those heart wrenching times. I have never shared this story with anyone, even my own children. After many years I convinced myself that I had released my anger and had forgiven those that sought to destroy me for their own personal gain. My close friends and family avoid any reference to those years, I suppose to spare me the personal pain and embarrassment. However, as time takes its toll, I've often wondered what I would leave behind and what would be remembered by those I love and care about. Does the truth matter? Is truth worth dredging up painful memories? When my life is over-said-and-done, the truth may be the only thing I can leave behind to my children and grandchildren.

And then, as life often does, I cross paths with Gordon Galloway. For over a year and a half I poured out notes and dozens of hours of tapes recounting the events and people that shaped my life. Then comes the endless accounts as Sheriff of Lake County. Admittedly, many of the stories seem far-fetched, never-could-happen tall tales. Not once did Gordon Galloway grin in disbelief or question me in regards to stretching the truth. Rather, Gordon went to work checking newspaper accounts, court records and interviewing dozens of people to validate and ensure this is a truthful account of what really happened.

I have to admit, doing this book brought back a lot of painful memories. I had to come to grips that much of the anger was still alive. I soon realized that I had not forgiven my enemies. I spent many sleepless nights and some of the "old" nightmares returned. It wasn't long before I realized that recounting and telling this story to Gordon was now proving to be something short of a miracle. For the first time in well over a decade I was actually getting it all out! I was feeling better, and now I can finally let the past go and hold my head up with pride to remember who I was and who I am.

Thank you Gordon Galloway for believing in me. Thank you for giving up some of your life to write my story. Most of all, thank you for showing me my life was not wasted, and I did make a difference.

"The Bulldog"
Bob Blevins

Gordon and Mary Ann Galloway

ABOUT THE AUTHOR

A graduate of Michigan State University, Gordon Galloway served in the Air Force with a Vietnam tour, then spent over 20 years flying fighters with the Michigan Air National Guard. He retired as a Colonel in 1988 and has since been employed as a Captain with Simmons Airlines (American Eagle) in Chicago. For over 30 years he has operated a farm near Morley and has discovered that he doesn't have to go farther than his own neighborhood to find fascinating stories to tell. He is dedicated to preserving those stories and letting you have an opportunity to meet the real celebrities that make up the fabric of this country.

Gordon's partner in bringing forth these stories is Mary Ann Galloway. A lifelong Michigan resident, Mary Ann obtained her education from Benzie County Normal and Western Michigan University. She taught grades Kdg.-8th for several years in the one room country schools of Ionia County.

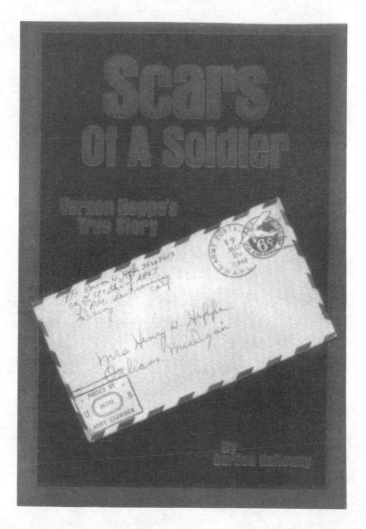

A Michigan farm boy with an eighth grade education is drafted into the Army during WWII, and becomes a man on the Pacific battlefields of Kwajalein, Leyte and Okinawa. Wounded four times, he received both the Silver and Bronze Star. With a keen sense of humor, Vernon Heppe tells about growing up on the farm and his adjustment to becoming a soldier. He describes combat in vivid detail and the memories that remain.

Hillbilly Poet

A True Story by Gordon Galloway

The boy was half Cherokee. He was raised by his blind grandmother until he was sent to an Episcopal boy's home. Ambitious and rebellious, he knew what he wanted out of life, and when the system wouldn't react to his needs, he simply bypassed it.

The mountains and rivers were his refuge and constructing power lines across this country became his vocation. This is a story filled with love, laughter and tears. The last chapter contains Ted Aldridge's poetry. Each poem remembers a special part of his life.